QUEUEING THEORY WITH APPLICATIONS TO PACKET TELECOMMUNICATION

QUEUEING THEORY WITH APPLICATIONS TO PACKET TELECOMMUNICATION

JOHN N. DAIGLE
Prof. of Electrical Engineering
The University of Mississippi
University, MS 38677

Springer

John N. Daigle
The University of Mississippi
Department of Electrical Engineering
University, MS 38677
wcdaigle@olemiss.edu

Queueing Theory with Applications to Packet Telecommunication

Library of Congress Cataloging-in-Publication Data

A C.I.P. Catalogue record for this book is available
from the Library of Congress.

ISBN 0-387-22857-8 e-ISBN 0-387-22859-4 Printed on acid-free paper.

Printed in the United States of America.

9 8 7 6 5 4 3 2 1 SPIN 11054061

springeronline.com

NOTE TO INSTRUCTORS

A complete solution manual has been prepared for use by those interested in using this book as the primary text in a course or for independent study. Interested persons should please contact the publisher or the author at http://www.olemiss.edu/~wcdaigle/QueueingText to obtain an electronic copy of the solution manual as well as other support materials, such as computer programs that implement many of the computational procedures described in this book.

Contents

List of Figures

List of Tables

Preface

Soon after Samuel Morse's telegraphing device led to a deployed electrical telecommunications system in 1843, waiting lines began to form by those wanting to use the system. At this writing queueing is still a significant factor in designing and operating communications services, whether they are provided over the Internet or by other means, such as circuit switched networks.

This book is intended to provide an efficient introduction to the fundamental concepts and principles underlying the study of queueing systems as they apply to telecommunications networks and systems. Our objective is to provide sufficient background to allow our readers to formulate and solve interesting queueing problems in the telecommunications area. The book contains a selection of material that provides the reader with a sufficient background to read much of the queueing theory-based literature on telecommunications and networking, understand their modeling assumptions and solution procedures, and assess the quality of their results.

This text is a revision and expansion of an earlier text. It has been used as a primary text for graduate courses in queueing theory in both Electrical Engineering and Operations Research departments. There is more than enough material for a one-semester course, and it can easily be used as the primary text for a two-semester course if supplemented by a small number of current journal articles.

Our goals are directed towards the development of an intuitive understanding of how queueing systems work and building the mathematical tools needed to formulate and solve problems in the most elementary setting possible. Numerous examples are included and exercises are provided with these goals in mind. These exercises are placed within the text so that they can be discussed at the appropriate time.

The instructor can easily vary the pace of the course according to the characteristics of individual classes. For example, the instructor can increase the pace by assigning virtually every exercise as homework, testing often, and covering topics from the literature in detail. The pace can be decreased to virtually any desired level by discussing the solutions to the exercises during the lecture periods. I have worked mostly with graduate students and have found that we achieve more in a course when the students work exercises on the blackboard during the lecture period. This tends to generate discussions that draw the students in and bring the material to life.

The minimum prerequisite for this course is an understanding of calculus and linear algebra. However, we have achieved much better results when the students have had at least an introductory course in probability. The best results have been obtained when the students have had a traditional electrical engineering background, including transform theory, an introductory course in stochastic processes, and a course in computer communications.

We now present an abbreviated summary of the technical content of this book. In Chapter 1, we introduce some general terminology from queueing systems and some elementary concepts and terminology from the general theory of stochastic processes, which will be useful in our study of queueing systems. The waiting time process for a single-server, first-come-first-serve (FCFS) queueing system, is discussed. We also demonstrate the application of queueing analysis to the design of wireless communication systems and IP switches. In the process, we demonstrate the importance of choosing queueing models that are sufficiently rich to capture the important properties of the problem under study.

In Chapter 2, we review some of the key results from the theory of random processes that are needed in the study of queueing systems. In the first section, we provide a brief review of probability. We begin with a definition of the elements of a statistical experiment and conclude with a discussion of computing event probabilities via conditioning. We then discuss random variables, their distributions, and manipulation of distributions. In the third and fourth sections, we develop some of the key properties of the exponential distribution and the Poisson process. In the fifth section, we review discrete- and continuous-parameter Markov chains defined on the nonnegative integers. Our goal is to review and reinforce a subset of the ideas and principles from the theory of stochastic processes that is needed for understanding queueing systems. As an example, we review in detail the relationship between discrete-time and discrete-parameter stochastic processes, which is very important to the understanding of queueing theory but often ignored in courses on stochastic processes. Similarly, the relationship between frequency-averaged and time-averaged probabilities is addressed in detail in Chapter 2.

In Chapter 3, we explore the analysis of several queueing models that are characterized as discrete-valued, continuous-time Markov chains (CTMCs). That is, the queueing systems examined Chapter 3 have a countable state space, and the dwell times in each state are drawn from exponential distributions whose parameters are possibly state-dependent. We begin by examining the well known M/M/1 queueing system, which has Poisson arrivals and identically distributed exponential service times. For this model, we consider both the time-dependent and equilibrium occupancy distributions, the stochastic equilibrium sojourn and waiting time distributions, and the stochastic equilibrium distribution of the length of the busy period. Several related processes, including the departure process, are introduced, and these are used to obtain equilibrium occupancy distributions for simple networks of queues.

After discussing the M/M/1 system, we consider the time-dependent behavior of finite-state general birth-death models. A reasonably complete derivation based upon classical methods is presented, and the rate of convergence of the system to stochastic equilibrium is discussed. Additionally, the process of randomization, or equivalently uniformization, is introduced. Randomization is described in general terms, and an example that illustrates its application is provided. We also discuss the balance equation approach to formulating equilibrium state probability equations for birth-death processes and other more general processes. Elementary traffic engineering models are introduced and blocking probabilities for these systems are discussed. Finally, we introduce the probability generating function technique for solving balance equations.

In Chapter 4, we continue our analysis of queueing models that are characterized by CTMCs. We discuss simple networks of exponential service stations of the feedforward, open, and closed varieties. We discuss the form of the joint state probability mass functions for such systems, which are of the so-called product form type. We discuss in detail a novel technique, due to Gordon [1990], for obtaining the normalizing constant for simple closed queueing networks in closed form. This technique makes use of generating functions and contour integration, which are so familiar to many engineers.

Next, we address the solution of a two-dimensional queueing model in which both the arrival and service rates are determined by the state of a single independent CTMC. This type of two-dimensional Markov chain is called a quasi-birth and death process (QBD), which is a vector version of the scalar birth-death process discussed previously. A number of techniques for solving such problems are developed. The first approach discussed uses the probability generating function approach. We make extensive use of eigenvector-based analysis to resolve unknown probabilities. Next, the matrix analytic technique is introduced and used to solve for the state probabilities. A technique based on solving eigensystems for finding the rate matrix of the matrix geometric method, which reveals the entire solution, is discussed next. Finally, a generalized state space approach, which seems to have been introduced first by Akar et. al [1998], is developed. We show how this technique can be used efficiently to obtain the rate matrix, thereby complementing the matrix analytic approach. We then introduce distributions of the phase (PH) type, and we provide the equilibrium occupancy distribution for the M/PH/1 system in matrix geometric form. We conclude the chapter with a set of supplementary exercises.

In Chapter 5, we introduce the M/G/1 queueing system. We begin with a classical development of the Pollaczek-Khintchine transform equation for the occupancy distribution. We also develop the Laplace-Stieltjes transforms for the ergodic waiting time, sojourn time, and busy period distributions.

We next address inversion of probability generating functions. Three methods are discussed. The first method is based upon Fourier analysis , the second approach is recursive, and the third approach is based on generalized state space methods, which were used earlier to determine the equilibrium probabilities for QBD processes. A number of practical issues regarding a variety of approximations are addressed using the generalized state space approach. For example, in the case of systems having deterministic service time, we obtain queue length distributions subject to batch arrivals for the cases where batch sizes are binomially distributed. We explore convergence of the queue length distribution to that of the M/D/1 system. We also explore the usefulness of the Pade approximation to deterministic service in a variety of contexts.

We next turn our attention to the direct computation of average waiting and sojourn times for the M/G/1 queueing system. Our development follows that for the M/M/1 system to the point at which the consequences of not having the Markovian property surfaces. At this point, a little renewal theory is introduced so that the analysis can be completed. Additional insight into the properties of the M/G/1 system are also introduced at this point. Following completion of the waiting- and sojourn-time development, we introduce alternating renewal theory and use a basic result of alternating renewal theory to compute the average length of the M/G/1 busy period directly. The results of this section play a key role in the analysis of queueing systems with priority, which we address in Chapter 6.

We begin Chapter 6 with an analysis of the M/G/1 system having the last come first serve service discipline. We show that the Pollaczek-Khintchine transform equations for the waiting and sojourn times can be expressed as geometrically weighted sums of random variables. Next, we analyze the M/G/1 queueing system with exceptional first service. We begin our development by deriving the Pollaczek-Khintchine transform equation of the occupancy distribution using the same argument by which Fuhrmann-Cooper decomposition was derived. This approach avoids the difficulties of writing and solving difference equations. We then use decomposition techniques liberally in the remainder of the chapter to study the M/G/1 queueing system with externally assigned priorities and head-of-the-line service. Transform equations are developed for the occupancy, waiting-time and sojourn-time distributions. Inversion of transform equations to obtain occupancy distribution is then discussed.

Finally, we develop expressions for the average waiting and sojourn times for the M/G/1 queueing system under both preemptive and nonpreemptive priority disciplines.

In Chapter 7 we introduce the G/M/1 and M/G/1 paradigms, which have been found to be useful in solving practical problems and have been discussed at length in Neuts' books. These paradigms are natural extensions of the ordinary M/G/1 and G/M/1 systems. In particular, the structure of the one-step transition probability matrices for the embedded Markov chains for these systems are simply matrix versions of the one-step transition probability matrices for the embedded Markov chains of the elementary systems.

In the initial part of the chapter, Markov chains of the M/G/1 and G/M/1 type are defined. The general solution procedure for models of the G/M/1 type and the M/G/1 with simple boundaries are discussed. The application of M/G/1 paradigm ideas to analysis of statistical multiplexing systems is then discussed by way of examples. Then, we extend our earlier development of the generalized state space methods to the case of the Markov chains of the M/G/1 type with complex boundary conditions. The methodology presented there is relatively new, and we believe our presentation is novel. Because generalized state-space procedures are relatively new, we attempt to provide a thorough introduction and reinforce the concepts through an example. Finally, additional environments where Markov chains of the G/M1 and M/G/1 types surface are discussed and pointers to descriptions of a variety of techniques are given.

We close in Chapter 8 with a brief discussion of a number of nontraditional techniques for gaining insights into the behavior of queueing systems. Among these are asymptotic methods and the statistical envelope approach introduced by Boorstyn and others.

JOHN N. DAIGLE

Acknowledgments

For their many valuable contributions, I am indebted to many people. John Mahoney of Bell Laboratories guided my early study of communications. Sheldon Ross of the University of California-Berkeley introduced me to the queueing theory and taught me how to think about queueing problems. Tom Robbins of Addison Wesley, Jim Meditch of The University of Washington, Ray Pickholtz of The George Washington University, and Bill Tranter of the Virginia Polytechnic Institute and State University initially encouraged me to write this book. Marty Wortman of Texas A&M University taught from early drafts and provided encouragement and criticism for several years. John Spragins of Clemson University and Dave Tipper of The University of Pittsburgh taught from the various draft forms and offered many suggestions for improvement. Discussions with Ralph Disney of Texas A&M University, Bob Cooper of Florida Atlantic University, Jim Meditch, and Paul Schweitzer of the University of Rochester also yielded many improvements in technical content.

Numerous students, most notably Nikhil Jain, John Kobza, Joe Langford, Marcos Magalhães, Naresh Rao, Stan Tang, and Steve Whitehead have asked insightful questions that have resulted in many of the exercises. Mary Jo Zukoski played a key role in developing the solution manual for the first edition.

Nail Akar of Bilikent University contributed much to my education on the generalized state space approach, and N. Cem Oğuz provided invaluable help in setting up and debugging a suitable programming environment for LAPACK routines. Ongoing discussions with Martin Reisslein, David Lucantoni, Ness Shroff, and Jorg Liebeherr have helped to keep me up-to-date on emerging developments. Alex Greene and Mellissa Sullivan of Kluwer Publishers have provided the appropriate encouragement to bring the manuscript to completion.

Finally, I am indebted to Katherine Daigle who read numerous drafts and provided many valuable suggestions to improve organization and clarity. For all errors and flaws in the presentation, I owe thanks only to myself.

Chapter 1

TERMINOLOGY AND EXAMPLES

Samuel Morse invented a telegraphing mechanism in 1837. He later invented a scheme for encoding messages, and then, under contract with the Government of the United States of America (USA), built the first telegraph system in 1843. Immediately, waiting lines began to form by those wanting to use the system. Thus, queueing problems in telecommunications began virtually simultaneously with the advent of electrical telecommunications.

The world-wide telecommunications infrastructure of today consists largely of two interrelated major infrastructures: the telephone network, which is a circuit-switched network, and the Internet, which is a packet-switched computer communications network. But, the lines are blurred; control of circuit switched systems has been accomplished using packet switching for almost three decades, and packet switching systems transport information over lines derived from circuit switching systems.

Today queueing theory is used extensively to address myriads of questions about quality of service, which has been a major concern of telecommunications systems from the beginning. Quality is measured in a variety of ways, including the delay in gaining access to a system itself, the time required to gain access to information, the amount of information lost, and the intelligibility of a voice signal. Usually the quantities in question are random variables and results are specified in terms of averages or distributions. A fundamental issue is the resource-quality trade-off; what quantities of resources, measured, say, in dollars, must be provided in order achieve a desired quality of service?

For at least three decades, there has been a trend towards ubiquitous service over packet-switched facilities, the primary motivation being cost reduction resulting from an increased capability to share resources. The primary obstacle has been the development of mechanisms that assure quality of service at a competitive costs. Queueing theory is one of the primary tools used to deal with questions involving trade-offs between the amount of resources allocated to provide a telecommunications service and the quality of service that will be experienced by the subscribers.

This chapter has three sections. In the first section, we present an overview of the terminology of queueing systems. Mathematical notation will be presented, but mathematical developments are deferred to later chapters. In the second section, we discuss a number of applications of queueing theory to system design. The primary objective is to provide the reader with basic information that can form the basis of thought about how queuing theory can be applied to telecommunications problems. The chapter concludes with a brief summary.

1.1 The Terminology of Queueing Systems

In this section, we introduce the reader to the terminology of queueing theory and to some definitions from the theory of stochastic processes that are needed in the study of queueing systems. We introduce some key random processes involved in queueing analysis, formally introduce the notion of an induced queueing process, and define some of the major quantities of interest.

In order to introduce notation and some of the dynamics of queueing systems, we consider the activities surrounding the use of a pay telephone, perhaps in an airport. Here, the telephone system itself is the *server*, and the customers who are waiting to use the telephone form the queue for the system. Figure 1.1 shows a schematic diagram of the queueing system.

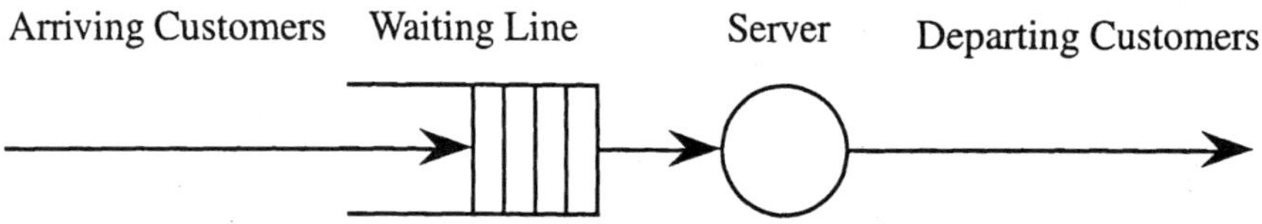

Figure 1.1. Schematic diagram of a single-server queueing system.

Assume that at time zero the telephone is *idle*; that is, no one is using the phone. Now, suppose that at time $\tilde{\tau}_1$ the first customer, whom we shall call C_1, arrives at the telephone and places a call. The system is now in a *busy* state, and the amount of time required to satisfy the customer's needs is dependent upon how many calls the customer makes, how long it takes to set up each call, and how long it takes the customer to conduct the business at hand. Define the total amount of time the telephone system is occupied by this customer as the *service-time requirement*, or simply the *service-time* of C_1 and denote this quantity by $\tilde{x}_1$. Then, C_1 leaves the system at time $\tilde{\tau}_1 + \tilde{x}_1$.

The waiting time, denoted by $\tilde{w}_1$, for C_1 is *zero* and the total time in the system for C_1 is $\tilde{x}_1$. We denote the total time in the system, which is sometimes called the *sojourn time*, by $\tilde{s}_1$. Thus, $\tilde{w}_1 = 0$ and $\tilde{s}_1 = \tilde{x}_1$. Figure 1.2 shows the sequence of events in this case. [1]

Now, suppose C_2 (the second customer) arrives at time $\tilde{\tau}_2$ and has service-time requirement $\tilde{x}_2$. Then C_2 will be ready to depart the system at time $\tilde{\tau}_2 + \tilde{x}_2 + \tilde{w}_2$, where $\tilde{w}_2$ is the amount of time C_2 waits for C_1 to finish using the telephone system; that is, the time between $\tilde{\tau}_2$ and $\tilde{\tau}_1 + \tilde{x}_1$, if any.

Clearly, if C_1 departs before $\tilde{\tau}_2$, then $\tilde{w}_2 = 0$; but if C_1 departs after $\tilde{\tau}_2$, then $\tilde{w}_2 = \tilde{\tau}_1 + \tilde{x}_1 - \tilde{\tau}_2$. Thus, we find $\tilde{w}_2 = \max\{0, \tilde{\tau}_1 + \tilde{x}_1 - \tilde{\tau}_2\}$. If we now define $(a)^+ = \max\{0, a\}$, then we find

$$\tilde{w}_2 = (\tilde{\tau}_1 + \tilde{x}_1 - \tilde{\tau}_2)^+.$$

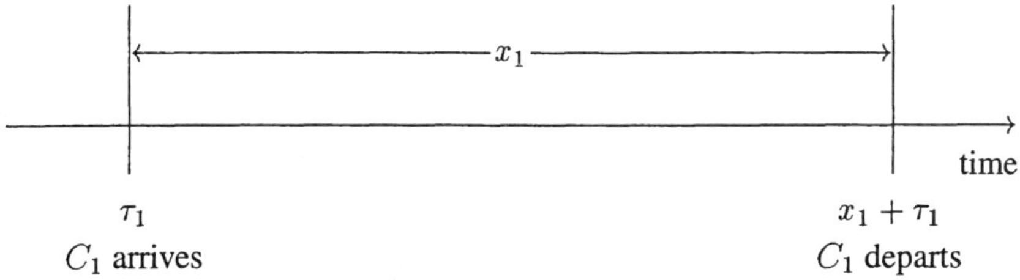

Figure 1.2. Sequence of events for first customer.

In general, the waiting time of the $(n+1)$st customer, C_{n+1}, is equal to the departure time of C_n minus the arrival time of C_{n+1} provided that difference is greater than zero. But, the departure time of C_n is $\tilde{\tau}_n + \tilde{x}_n + \tilde{w}_n$, so,

$$\tilde{w}_{n+1} = (\tilde{\tau}_n + \tilde{x}_n + \tilde{w}_n - \tilde{\tau}_{n+1})^+$$

or, equivalently,

$$\tilde{w}_{n+1} = (\tilde{w}_n + \tilde{x}_n - \tilde{t}_{n+1})^+ \tag{1.1}$$

where $\tilde{t}_{n+1} = \tilde{\tau}_{n+1} - \tilde{\tau}_n$ is called the *interarrival time* for C_{n+1}.

Figure 1.3 gives a graphic description of the sequence of events experienced by the general customer.

We note that $\{\tilde{w}_n, n = 1, 2, \cdots\}$, $\{\tilde{x}_n, n = 1, 2, \cdots\}$, and $\{\tilde{t}_n, n = 1, 2, \cdots\}$ are all discrete-parameter stochastic processes. The distribution of the random variables $\tilde{x}_n$ and $\tilde{t}_n$ may be discrete, continuous, or mixed, depending upon the particular system under study. The complexity of these distributions influences the difficulty of solving a particular problem. For continuity, we remind the reader of the following definition.

[1]Note that random variables are designated by tildes and their values by the same variables without tildes. For example, $\tilde{x}_1$ denotes a random variable and x_1 denotes its value.

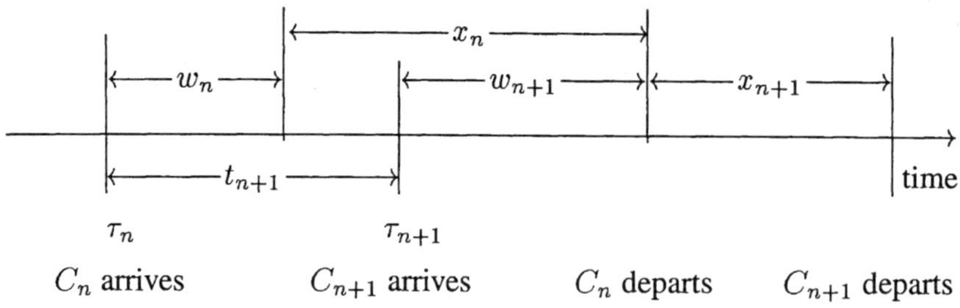

Figure 1.3. Sequence of events for general customer.

DEFINITION 1.1 **Stochastic process** (Ross [1983]) . A stochastic process (SP) $\{\tilde{x}(t), t \in \mathcal{T}\}$ is a collection of random variables, $\tilde{x}(t)$, indexed on t, $t \in \mathcal{T}$. That is, for each $t \in \mathcal{T}$, $\tilde{x}(t)$ is a random variable.

We now turn to a more formal definition of a queueing process. Before proceeding, we need the definitions for statistical independence and common distributions.

DEFINITION 1.2 **Common distribution.** A random variable $\tilde{x}$ having a distribution F means $F_{\tilde{x}}(x) \stackrel{\Delta}{=} P\{\tilde{x} \leq x\}$. If $\tilde{x}_1, \tilde{x}_2, \cdots$ have a common distribution F, then

$$P\{\tilde{x}_1 \leq x\} = P\{\tilde{x}_2 \leq x\} = \cdots P\{\tilde{x}_n \leq x\} \cdots = F_{\tilde{x}}(x).$$

That is, the random variables, $\tilde{x}_1$, $\tilde{x}_2$, ... all have the same distribution, which is $F_{\tilde{x}}(x)$.

DEFINITION 1.3 **Statistical independence.** A set, $\tilde{x}_1, \tilde{x}_2, \cdots, \tilde{x}_n$, of random variables are said to be statistically independent, or simply independent, if for all real $x_1, x_2, \cdots, x_n$,

$$P\{\tilde{x}_1 \leq x_1, \cdots, \tilde{x}_n \leq x_n\} = P\{\tilde{x}_1 \leq x_1\} \cdots P\{\tilde{x}_n \leq x_n\}.$$

A sequence of random variables $\{\tilde{x}_i, i = 1, 2, \cdots\}$ is said to be an independent sequence if every finite collection from the sequence is independent. In either case, the random variables are also said to be mutually independent.

DEFINITION 1.4 **Induced queueing process** (Feller [1971], pp. 194-195). Let $\tilde{z}_1, \tilde{z}_2, \cdots$ be mutually independent random variables with common distribution F. Then the induced queueing process is the sequence of random variables $\tilde{w}_0, \tilde{w}_1, \cdots$ defined recursively by $\tilde{w}_0 = 0, \tilde{z}_0 = 0$ and

$$\tilde{w}_{n+1} = (\tilde{w}_n + \tilde{z}_n)^+. \tag{1.2}$$

By analogy with the process defined by (1.1), we see that

$$\tilde{z}_n = \tilde{x}_n - \tilde{t}_{n+1}. \tag{1.3}$$

Intuitively, one might argue that the difference between the *service-time* of the nth customer and the *interarrival time* of the $(n+1)$st customer induces a delay for the customers that follow. If the difference is positive, the effect is to increase the waiting times of the customers that follow; if the difference is negative, the waiting times of the customers that follow tends to be decreased.

Now, suppose that for every realization of the queueing process and for every n, it turns out that $x_n < t_{n+1}$. Then there would never be any customers waiting because C_n would have completed service before C_{n+1} arrived for every n. On the other hand, if $x_n > t_{n+1}$ for every n, then the server would get further behind on every customer. Thus, the waiting time would build to infinity as time increased beyond bound. But, in the general case, for a given value of n, $\tilde{z}_n = \tilde{x}_n - \tilde{t}_{n+1}$ may be negative, zero, or positive, and $\tilde{u}_n$ is a measure of the *elbow room*.

We note that it is sometimes, but not usually, convenient to work with (1.1) when solving a queueing problem for reasons that will be considered later. The reader is referred to Ackroyd [1980] for a description of a method dealing directly with (1.1) and to Akar [2004] for a modern treatment. More often than not, however, initial results are obtained in terms of queue length distributions, and other results are derived from the results of the queue length analysis. In Section 1.2, we present examples in which the dynamical equations that are solved are expressed directly in terms of queue lengths.

We now introduce the concept of unfinished work. This is a continuous-time, continuous-valued stochastic process that is sometimes extremely useful in the analysis of queueing systems operating under complicated service disciplines such as those employing service priority.

DEFINITION 1.5 **Unfinished work.** Let $\tilde{u}(t)$ denote the amount of time it would take the server to empty the system starting at time t if no new arrivals occur after time t. Then $\tilde{u}(t)$, which excludes any arrival that might occur at time t, is called the *unfinished work.* The unfinished work is, then, a measure of the server's backlog at time t, $t \geq 0$.

Sometimes $\tilde{u}(t)$ is called the *virtual waiting time* because $\tilde{u}(t)$ is the length of time a customer would have to wait in a first-come-first-serve (FCFS) queueing system if the customer arrived at time t. A typical realization for $\tilde{u}(t)$ is shown in Figure 1.4. Completing Exercise 1.1 will help the reader to understand the concept more fully and to see the relationship between waiting time and unfinished work.

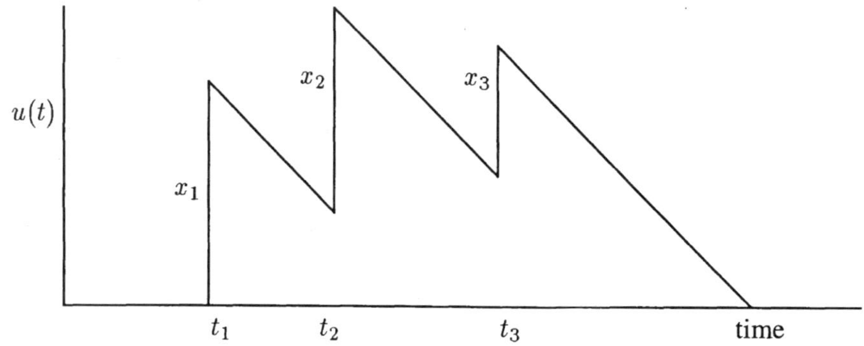

Figure 1.4. Typical realization for unfinished work.

EXERCISE 1.1 Assume values of $\tilde{x}$ and $\tilde{t}$ are drawn from truncated geometric distributions. In particular, let $P\{\tilde{x} = n\} = 0$ and $P\{\tilde{t} = n\} = 0$ except for $1 \leq n \leq 10$, and let $P\{\tilde{x} = n\} = \alpha p_x(1 - p_x)^n$ and $P\{\tilde{t} = n\} = \beta p_t(1 - p_t)^n$ for $1 \leq n \leq 10$ with $p_x = 0.292578$ and $p_t = 0.14358$.

1. Using your favorite programming language or a spreadsheet, generate a sequence of 100 random variates each for $\tilde{x}$ and $\tilde{t}$.

2. Plot $\tilde{z}(t)$ as a function of t, compute $\tilde{z}_n$ from (1.3) and $\tilde{w}_n$ from (1.2) for $1 \leq n \leq 20$.

3. Compute $\tilde{\tau}_n$ for $1 \leq n \leq 20$ and verify that $\tilde{w}_n$ can be obtained from $\tilde{z}(t)$.

4. Compute $\tilde{z}_n$ from (1.3) and $\tilde{w}_n$ from (1.2) for $1 \leq n \leq 100$ and compute the average waiting times for the 100 customers.

DEFINITION 1.6 **Busy period.** With reference to Figure 1.4, it is seen that the unfinished work is 0 prior to t_1 and that the level of unfinished work returns to zero after customer C_3 is served. The period of time between a transition from zero to a positive level of unfinished work and a transition from a positive to zero level of unfinished work is called a *busy period*. The sequence of busy periods is a stochastic process, which is usually denoted by $\{\tilde{y}_i, i = 0, 1, \ldots\}$.

EXERCISE 1.2 Assume values of $\tilde{x}$ and $\tilde{t}$ are drawn from truncated geometric distributions as given in Exercise 1.1.

1. Using the data obtained in Exercise 1.1, determine the lengths of all busy periods that occur during the interval.

2. Determine the average length of the busy period.

3. Compare the average length of the busy period obtained in the previous step to the average waiting time computed in Exercise 1.1. Based on the results of this comparison, speculate about whether or not the average length of the busy period and the average waiting time are related.

In general, queueing systems are classified according to their properties. Some of these properties are now given:

1. The form of the interarrival distribution $F_{\tilde{t}}(t) \stackrel{\Delta}{=} P\{\tilde{t} \leq t\}$ where $\tilde{t}$ represents a generic $\tilde{t}_j$;

2. The form of the service-time distribution $F_{\tilde{x}}(x) \stackrel{\Delta}{=} P\{\tilde{x} \leq x\}$ where $\tilde{x}$ represents a generic $\tilde{x}_j$;

3. The number of arrivals in a batch;

4. The number of servers;

5. The service discipline - the order in which service is rendered, the manner in which service is rendered (time shared, etc.), whether the system has priority;

6. The number of customers allowed to wait;

7. The number of customers in the population (usually denoted only if the population is finite).

A queueing system is usually described using a shorthand notation (due to D.G. Kendall) of the form G/G/s/K. In this notation, the first G denotes the form of the interarrival time distribution, the second G denotes the form of the service-time distribution, the value of s denotes the number of servers, and the value of K denotes the number of customers allowed to wait. Sometimes the notation GI is used in place of G to emphasize independence, as, for example, in the notation GI/M/1/K to denote the queueing system having general and independent interarrivals, a single exponential server, and a finite waiting room of capacity K. [2]

Remark. The induced queueing process is defined in terms of an independent sequence of random variables, $\{\tilde{z}_1, \tilde{z}_2, \cdots\}$. In many cases, analysis of a simple queueing model based on that assumption is sufficient to address a system design question. But, most often a model that captures some aspects of dependence among the system's random variables is needed to gain an understanding of an issue. Indeed, the objective of an analysis is often to explain such dependence. A significant portion of this text is devoted to the topic of developing specialized models that capture key properties of real systems. It is also true that (1.2) holds whether or not $\{\tilde{z}_1, \tilde{z}_2, \cdots\}$ is an independent sequence. However, the difficulty of the solving (1.2) or an alternate formulation is certainly dependent upon whether or not that sequence is independent.

Some of the quantities of interest in the study of queueing systems include the waiting-time distribution, the system-time distribution, the distribution of number of customers in the system, the probability that the server is busy (idle), the distribution of the length of a busy period, the distribution of the number of customers served during a busy period, averages for waiting-time, time in system, number in system, and the number served in busy period.

Remark. For a particular problem, all of these quantities are not necessarily of interest in themselves, but they are useful tools through which other more interesting quantities can be determined. For example, busy period analysis is a useful tool in the study of priority queueing systems, as we shall see later.

Remark. It's easy to state queueing problems that defy analysis, and it's easy to mistake one queueing problem for another. The reader is encouraged to think very carefully and rigorously before settling on assumptions and before using off-the-shelf results of questionable relevance. It is equally important to take special care not to define a queueing model that is overly complicated for a given application; the specific question being addressed should constantly be kept in mind when the analytical model is defined.

Remark. Usually, it is not feasible, and sometimes it is impossible, to obtain an accurate description of a system under study. Thus, the specific numerical results from a queueing analysis, in and of themselves, are not usually very useful. The useful part of a queueing analysis usually derives from the analysts' ability to determine trends and sensitivities. For example, "Does the system degrade gradually or catastrophically as load is increased?" The result

[2] We note that an exponential random variable has the distribution $F_{\tilde{x}}(x) = 1 - e^{-\mu x}$ where μ is called the rate parameter. The exponential distribution and its properties will be discussed in detail in Chapter 2.

is that a substantial factor in the value of a queueing analysis is the care taken to define the problem.

1.2 Examples of Application to System Design

In this section, we present three examples that illustrate the application of queueing theory to practical problems in the design of telecommunications systems. Each example is covered in a subsection. Our examples present only the problem and its solution, the method of solution being a topic for discussion in later chapters. Pointers to sections where solution methodologies are discussed within the text are given in each subsection.

The first example applies concepts from classical traffic engineering to the problem of designing cellular telephone systems. The specific example given addresses analog cellular systems, but the same problems exist is both time division multiple access (TDMA) and code division multiple access (CDMA) -based cellular systems, and they are addressed in the same way as described in our example.

The second example is related to the design of modern IP switching systems. At issue is the impact that correlation in the arrival process has in the backlog at the output ports of the switch. The backlog is related to the delay that will be experienced at the output port of the switch, which may be an important component of the total delay experienced as traffic traverses the switch.

The third example considers the backlog at the intersection of the traditional Internet and a high data rate cellular data transmission system. The primary feature considered in this example is the variability in the service capacity of the forward wireless link, which is due to variation in path loss and fading as the mobile travels around within the coverage area.

1.2.1 Cellular Telephony

In an analog cellular communication system, there are a total of 832 available frequencies, or channels. These are typically divided between two service vendors so that each vendor has 416 channels. Of these 416 channels, 21 are set aside for signalling. A cellular system is tessellated, meaning that the channels are shared among a number of cells, typically seven. Thus, each cell has about 56 channels. In order to get a feel for where cell cites should be placed, the vendor would like to estimate the call blocking probability as a function of the total population of customers using the system.

The call blocking probability is defined as follows. Suppose a customer would like to make a call. The customer enters the number and attempts the call. If the systems responds that no service is available at the time, then the call attempt is said to be *blocked.* The ratio of the total number of call attempts blocked to the total number of calls attempted by all customers over a given period of time is defined as the call blocking ratio or call blocking probability.

DEFINITION 1.7 **Frequency-averaged metric**. Suppose a probability is defined as the limiting proportion of the number of occurrences of a specific event to the total number of occurrences of an event of which the former event is a subset. Then that probability is said to be a *frequency-averaged metric* or *frequency-averaged probability*.

Remark. Blocking probability is a frequency-averaged probability. Under certain conditions, this frequency-averaged metric is equivalent to a time-averaged probability. However, too frequently in the literature, a time-averaged metric is incorrectly reported as a blocking probability. Fortunately, it is usually straightforward to convert to a frequency averaged probability from a time-averaged probability. Frequency and time-average probabilities are discussed in Chapter 3.

There are a number of important issues involved in completing the problem definition. Obvious questions are "How many calls does a typical customer make?" and "What is the duration of a call?"

Each of these questions might be answered by specifying the distribution of a random variable, that is, by providing the distribution of the number of calls made by a typical customer during the busiest hour of the busiest day of the week and the distribution of the length of a call of a typical user during that busiest period. From such distributions, elementary parameters of the system such as the average call generation rate per customer and the average holding time per call can be estimated. Alternatively, these parameters could be estimated directly.

Engineering of a system is virtually always based on traffic loads placed on the system during the busiest times, which is frequently referred to as *the busy hour*. Define λ_C to be the call generation rate per customer during the busy hour, $E[\tilde{x}]$ to be the average call holding time during the busy hour, and $\rho_C = \lambda_C E[\tilde{x}]$, which quantity represents the utilization per channel per customer during the busy hour. For example, if a typical customer attempts an average of two calls per hour during the busy hour and average call holding time is 3 minutes, then $\rho_C = 0.1$.

This information, together with a few additional assumptions is sufficient for obtaining a first cut at the blocking probability. In particular, we assume that the sequence of interarrival times is a sequence of independent, identically distributed exponential random variables, and the sequence of holding times is a sequence of independent, identically distributed random variables having an arbitrary distribution.

The results are show in Figure 1.5. Typically, a system is designed so that the blocking probability meets an objective of 0.01 or less. From the graph, it is seen that the objective would be met if the population is less than about 430. But, we also notice that the blocking probability increases very quickly with increased population in the neighborhood of a 0.01 blocking probability. Indeed, the blocking probability increases to 0.1 with a population increase to only 560. This raises a number of *sensitivity issues*, such as the effect of changes in average call holding time on the blocking probability. Machinery for dealing with these types of problems will be developed in Chapter 3.

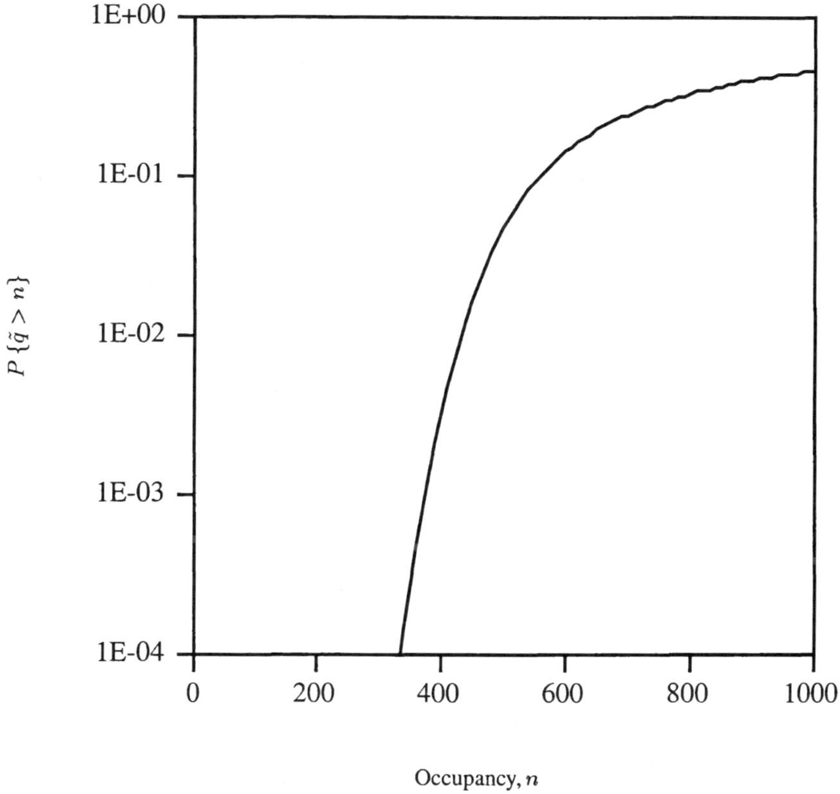

Figure 1.5. Blocking probability as a function of population size at a load of $\rho_C = 0.1$.

1.2.2 Multiplexing Packets at a Switch

As data traverses the Internet, it is multiplexed onto and demultiplexed from data communications lines at a number of switches, the interconnection of which forms an end-to-end path. Queues form at many points along the path, but this example focusses on the queue that forms at the communication lines that are attached to the output ports of a switch.

In general, a switch has N input and output ports at which input and output communications lines are connected. Upon arrival to the input processor of a switch, packets are usually partitioned into fixed-length data blocks, and it is these data blocks that are actually switched. We wish to determine the effect of N upon the queue length distribution at a typical output line at a given traffic load. We also wish to know if the queue length is affected by the form of the arrival process.

An elementary abstraction of the queueing problem is as follows. We suppose that our system is time-slot oriented, where one time slot is the time required for one packet to enter or leave the switch on each communication line. Since the output port of the switch serves N incoming lines, as many as N packets destined to a particular output port may arrive to the switch during one time slot. But, only one packet may depart from the switch during any given time slot. Therefore a queue forms. For the present, we assume an infinite buffer size so that the queue may grow without bound.

Define $\tilde{q}_k$ to be the number of units in the queue at the end of the kth slot, $k \in \{0, 1, \ldots\}$. Then, $\{\tilde{q}_k, k = 0, 1, \ldots\}$ is a discrete valued, discrete parameter stochastic process. Later in the book, it will be shown that time-slot oriented queueing systems behave according to the following dynamical equation:

$$\tilde{q}_{k+1} = (\tilde{q}_k - 1)^+ + \tilde{v}_{k+1}, \tag{1.4}$$

where $\tilde{v}_k$ denotes the number of items that arrive (are added to the queue) during the kth slot.

The sequence $\{\tilde{v}_k, k = 0, 1, \ldots\}$, which is the arrival process, is, itself, a discrete valued, discrete parameter, stochastic process. The degree of difficulty in solving the queueing equation, in fact, depends upon the complexity of the arrival process.

Under an appropriate system load, the distribution of the random variable $\tilde{q}_k$, which we denote by $F_{\tilde{q}_k}(\cdot)$ converges to an equilibrium distribution, which we will denote by $F_{\tilde{q}}(\cdot)$; that is, $F_{\tilde{q}}(x) = \lim_{k\to\infty} F_{\tilde{q}_k}(x)$. We wish to determine $F_{\tilde{q}}(x)$ under a given set of conditions.

With respect to the arrival process, we wish to consider two cases. In the first case, we take the simplest possible assumption for the arrival process; in each time slot, each incoming line, independent of everything, has a packet destined for the target output line with a fixed probability p. This assumption then leads to the fact that $\{\tilde{v}_k, k = 0, 1, \ldots\}$ is a sequence of independent, identically distributed binomial random variables with parameters N and p. Further, we set the value of p such that, on average, a packet is transmitted on the target output line in 90% of the time slots. Thus, $p = 0.9/N$.

In the second case, we assume the arrival processes due to different lines are independent, but the arrival steam from any given line is correlated. In particular, on each given line, packets come in bursts so that there is a run of slots having packets followed by a run of slots not having packets. Such an arrival process is called an *on-off* process. In the simplest case, if the process is in the *on* state during a slot, a packet arrives to the system, else no packet arrives. At the end of each time slot, given that the system is in the on state, the system transitions back into the on state with probability p_{11} or into the off state with probability $p_{10} = 1 - p_{11}$. Similarly, given that the system is in the off state, the system transitions back into the off state with probability p_{00} and into the on state with probability $p_{01} = 1 - p_{00}$. The length of a run then has the geometric distribution with parameter p_{10}; that is, the probability that the run length is n is $p_{10}p_{11}^{n-1}$.

Again, we take the simplest possible nontrivial case wherein the run lengths on all of the incoming lines have identical geometric distributions. To obtain the desired utilization, the proportion of slots having packets is set to $p = 0.9/N$ as before. In this case, it turns out that the process $\{\tilde{v}_k, k = 0, 1, \ldots\}$ is a discrete-valued, discrete parameter Markov chain, which will be discussed later.

EXERCISE 1.3 For the general case where the packet arrival process has run lengths, it will be seen that the survivor functions decrease with decreasing run lengths. Determine whether or not there exists an average run length at which the packet arrival process becomes a sequence of independent Bernoulli trials. If such a choice is possible, find the value of run length at which independence occurs. Discuss the result of reducing the run length below that point. [Hint: A packet arrival occurs whenever the system transitions into the on state. Thus, if the probability of transitioning into the on state is independent of the current state, the arrival process becomes a sequence of independent Bernoulli trials.]

Figure 1.6 shows the survivor functions that result with $N = 4$, $N = 16$, and $N = 64$ for the case of independent arrivals. From Figure 1.6, it can be seen that the number of multiplexed lines does have some effect upon the queue length distribution. For example, the probability that the queue length exceeds 30 packets is about 1.7×10^{-3} with $N = 64$, but only about 2.6×10^{-4} with $N = 4$. From the graph it is also seen that the change in the queue length distribution decreases as N increases. For example, the change from $N = 4$ to $N = 16$ is much larger than the change from $N = 16$ to $N = 64$.

Figure 1.7 shows the effect of changes of average run length on the survivor function with the number of input lines held constant at 8. From Figure 1.7 it is readily seen that the queue length distribution is fairly sensitive to run length even for modest values. In fact, from the data used to plot this figure it can be found that the probability that the queue length exceeds 40 packets increases

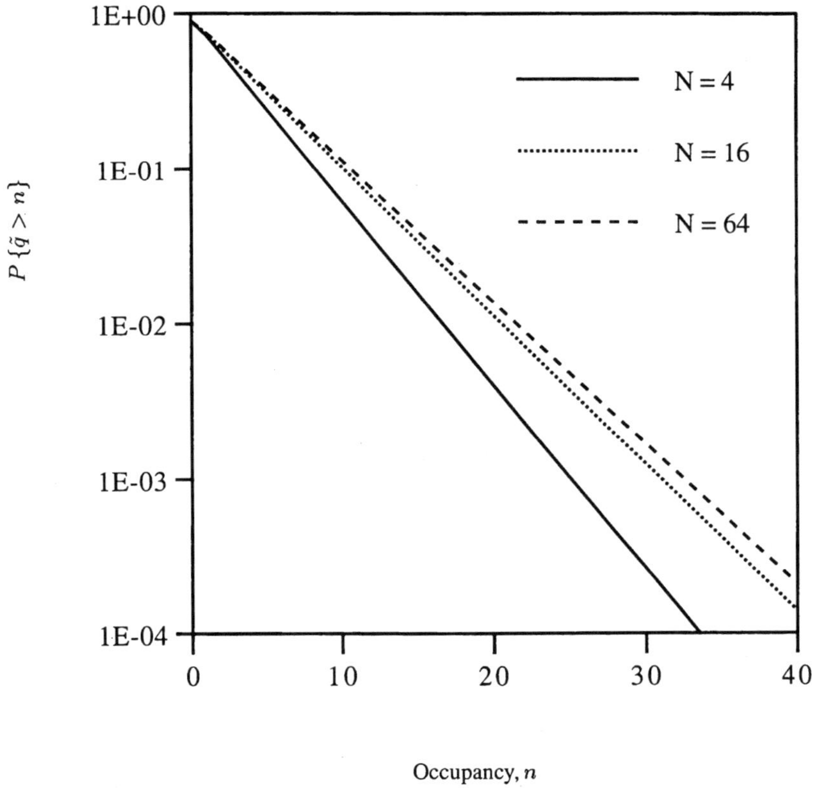

Figure 1.6. Queue length survivor function for an N-to-1 multiplexing system at a traffic intensity of 0.9 with N as a parameter and with independent, identically distributed arrivals.

from about 7.6×10^{-05} for independent arrivals to about 1.7×10^{-03} at an average run length of 1.4 to about 2.3×10^{-02} at an average run length of 2.0. These increases in the probability of exceeding 40 are factors of 22 and 300 at run lengths of 1.4 and 2.0, respectively. From this it is clear that the form of the arrival processes can have a significant effect upon queueing within a system.

We will develop modeling machinery to produce curves such as those shown in Figures 1.6 and 1.7 in Chapters 5 and 7.

1.2.3 CDMA-Based Cellular Data

High data rate transmission based on frame-oriented time division multiplexing has been proposed as a paradigm for forward-link transmission in CMDA-based cellular systems (Bender [2000]). In such a system, the capac-

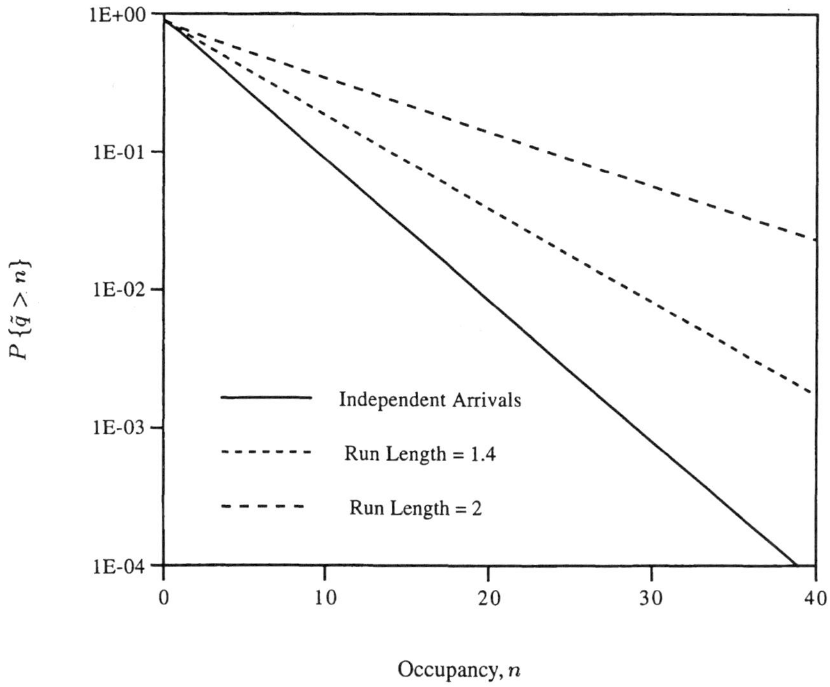

Figure 1.7. Queue length survivor function for an 8-to-1 multiplexing system at a traffic intensity of 0.9 with average run length as a parameter.

ity of a frame depends upon the signal plus noise to interference ratio (SINR) of the target mobile receiver, which is dependent upon the path losses between all transmitting cell sites and the target receiver as well as fading conditions. Available capacities range from one data block of 1024 bits to sixteen data blocks at 4096 bits, the latter of which is equivalent to 64 blocks at 1024 bits.

Similar to the previous section, define $\tilde{q}_k$ to the number of units in the queue at the end of the kth frame, $k\,\{0, 1, \ldots\}$. Then, again, $\{\tilde{q}_k, k = 0, 1, \ldots\}$ is a discrete valued, discrete parameter stochastic process.

Later in the book, it will be shown that frame-oriented queueing systems generally behave according to the following dynamical equation:

$$\tilde{q}_{k+1} = (\tilde{q}_k - \tilde{c}_{k+1})^+ + \tilde{v}_{k+1}, \tag{1.5}$$

where $\tilde{c}_k$ denotes the number of items served (removed from the queue) during the kth frame and $\tilde{v}_k$ denotes the number of items that arrive (are added to the queue) during the kth frame.

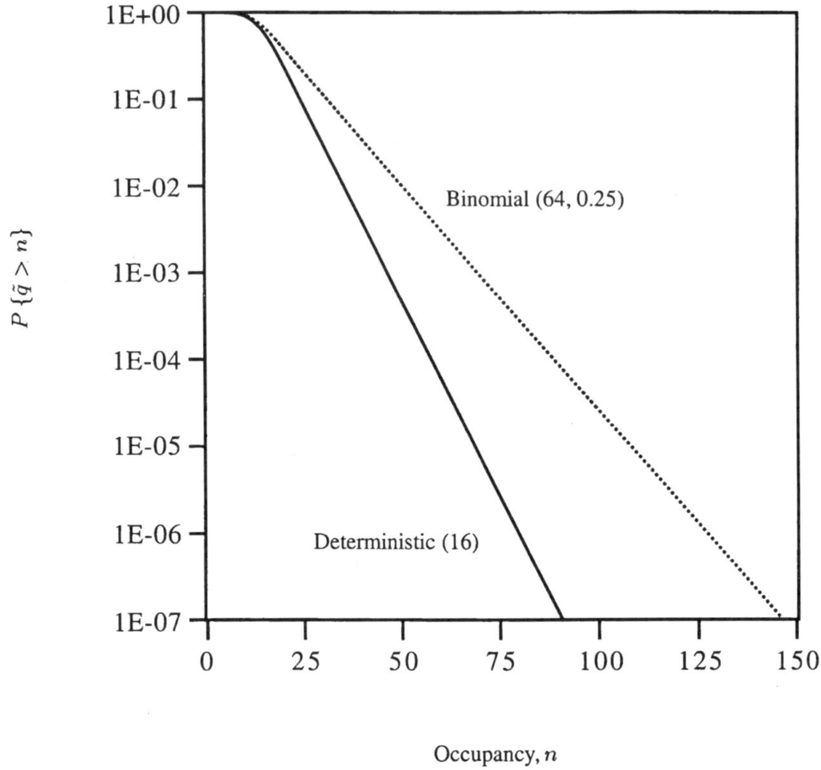

Figure 1.8. Comparison between a system serving a fixed number of 16 units per frame and a system serving a binomial number of units with an average of 16 at a traffic intensity of 0.9.

One way to begin to understand the queueing behavior of the actual forward link is to perform an analysis that compares a system that serves a random number of units per frame to a system that serves a fixed number of units per frame. As an example, we might compare the case where the random process $\{\tilde{c}_k, k = 0, 1, \ldots\}$ is a sequence of independent, identically distributed binomial random variables with mean 16 to the case where $\tilde{c}_k = 16$ with probability 1 for all $k \in \{1, 2, \ldots\}$. We can think of the system as having *bulk* or *batch* services, where $\tilde{c}_k$ denotes the size of the batch served during frame k.

As before, we denote the equilibrium queue length distribution by $F_{\tilde{q}}(x) = \lim_{k\to\infty} F_{\tilde{q}_k}(x)$.

Figure 1.8 shows survivor functions for the binomially and deterministically distributed batch sizes. From this figure, it is quite obvious that there is a significant difference between the queueing behavior of the two systems. For example, for the random batch size case, the probability that there will be more than 75 packets in the queue is approximately 5×10^{-4} while the probability that the queue size exceeds 75 in the deterministic batch-size case is only about 2.5×10^{-6}. Thus, it is about 200 times more likely to find a queue length exceeding 75 in the case of binomially distributed batch sizes with a mean of 16 than it is in the case of a system serving fixed batches of size 16.

1.3 Summary

In this chapter we have provided a brief introduction to the language of queueing theory, and we have given a number of examples that illustrate the application of queueing theory to system design. Although we have avoided mathematical development, we have introduced some key concepts that are useful in understanding the nature of queueing problems.

From the examples in this chapter, it is quite clear that correlation in the arrival process and variability in the service process can have a significant impact upon the performance of a system and, consequently, on the quality of service delivered by a system. Failure to recognize this fact can lead to an erroneous prediction of the amount of resources required to support a service at a desired level of quality.

A significant proportion of this text is devoted to developing the machinery required to develop solid problem definitions, understand the current literature on applied queueing systems, and formulate new approaches for solving queueing problems that may arise in the design of real systems.

Chapter 2

REVIEW OF RANDOM PROCESSES

In this chapter, we review some of the key results from the theory of random processes that are needed in the study of queueing systems. In the first section, we provide a brief review of probability. We begin with a definition of the elements of a statistical experiment and conclude with a discussion of computing event probabilities via conditioning. In the second section, we discuss random variables, their distributions, and manipulation of distributions. In the third and fourth sections, we discuss the exponential distribution and the Poisson process, respectively, which play a key role in queueing analysis, and we develop some of their key properties. In the fifth section, we provide a brief review of discrete and continuous parameter Markov chains defined on the nonnegative integers.

While the materials presented here are, for the most part, self-contained and a mastery of the materials presented here would provide an adequate basis for understanding queueing systems, our experience is that these materials cannot be used as a substitute for good courses on probability theory and random processes. Rather, our presentation is intended primarily as review and reinforcement of a subset of the ideas and principles from probability theory that are useful in understanding queueing systems. As as example, in courses on stochastic processes the distinction between discrete time and discrete parameter stochastic processes is often mentioned briefly and then ignored. But, in the study of queueing systems, this difference is significant, and we reinforce that fact herein. Similarly, the relationship between frequency-averaged probabilities and time-averaged probabilities is addressed in detail at the end of this chapter.

2.1 Statistical Experiments and Probability

Possibly the most difficult aspects of any applied probability problems are to properly formulate the problem and to properly specify the parameters of the problem. In the case of queueing problems, this often requires very careful definitions of statistical experiments and manipulation of the laws of probability. In some cases, experiments are very complicated and direct computation of event probabilities is very difficult. In such cases, it is often helpful to compute certain event probabilities by conditioning on the occurrence of other events, which is really a process of breaking the experiment down into a set of more easily understood sub-experiments. In the first subsection, we discuss statistical experiments and their properties and in the second subsection we discuss computation of probabilities via conditioning.

2.1.1 Statistical Experiments

In this section, we define the properties of a statistical experiment, introduce the laws of probability, and show how the laws of probability are used to compute event probabilities.

DEFINITION 2.1 **Statistical experiment** A statistical experiment is an experiment whose outcome is not known in advance. A statistical experiment has three major characteristics:

1. The **sample space**, which is the set of all possible outcomes of the experiment. The sample space is denoted by $\mathcal{S}$.

2. The **event space**, which is the set of all possible subsets of the sample space, an event being defined as any specific subset of the sample space. The event space is denoted by Ω.

3. The **probability measure of the events**. In general, to each $\omega \in \Omega$, we assign a number $P\{\omega\}$, which represents the probability that the event ω occurs.

Note that the sample space of an experiment contains *all* of its possible outcomes. Exactly one element of the sample space results whenever the experiment is conducted.

Note also that an event is a set. As such, mathematical operations on events follow the same rules as mathematical operations on sets. If any set is a subset of a sample space, then that set is an event; otherwise the subset is not an event. Since the empty set, denoted by $\emptyset$, is always a subset of any set, $\emptyset$ is always an event; likewise, a set is always a subset of itself so that $\mathcal{S}$ is an event. An event that contains exactly one element of the sample space is called an *elementary event*, and an event formed by taking unions of elementary events is called a *compound event*.

DEFINITION 2.2 **Mutually exclusive events.** Two events, $\omega_1, \omega_2 \in \Omega$, are said to be mutually exclusive if they have no elements in common; that is, ω_1 and ω_2 are said to be mutually exclusive if $\omega_1 \cap \omega_2 = \emptyset$.

The notion of an event probability is quite abstract, and, indeed, the assignments need not make any sense in the real world. In the world of engineering, of course, the assignment of probabilities should make sense in the practical world. In theory, we are at liberty to assign probabilities to events as we please, but whatever assignment we do make must be consistent with the laws of probability, which are now defined.

DEFINITION 2.3 **Laws of probability**. The laws of probability are as follows:

1. $P\{\emptyset\} = 0$.
2. For any $\omega \in \Omega$, $0 \le P\{\omega\} \le 1$.
3. Suppose $\omega_1, \omega_2 \in \Omega$ are mutually exclusive. Then $P\{\omega_1 \cup \omega_2\} = P\{\omega_1\} + P\{\omega_2\}$.
4. $P\{\mathcal{S}\} = 1$.

We note that the elementary events are mutually exclusive of each other and their union is the sample space. Therefore the probabilities of the elementary events must sum to unity.

EXAMPLE 2.1 Consider choosing a mode of accessing the Internet. Suppose there are exactly four possible choices: ordinary telephone line denoted by T; cable modem, denoted by C; satellite, denoted by S; and no access at all, denoted by N. Suppose further that each individual has made exactly one of those choices. The experiment is to choose an individual at random and ascertain that individual's choice.

The sample space of the experiment is then $\mathcal{S} = \{T, C, S, N\}$. Note that because $\mathcal{S}$ is a set, the order in which its elements are listed is immaterial.

The event space of the experiment has 16 elements. These elements are as follows, , $\{N\}$, $\{S\}$, $\{C\}$, $\{T\}$, $\{S, N\}$, $\{C, N\}$, $\{T, N\}$, $\{C, S\}$, $\{T, S\}$, $\{T, C\}$, $\{C, S, N\}$, $\{T, S, N\}$, $\{T, C, N\}$, $\{T, C, S\}$, $\{T, C, S, N\}$.

Since we are free to assign the probabilities as we see fit, we assign the probability of an event to be the proportion of all individuals who selected each of the four possible choices. We assume we have perfect knowledge of the choices and the proportions are according to the following: $P\{T\} = 0.11$, $P\{C\} = 0.66$, $P\{S\} = 0.19$, and $P\{N\} = 0.04$. From these assignments, we may then find the probabilities of the remaining events by following the laws of probability. For example, $\{S, N\} = \{S\} \cup \{N\}$ and $\{S\} \cap \{N\} = \emptyset$. Therefore from the second law of probability, $P\{S, N\} = P\{S\} + P\{N\}$.

Suppose an experiment is conducted and we find that the event ω_1 occurs. Since we know that the outcome of any experiment is exactly one element of the sample space, suppose the actual outcome of the experiment is s_1. Then, the statement ω_1 occurs means that $s_1 \in \omega_1$. Alternatively, suppose $\omega_1 = \{s_1, s_2, s_3\}$. Then the statement ω_1 occurs means that the outcome of the experiment was either s_1, s_2, or s_3. For example, occurrence of the event $\{C, S, N\}$ means that the individual chosen at random for the experiment may or may not have access to the Internet. If the individual does have internet access, then it is either by cable modem or satellite.

EXERCISE 2.1 For the experiment described in Example 2.1, specify all of the event probabilities.

EXERCISE 2.2 For the experiment described in Example 2.1, there are a total of four possible outcomes and the number of events is 16. Show that it is always true that card $(\Omega) = 2^{\text{card }(\mathcal{S})}$, where card $(\mathcal{A})$ denotes the cardinality of the set $\mathcal{A}$, which, in turn, is the number of elements of $\mathcal{A}$. [Hint: The events can be put in one-to-one correspondence with the card $(\mathcal{S})$-bit binary numbers.]

2.1.2 Conditioning Experiments

In many practical situations, statistical experiments are very complicated. In some cases, the experiments are so complicated that it is virtually impossible to understand the entire experiment without breaking the experiment down into a set of smaller experiments that are more easily understood. The act of breaking down an experiment into a set of smaller experiments is called conditioning.

First consider the simple experiment of rolling a fair die and observing the number of dots on the side facing up. The sample space for this experiment is $\mathcal{S} = \{1, 2, 3, 4, 5, 6\}$. Suppose we define the events $\mathcal{O} = \{1, 3, 5\}$ and $\mathcal{E}\,\{2, 4, 6\}$ to be the events of an odd and an even outcome, respectively. Now suppose we want to know the probability of rolling 5 given that the event $\mathcal{E}$ has occurred. Since we know $\mathcal{E}$ has occurred, we know that either 2, 4, or 6 was the outcome of the roll. Therefore, the probability that the event $\{5\}$ occurred given that the event $\mathcal{E}$ has occurred is zero.

Now, suppose we want to know the probability that a 6 was rolled given that $\mathcal{E}$ has occurred. We note that $\mathcal{E}$ has three possible outcomes, each of which is equally likely to occur. Because the number 6 is one of those three equally likely possible outcomes, the probability that the event $\{6\}$ occurred given that the event $\mathcal{E}$ has occurred is 1/3. Thus, the unconditional probability of rolling a 6 is 1/6 but the conditional probability given an even number was rolled is 1/3.

Similarly, the probability that the event $\{5, 6\}$ occurred given that the event $\mathcal{E}$ has occurred is 1/3 since given that the event $\mathcal{E}$ has occurred, the only way for the event $\{5, 6\}$ to occur is that the event $\{6\}$ occurs. Thus we see that only elementary events that are subsets of $\mathcal{E}$ contribute to probabilities that are conditioned on the event $\mathcal{E}$.

In computing conditional probability of an event, say ω, given that an event, say $\mathcal{E}$, has occurred the irrelevant elementary events are eliminated from consideration by taking the intersection of ω with $\mathcal{E}$. This intersection operation yields a new event, $\omega^* = \omega \cap \mathcal{E}$. Then, either $\omega^* = \emptyset$, in which case its probability of occurrence is zero, or ω^* is the union of elementary events, in which case its probability can be calculated using the laws of probability. In order to complete the computation of the conditional probability of ω given $\mathcal{E}$, we simply divide $P\{\omega^*\}$ by $P\{\mathcal{E}\}$.

The conditional probability of ω given $\mathcal{E}$ is denoted by $P\{\omega|\mathcal{E}\}$, and based on the arguments of the previous paragraph, we have the following computational formula:

$$P\{\omega|\mathcal{E}\} = \frac{P\{\omega \cap \mathcal{E}\}}{P\{\mathcal{E}\}}. \tag{2.1}$$

Basically, there are two approaches to calculating a conditional probability. Sometimes, it is straightforward to formulate a new experiment, where the sample space is replaced by the event and then to calculate the conditional probability directly from that. For example, suppose the experiment is to roll a fair die until an even number of dots appears, then observe the outcome. The sample space for this experiment is $\mathcal{S} = \{2, 4, 6\}$. The probability of rolling a 2 or a 5 is then obviously the same as rolling a 2 since 5 is not a possible outcome. Hence, $P\{2 \text{ or } 5|\text{even}\}$ is the same as $P\{2\}$ in the current experiment, which is 1/3.

The second approach is to work directly with the original experiment, and this approach uses the following steps to find $P\{\omega|\mathcal{E}\}$:

1. Find $P\{\mathcal{E}\}$ using the laws of probability.
2. Find $\omega^* = \omega \cap \mathcal{E}$.
3. Find $P\{\omega^*\}$ using the laws of probability.
4. Divide $P\{\omega^*\}$ by $P\{\mathcal{E}\}$ to find $P\{\omega|\mathcal{E}\}$.

We note that in practice, we never intentionally condition on the null event because that event can never occur in an actual experiment; that is, the null event occurs only if the experiment is not conducted. In addition, given the information that the null event has occurred, we know that no other event can occur because the experiment has not been conducted. In addition, conditioning on the null event would create a division-by-zero problem with the third step in the procedure for computing conditional probabilities.

EXAMPLE 2.2 Suppose we are an international company and we offer a total of 8 options, where each of our customers chooses exactly one option. Our experiment is to choose a customer at random, consult that customer's records and determine which option the customer has chosen. Define $\mathcal{S} = \{s_1, s_2, \ldots, s_8\}$ and assume that we provide services in countries A and B only. In country A we do not offer options s_3, s_5 or s_7, but in country B we offer options s_3, s_5 and s_7 only. Then, there would be a natural *partition* of $\mathcal{S}$, namely $\mathcal{S} = \mathcal{S}_A \cup \mathcal{S}_B$, where $\mathcal{S}_A = \{s_1, s_2, s_4, s_6, s_8\}$ and $\mathcal{S}_B = \{s_3, s_5, s_7\}$. Suppose the elementary events are assigned the probabilities $P\{s_i\} = i/36$, $i = 1, 2, \ldots, 8$. Define $\omega_1 = \{s_1, s_2, s_3\}$. Compute $P\{\omega_1\}$, $P\{\omega_1|\mathcal{S}_A\}$, and $P\{\omega_1|\mathcal{S}_B\}$.

Solution. Since ω_1 is a union of (disjoint) elementary events, we can find $P\{\omega_1\}$ by simply summing the probabilities of its constituent elementary events. Thus,

$$P\{\omega_1\} = P\left\{\cup_{i\in\{1,2,3\}} s_i\right\} = \sum_{i\in\{1,2,3\}} P\{s_i\} = \frac{6}{36} = \frac{1}{6}.$$

1. Find $P\{\mathcal{S}_A\}$ using the laws of probability. Since $\mathcal{S}_A$ is specified as a union of disjoint events, we can determine the probability of the union as the sum of the event probabilities. Therefore,

$$P\{\mathcal{S}_A\} = P\left\{\cup_{i\in\{1,2,4,6,8\}} s_i\right\} = \sum_{i\in\{1,2,4,6,8\}} P\{s_i\} = \frac{21}{36}.$$

2. Find $\omega^* = \omega_1 \cap \mathcal{S}_A$.

$$\omega^* = \{s_1, s_2, s_3\} \cap \{s_1, s_2, s_4, s_6, s_8\} = \{s_1, s_2\}.$$

3. Find $P\{\omega^*\}$ using the laws of probability.

$$P\{\omega^*\} = P\left\{\cup_{i\in\{1,2\}} s_i\right\} = \sum_{i\in\{1,2\}} P\{s_i\} = \frac{3}{36}.$$

4. Divide $P\{\omega^*\}$ by $P\{\mathcal{E}\}$ to find $P\{\omega|\mathcal{E}\}$.

$$P\{\omega_1|\mathcal{S}_A\} = \frac{P\{\omega^*\}}{P\{\mathcal{E}\}} = \frac{1}{7}.$$

To compute $P\{\omega_1|\mathcal{S}_B\}$, we follow the same procedures to find $P\{\mathcal{S}_B\} = 15/26, \omega_1 \cap \mathcal{S}_B = \{s_3\}, P\{\omega_1 \cap \mathcal{S}_B\} = P\{s_3\} = 3/36$, and finally

$$P\{\omega_1|\mathcal{S}_B\} = \frac{P\{\omega_1 \cap \mathcal{S}_B\}}{P\{\mathcal{S}_B\}} = \frac{3/36}{15/36} = \frac{1}{5}.$$

In summary we find that the unconditional probability and the two conditional probabilities all have different values, which are

$$P\{\omega_1\} = \frac{1}{6}, \quad P\{\omega_1|\mathcal{S}_A\} = \frac{1}{7}, \quad \text{and} \quad P\{\omega_1|\mathcal{S}_B\} = \frac{1}{5}.$$

EXERCISE 2.3 Repeat the computations of Example 2.2 by constructing restricted experiments based on $\mathcal{S}_A$ and $\mathcal{S}_B$. [Hint: The probabilities of the elementary events in the restricted experiments must be normalized by dividing the probabilities of the individual elementary events of the restricted experiment by the sum of the probabilities of the constituent elementary events of the restricted experiment.]

DEFINITION 2.4 **Joint probability** Suppose $\omega_1, \omega_2 \in \Omega$. Then $P\{\omega_1 \cap \omega_2\}$ is called the *joint probability* of the events ω_1 and ω_2. We sometimes express $P\{\omega_1 \cap \omega_2\}$ as $P\{\omega_1\omega_2\}$ or $P\{\omega_1, \omega_2\}$.

From (2.1), we readily find that

$$P\{\omega \cap \mathcal{E}\} = P\{\omega|\mathcal{E}\} P\{\mathcal{E}\}. \tag{2.2}$$

Now, if $\mathcal{E}^c$ denotes the complement of $\mathcal{E}$, we find $\mathcal{E} \cap \mathcal{E}^c = \emptyset$. Therefore $\{\omega \cap \mathcal{E}\} \cap \{\omega \cap \mathcal{E}^c\} = \emptyset$. Also, $\omega = \{\omega \cap \mathcal{E}\} \cup \{\omega \cap \mathcal{E}^c\}$. Therefore by the second law of probability,

$$P\{\omega\} = P\{\omega \cap \mathcal{E}\} + P\{\omega \cap \mathcal{E}^c\}.$$

It then follows from (2.2) that

$$P\{\omega\} = P\{\omega|\mathcal{E}\} P\{\mathcal{E}\} + P\{\omega|\mathcal{E}^c\} P\{\mathcal{E}^c\}. \tag{2.3}$$

In the previous example, because $\mathcal{S}_B = \mathcal{S}_A^c$, we can find $P\{\omega_1\}$ as follows:

$$P\{\omega_1\} = P\{\omega_1|\mathcal{S}_A\} P\{\mathcal{S}_A\} P\{\omega_1|\mathcal{S}_B\} P\{\mathcal{S}_B\} = \frac{1}{7}\frac{21}{36} + \frac{1}{5}\frac{15}{36} = \frac{1}{6}.$$

DEFINITION 2.5 **Partition.** Define $\{\mathcal{E}_1, \mathcal{E}_2, \ldots, \mathcal{E}_N\}$ to be a set of mutually disjoint events such that $\cup_{i=1}^N \mathcal{E}_i = \mathcal{S}$. Then $\{\mathcal{E}_1, \mathcal{E}_2, \ldots, \mathcal{E}_N\}$ is called a partition of $\mathcal{S}$.

If $\{\mathcal{E}_1, \mathcal{E}_2, \ldots, \mathcal{E}_N\}$ is a partition of $\mathcal{S}$ and $\omega \in \Omega$, it is readily deduced that $\{\omega \cap \mathcal{E}_1, \omega \cap \mathcal{E}_2, \ldots, \omega \cap \mathcal{E}_N\}$ is a collection of mutually disjoint sets such that $\cup_{i=1}^{N} (\omega \cap \mathcal{E}_i) = \omega$. Therefore, from the second law of probability and (2.2), we have

$$P\{\omega\} = \sum_{i=1}^{N} P\{\omega \cap \mathcal{E}_i\} = \sum_{i=1}^{N} P\{\omega|\mathcal{E}_i\} P\{\mathcal{E}_i\}. \tag{2.4}$$

We note that a conditional probability is rarely computed from a joint probability in practice. The more likely case is that an unconditional probability is ultimately required, conditioning is used to find a set of joint probabilities, and then the unconditional probability is computed by using (2.4).

Suppose $\mathcal{A}$ and $\mathcal{B}$ are any two sets. Then, it is always true that $\mathcal{A} \cap \mathcal{B}$ and $\mathcal{B} \cap \mathcal{A}$ are the same set. Thus, since events are sets, suppose $\omega_1, \omega_2 \in \Omega$ with $P\{\omega_1\} \neq 0$ and $P\{\omega_2\} \neq 0$. Then from (2.2), we readily find

$$P\{\omega_1 \cap \omega_2\} = P\{\omega_1|\omega_2\} P\{\omega_2\} = P\{\omega_2|\omega_1\} P\{\omega_1\}.$$

Upon solving for $P\{\omega_2|\omega_1\}$, we find

$$P\{\omega_2|\omega_1\} = \frac{P\{\omega_2|\omega_1\} P\{\omega_1\}}{P\{\omega_2\}}. \tag{2.5}$$

Equation (2.5) is called *Bayes' rule.* Note that the numerator of (2.5) is just the joint probability of the events ω_1 and ω_2, but, as we have said earlier, we usually compute joint probabilities by first computing a conditional probability and then using (2.2). Bayes' rule is useful in cases where an experiment based on one event may be easy to visualize, but an experiment based on a second event may be difficult to visualize.

EXAMPLE 2.3 In Example 2.2, we computed the conditional probabilities that a customer chosen at random has one of three options given that a customer was selected in each of the countries where we provide service; that is, we computed $P\{\omega_1|\mathcal{S}_A\}$, and $P\{\omega_1|\mathcal{S}_B\}$. Now, suppose we want to know the probability that customer chosen at random is from country A given that the selected customer has one of the three options. We would then find

$$P\{\mathcal{S}_A|\omega_1\} = \frac{P\{\omega_1|\mathcal{S}_A\} P\{\mathcal{S}_A\}}{P\{\omega_1\}} = \frac{1}{7}\frac{7}{21} \div \frac{1}{6} = \frac{1}{2}.$$

Notice conducting an experiment on a country-by-country basis makes intuitive sense, but conducting an experiment on each of the possible subsets of the options is harder to visualize.

An alternate form of Bayes' rule is available for cases where the probability of an event is obtained from a partition. Suppose $\{\mathcal{E}_1, \mathcal{E}_2, \ldots, \mathcal{E}_N\}$ is a partition of $\mathcal{S}$ and $\omega \in \Omega$, then

$$P\{\mathcal{E}_i|\omega\} = \frac{P\{\omega|\mathcal{E}_j\}}{P\{\omega\}} = \frac{P\{\omega|\mathcal{E}_j\}}{\sum_{i=1}^{N} P\{\omega|\mathcal{E}_i\} P\{\mathcal{E}_i\}}. \tag{2.6}$$

We turn now to a discussion of statistical independence among events.

DEFINITION 2.6 **Statistical independence.** Suppose $\omega_1, \omega_2 \in \Omega$ and that $P\{\omega_2\} \neq 0$. Then, the event ω_1 is said to be *statistically independent* of the event ω_2 if, and only if, $P\{\omega_1|\omega_2\} = P\{\omega_1\}$. That is, ω_1 is statistically independent of ω_2 if, and only if, its unconditional probability of occurrence and its conditional probability of occurrence given ω_2 have the same value.

From (2.2), we readily find that if ω_1 and ω_2 are statistically independent events, then, since $P\{\omega_1|\omega_2\} = P\{\omega_1\}$,

$$P\{\omega_1 \cap \omega_2\} = P\{\omega_1\} P\{\omega_2\}. \tag{2.7}$$

That is, if two events, ω_1 and ω_2, are statistically independent, then their joint probability is the product of their individual unconditional probabilities.

EXERCISE 2.4 Suppose $\omega_1, \omega_2 \in \Omega$ with $P\{\omega_1\} \neq 0$ and $P\{\omega_2\} \neq 0$. Show that if ω_1 is statistically independent of ω_2, then necessarily, ω_2 is statistically independent of ω_1. In other words, show that statistical independence between events is a mutual property.

EXERCISE 2.5 Suppose $\omega_1, \omega_2 \in \Omega$ but $P\{\omega_1\} = 0$ or $P\{\omega_2\} = 0$. Discuss the concept of independence between ω_1 and ω_2.

At the other extreme of independence between events is the concept of mutual exclusivity, which we formerly defined and now revisit.

DEFINITION 2.7 **Mutually exclusive events (revisited).** Suppose $\omega_1, \omega_2 \in \Omega$ with $P\{\omega_1\} \neq 0$ and $P\{\omega_2\} \neq 0$. Then, the events ω_1 and ω_2 are said to be *mutually exclusive* if, and only if, $P\{\omega_1|\omega_2\} = 0$ (and, therefore, $P\{\omega_2|\omega_1\} = 0$). That is, mutual exclusivity between two events means the occurrence of one of the events literally excludes the possibility that the other event occurs.

Some examples of mutually exclusive events are the events of a partition and the elementary events of any experiment.

2.2 Random Variables

In this section, we first provide a formal definition of the term *random variable*, and then we discuss distributions of random variables. Next, we discuss characterization of random variables according to the form of their distributions. We then discuss probability mass functions and probability density functions for random variables. Next we discuss computation of expectation of functions of random variables. Finally, we discuss computation of the distribution of sums of random variables.

DEFINITION 2.8 **Random variable.** A random variable is a function that maps a sample space into the real numbers. Let $\mathcal{R}$ denote the set of real numbers, and suppose $\tilde{x}$ is a random variable defined on $\mathcal{S}$. Then, for each $s \in \mathcal{S}$, $\tilde{x}(s) \in \mathcal{R}$. Alternatively stated, $\tilde{x} : \mathcal{S} \to \mathcal{R}$.

EXAMPLE 2.4 Consider the experiment defined in Example 2.2. Suppose we want to define a random variable that reflects the monthly cost of a customer's service. We might then define the random variable $\tilde{c}$ to represent the cost specified in \$US. In order to define $\tilde{c}$, we would then define a specific function that maps the elements of $\mathcal{S}$ into the real numbers. For example, we may have $\tilde{c}(s_1) = 132.22$, $\tilde{c}(s_2) = 104.54$, $\tilde{c}(s_3) = 99.99$, $\tilde{c}(s_4) = 96.75$, $\tilde{c}(s_5) = 84.35$, $\tilde{c}(s_6) = 75.29$, $\tilde{c}(s_7) = 70.07$, and $\tilde{c}(s_8) = 56.48$.

Once a random variable is defined on a sample space, events can be defined in terms of the random variable. For example, we can define an event such as $\{\tilde{c} \in (80, 110)\}$ for the random variable defined in Example 2.4. Upon consulting our definition of $\tilde{c}$, we find that $\{\tilde{c} \in (80, 110)\} = \{s_2, s_3, s_4, s_5\}$. Thus, if we want to know $P\{\tilde{c} \in (80, 110)\}$, we can compute $P\{s_2, s_3, s_4, s_5\}$.

In order to organize computations of event probabilities involving random variables, a distribution that captures all of the event probabilities function is defined for each random variable. This distribution function is variously referred to as the *distribution, distribution function, cumulative distribution function*, or *probability distribution function*, and is defined as follows:

DEFINITION 2.9 **Distribution function of a random variable**. Let $F_{\tilde{x}}(x)$ denote the *distribution function* of a random variable, $\tilde{x}$. Then,

$$F_{\tilde{x}}(x) \triangleq P\{\tilde{x} \leq x\} \quad \text{for} -\infty < x < \infty.$$

Figure 2.1 shows the distribution function for the random variable $\tilde{c}$ defined in Example 2.4.

All distribution functions share a common set of properties, which are as follows:

1. $F_{\tilde{x}}(x)$ is a nondecreasing function of x,

2. $\lim_{x \to -\infty} F_{\tilde{x}}(x) = 0$, and

3. $\lim_{x \to \infty} F_{\tilde{x}}(x) = 1$.

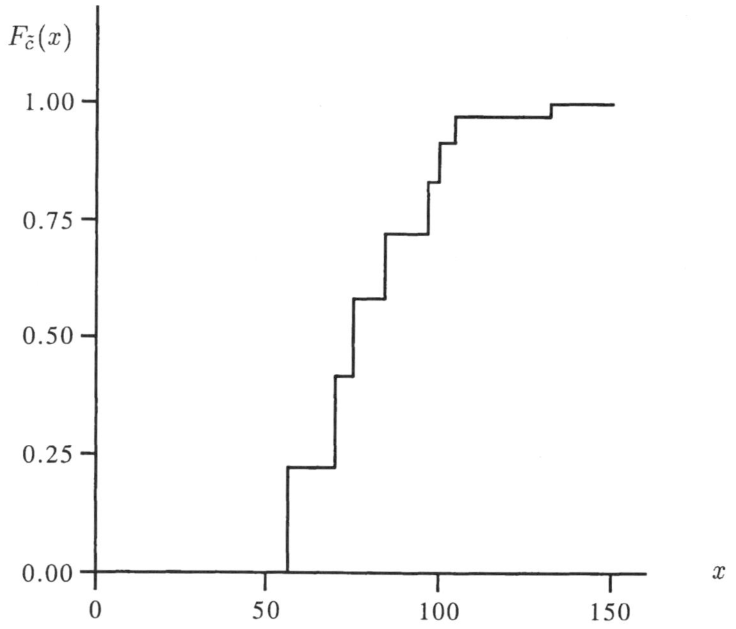

Figure 2.1. Distribution function for the random variable $\tilde{c}$ defined in Example 2.4.

Random variables are classified as either *discrete*, *continuous*, or *mixed*, depending on the form of their distribution function. If a random variable's distribution function has a derivative of zero except for a countable collection of points, then the random variable is said to be *discrete*. That is, a random variable $\tilde{x}$ is said to be discrete if

$$\frac{d}{dx} F_{\tilde{x}}(x) = 0, x \ni \{x_1, x_2, \ldots\} .$$

Further, the set of values of x for which $\frac{d}{dx} F_{\tilde{x}}(x) > 0$ is called the support set of x, and we usually denote the support set of $\tilde{x}$ by $\mathcal{X}$.

If $F_{\tilde{x}}(x)$ is continuous at all points $x \in (-\infty, \infty)$, then the random variable $\tilde{x}$ is said to be *continuous*. The set of values of x over which $\frac{d}{dx} F_{\tilde{x}}(x) > 0$ is, as in the case of a discrete random variable, called the support set of x.

If a random variable is neither discrete nor continuous, then that random variable is said to be *mixed*. As an example, from the distribution function shown in Figure 2.1, it is clear that the random variable $\tilde{c}$ is discrete.

With respect to the primary random variables involved in queueing systems, queue lengths are usually discrete random variables while waiting times are usually mixed random variables because they usually have a probability mass at zero, and the lengths of busy periods are usually continuous.

For any random variable $\tilde{x}$, the *probability mass* at any point x_0 can be computed as follows:

$$P\{\tilde{x} = x_0\} = \lim_{\epsilon \to 0} F_{\tilde{x}}(x_0 + \epsilon) - \lim_{\epsilon \to 0} F_{\tilde{x}}(x_0 - \epsilon); \tag{2.8}$$

that is, the probability mass at x_0 is determined by subtracting the limit of $F_{\tilde{x}}(x)$ as $x \to x_0$ from the left from the limit of $F_{\tilde{x}}(x)$ as $x \to x_0$ from the right. Clearly, since a continuous random variable has a continuous distribution function, it is always true that a continuous random variable has no probability mass anywhere. That is, if $\tilde{x}$ is continuous, then

$$P\{\tilde{x} = x\} = 0 \quad \text{for all} \quad x \in (-\infty, \infty).$$

DEFINITION 2.10 **Probability mass function.** Suppose $\tilde{x}$ is discrete. Then $\mathcal{X}$ is a countable set, and $P\{\tilde{x} = x_i\} > 0$ for all $x_i \in \mathcal{X}$. In the case of a discrete random variable, the discrete function $p_{\tilde{x}}(x) = P\{\tilde{x} = x\}$ for $x \in \mathcal{R}$ is called the *probability mass function* of $\tilde{x}$.

DEFINITION 2.11 **Probability density function.** If $\tilde{x}$ is continuous, then the function $f_{\tilde{x}}(x) = \frac{d}{dx}F_{\tilde{x}}(x)$ is called the *probability density function* of $\tilde{x}$.

Many operations on the distributions of random variables involve integration of the product of functions of the random variable and the differential of its distribution. In order to unify discussion, the differential of the distribution function of a discrete or mixed random variable can be represented in terms of *Dirac delta functions*. The Dirac delta function is defined as follows:

DEFINITION 2.12 **Dirac delta function.** Let $v(x)$ represent any function that is continuous at x_0. A Dirac delta function is defined as a function, $\delta(x)$ such that

$$\int_{x_1}^{x_2} v(x)\delta(x - x_0)dx = v(x_0) \quad \text{provided} \quad x_1 < x_0 < x_2.$$

Suppose $F_{\tilde{x}}(x)$, whether $\tilde{x}$ is discrete or mixed, has probability mass at x_0. If we define

$$\frac{d}{dx}F_{\tilde{x}}(x) = P\{\tilde{x} = x_0\}\,\delta(x - x_0),$$

we can represent the differential of $F_{\tilde{x}}(x)$ as follows:

$$dF_{\tilde{x}}(x) = \begin{cases} \sum_{x_i \in \mathcal{X}} P\{\tilde{x} = x_i\}\,\delta(x - x_i)dx, & \text{if } \tilde{x} \text{ is discrete,} \\ f_{\tilde{x}}(x)dx, & \text{if } \tilde{x} \text{ is continuous,} \\ \sum_{x_i \in \mathcal{X}} P\{\tilde{x} = x_i\}\,\delta(x - x_i)dx + f_{\tilde{x}}(x)dx, & \text{if } \tilde{x} \text{ mixed.} \end{cases} \tag{2.9}$$

We would than have, for example, for any random variable,

$$F_{\tilde{x}}(x) = \int_{-\infty}^{x^+} dF_{\tilde{x}}(x).$$

The above integral is called a *Riemann-Stieltjes integral.*

EXERCISE 2.6 Develop the expression for $dF_{\tilde{c}}(x)$ for the random variable $\tilde{c}$ defined in Example 2.4.

DEFINITION 2.13 **Expectation of a random variable.** The expectation of a random variable $\tilde{x}$ is denoted by $E[\tilde{x}]$ and it is defined as

$$E[\tilde{x}] = \int_{-\infty}^{\infty} x dF_{\tilde{x}}(x). \tag{2.10}$$

The value of $E[\tilde{x}]$ is called the *mean of* $\tilde{x}$.

From the definition of a random variable, we know that a random variable is simply a function that maps an experiment into a real number. Therefore, any real function of a random variable is also a random variable. Let $h(\tilde{x})$ denote an arbitrary real function of a random variable, $\tilde{x}$. Then, by definition,

$$E[h(\tilde{x})] = \int_{-\infty}^{\infty} x dF_{h(\tilde{x})}.$$

That is, the expectation is about the distribution of the random variable, not the random variable itself. However, it turns out that the expectation of $h(\tilde{x})$ can be obtained without first obtaining the distribution of $h(\tilde{x})$. In fact, a computational formula is as follows:

$$E[h(\tilde{x})] = \int_{-\infty}^{\infty} h(x) dF_{\tilde{x}}(x). \tag{2.11}$$

Sometimes (2.11) is referred to as the *law of the unconscious statistician* because the computational formula is used as though it were actually the definition of $E[h(\tilde{x})]$, which it is not.

Many special forms of $h(\tilde{x})$ have special names. Among these is the following:

DEFINITION 2.14 **nth moment of $\tilde{x}$.** Let $h(\tilde{x}) = \tilde{x}^n$. Then, the expectation $E[x^n]$ is called the nth moment of $\tilde{x}$. Using (2.11) with $h(\tilde{x})$ replaced by $\tilde{x}^n$, we find

$$E[\tilde{x}^n] = \int_{-\infty}^{\infty} x^n dF_{\tilde{x}}(x).$$

DEFINITION 2.15 **Variance of $\tilde{x}$.** The expectation $E[(\tilde{x} - E[\tilde{x}])^2]$ is called the *variance of* $\tilde{x}$ and is denoted by var $(\tilde{x})$.

EXERCISE 2.7 Find $E[\tilde{c}]$ and var $(\tilde{c})$ for the random variable $\tilde{c}$ defined in Example 2.4.

In queueing analysis it is often necessary to find the distribution of a random variable when the random variable of interest is expressed as a function of other random variables whose distributions are known. For example, it is often necessary to compute the probability mass function for the sum of two nonnegative integer-valued random variables or the probability density function of the sum of two nonnegative continuous random variables. We first consider the discrete case. Suppose $\tilde{x}_1$ and $\tilde{x}_2$ are two nonnegative integer-valued random variables, and suppose we define $y = \tilde{x}_1 + \tilde{x}_2$. Then, clearly, $\tilde{x}$ is a nonnegative integer-valued random variable. Suppose we want to know $P\{\tilde{x} = 3\}$. If $\tilde{x} = 3$, then we must have $(\tilde{x}_1, \tilde{x}_2) = (0, 3)$, $(\tilde{x}_1, \tilde{x}_2) = (1, 2)$, $(\tilde{x}_1, \tilde{x}_2) = (2, 1)$, or $(\tilde{x}_1, \tilde{x}_2) = (3, 0)$. Thus,

$$P\{\tilde{x} = 3\} = \sum_{i=0}^{3} P\{\tilde{x}_1 = i, \tilde{x}_2 = 3 - i\}.$$

Now, from (2.2), we know that

$$P\{\tilde{x}_1 = i, \tilde{x}_2 = 3 - i\} = P\{\tilde{x}_2 = 3 - i \mid \tilde{x}_1 = i\} P\{\tilde{x}_1 = i\}.$$

Thus,

$$P\{\tilde{x} = 3\} = \sum_{i=0}^{3} P\{\tilde{x}_2 = 3 - i \mid \tilde{x}_1 = i\} P\{\tilde{x}_1 = i\}.$$

Similarly, for any integer value, n, and replacing $\tilde{x}$ by $\tilde{x}_1 + \tilde{x}_2$, we find

$$P\{\tilde{x}_1 + \tilde{x}_2 = n\} = \sum_{i=0}^{n} P\{\tilde{x}_2 = n - i \mid \tilde{x}_1 = i\} P\{\tilde{x}_1 = i\}. \tag{2.12}$$

If $\tilde{x}_1$ and $\tilde{x}_2$ are independent, then $P\{\tilde{x}_2 = n - i \mid \tilde{x}_1 = i\} = P\{\tilde{x}_2 = n - i\}$. Thus, in the specific case that $\tilde{x}_1$ and $\tilde{x}_2$ are independent, (2.12) reduces to

$$P\{\tilde{x}_1 + \tilde{x}_2 = n\} = \sum_{i=0}^{n} P\{\tilde{x}_2 = n - i \mid \tilde{x}_1 = i\} P\{\tilde{x}_1 = i\}. \tag{2.13}$$

The right hand side of (2.13) is readily recognized as the discrete convolution of $p_{\tilde{x}_1}(x)$ and $p_{\tilde{x}_2}(x)$. Thus, when $\tilde{x}_1$ and $\tilde{x}_2$ are independent, we have $p_{\tilde{x}_1+\tilde{x}_2}(x) = p_{\tilde{x}_1}(x) \otimes p_{\tilde{x}_2}(x)$.

As an example, in Section 1.2.1, we discussed multiplexing of traffic at the output of an IP switch. The queue length, as seen at the end of a time slot, was described as a discrete valued, discrete parameter stochastic process, and its evolution was described by the dynamical equations

$$\tilde{q}_{k+1} = (\tilde{q}_k - 1)^+ + \tilde{v}_{k+1}.$$

The random variable on the left hand side, $\tilde{q}_{k+1}$, is the sum of two random variables, $(\tilde{q}_k - 1)^+$ and $\tilde{v}_{k+1}$. Thus we could compute the probability mass function for $\tilde{q}_{k+1}$ by using (2.12) with $\tilde{x}_1 = (\tilde{q}_k - 1)^+$ and $\tilde{x}_1 = \tilde{v}_{k+1}$.

EXERCISE 2.8 Let $\mathcal{X}_1$ and $c X_2$ denote the support sets for $\tilde{x}_1$ and $\tilde{x}_2$, respectively. Specialize (2.12) where $\mathcal{X}_1 = \{5, 6, \ldots, 14\}$ and $\mathcal{X}_2 = \{11, 12, \ldots, 22\}$.

For the continuous case, the following are computational formulas for computing the probability density function for the sum of two nonnegative continuous random variables:

$$f_{\tilde{x}_1+\tilde{x}_2}(x) = \int_0^x f_{\tilde{x}_2|\tilde{x}_1}(x - y \mid y) f_{\tilde{x}_1}(y) dy, \tag{2.14}$$

where $f_{\tilde{x}_2|\tilde{x}_1}(x - y)$ is defined as *the conditional probability function of* $\tilde{x}_2$ *given that* $\tilde{x}_1 = y$. Such conditional density functions can be computed in a manner similar to that demonstrated above for the case of discrete random variables. The main difference is that we have to define probabilities in limiting forms. The usual approach is to use expressions such as $P\{\tilde{x} \in (x, x + dx)\} \approx f_x(x)dx$, make all the probability arguments in terms of probabilities rather than densities, and then take limits to obtain the desired results.

EXERCISE 2.9 Derive (2.14) as indicated in the previous paragraph.

If $\tilde{x}_1$ and $\tilde{x}_2$ are independent, then $f_{\tilde{x}_2|\tilde{x}_1}(x - y) = f_{\tilde{x}_2}(x - y)$, and (2.14) reduces to

$$f_{\tilde{x}_1+\tilde{x}_2}(x) = \int_0^x f_{\tilde{x}_2}(x - y) f_{\tilde{x}_1}(y) dy. \tag{2.15}$$

Thus, as in the discrete case, when $\tilde{x}_1$ and $\tilde{x}_2$ are independent, the probability density function for their sum is given by the convolution of the density functions of $\tilde{x}_1$ and $\tilde{x}_2$;

$$f_{\tilde{x}_1+\tilde{x}_2}(x) = f_{\tilde{x}_1}(x) \otimes f_{\tilde{x}_2}(x).$$

2.3 Exponential Distribution

Certain ideas and concepts from the theory of stochastic processes are basic in the study of elementary queueing systems. Perhaps the most important of these are the properties of the exponential distribution and the Poisson process. The purpose of this and the next section is to discuss these and related concepts. We begin with a definition of the memoryless property of a random variable and then relate this to the exponential distribution.

Much of the literature and results in stochastic analysis are based upon the assumption that the times between events in the stochastic processes under study are drawn from exponential distributions. These assumptions are normally made for purposes of analytical tractability; the analyst chooses a simplified analysis in preference to no analytical results. In this section, we recognize the importance of making simplifying assumptions, but we introduce important concepts so that the implications of the assumptions are better understood.

Exponential distributions have the *memoryless property*, which is defined as follows:

DEFINITION 2.16 **Memoryless property.** A random variable $\tilde{x}$ is said to be memoryless if, and only if, for every $\alpha, \beta \geq 0$,

$$P\{\tilde{x} > \alpha + \beta | \tilde{x} > \beta\} = P\{\tilde{x} > \alpha\}.$$

The implication of the memoryless property is that the lifetime of the process in question begins all over again at every single point in time. Thus, if for example, $\tilde{x}$ represents the lifetime of a light bulb, and $\tilde{x}$ is memoryless, then at every single point in time, the light bulb is as good as new.

In general, from the definition of conditional probability, we know that

$$\begin{aligned} P\{\tilde{x} > \alpha + \beta | \tilde{x} > \beta\} &= \frac{P\{\tilde{x} > \alpha + \beta, \tilde{x} > \beta\}}{P\{\tilde{x} > \beta\}} \\ &= \frac{P\{\tilde{x} > \alpha + \beta\}}{P\{\tilde{x} > \beta\}}. \end{aligned}$$

But if $\tilde{x}$ is memoryless, then

$$P\{\tilde{x} > \alpha + \beta \mid \tilde{x} > \beta\} = P\{\tilde{x} > \alpha\}.$$

Thus, for $\tilde{x}$ memoryless, we have

$$P\{\tilde{x} > \alpha + \beta\} = P\{\tilde{x} > \alpha\}P\{\tilde{x} > \beta\}.$$

DEFINITION 2.17 **Exponentially distributed.** A random variable $\tilde{x}$ is said to be exponentially distributed if for some finite, positive λ, $P\{\tilde{x} > x\} = e^{-\lambda x}$ for $x \geq 0$.

With regard to the memoryless property, we state the following two lemmas, the proofs of which are deferred to the exercises.

LEMMA 2.1 If $\tilde{x}$ is exponentially distributed, then $\tilde{x}$ is memoryless.

EXERCISE 2.10 Prove Lemma 2.1.

LEMMA 2.2 Let g be a nonnegative right-continuous function with $g(t+s) = g(t)g(s)$ for all $s, t > 0$. Then either $g(t) = 0$ for $t > 0$ or $g(t) = e^{-\lambda t}$ for some positive $\lambda < \infty$.

EXERCISE 2.11 Prove Lemma 2.2. [*Hint*: Start with rational arguments. Extend to the real line using a continuity argument. The proof is given in Feller, [1968] pp. 458 - 460, but it is strongly recommended that the exercise be attempted without going to the reference.]

From Lemmas 2.1 and 2.2, we have the following theorem.

THEOREM 2.1 *A continuous random variable, $\tilde{x}$, is exponentially distributed if and only if, $\tilde{x}$ is memoryless. That is, the memoryless property is unique to the exponential random variable.* □

Now, from Theorem 2.1 we find that for $\tilde{x}$ memoryless,

$$P\{\tilde{x} > x\} = e^{-\lambda x}, \qquad x \geq 0.$$

Thus,

$$P\{\tilde{x} \leq x\} = 1 - e^{-\lambda x}, \qquad x \geq 0,$$

and

$$\frac{d}{dx} P\{\tilde{x} \leq x\} = \lambda e^{-\lambda x}, \qquad \text{for } x \geq 0.$$

The parameter λ is sometimes called the rate, and we say "$\tilde{x}$ is exponentially distributed with rate λ," and we write $\tilde{x} \sim E(\lambda)$.

EXAMPLE 2.5 An office shared by a number of graduate students has two telephones. When Alice decides to use a telephone, she sees that Bob and Charlie are using them, but no one else is waiting. Alice knows she can use the phone as soon as either Bob or Charlie completes his call. Suppose the holding time of each call are drawn independently from an exponential distribution with parameter μ. What is the probability that Alice completes her call before Charlie?

Solution: Since service is exponential, and therefore memoryless, when Alice enters, the remaining time for Bob's and Charlie's calls are independent exponential random variables with parameter μ. Thus Bob and Charlie are equally likely to finish last, and $P\{\text{Bob before Charlie}\} = 1/2$. If Bob completes his call before Charlie, then from the point when Bob finishes, Charlie and Alice will use the phones an amount of time drawn independently from an exponential distribution with rate μ. Hence $P\{\text{Alice before Charlie} \mid \text{Bob before Charlie}\} = 1/2$. Thus, $P\{\text{Alice before Charlie}\} = 1/4$.

If the holding times in the above example were deterministic rather than exponential, then the result would have been quite different. Comparison between exponential and deterministic assumptions are explored later in the text, but an initial comparison is encouraged in the next exercise.

EXERCISE 2.12 Repeat Exercise 2.5, assuming all students have a deterministic holding time of one unit. How do the results compare? Would an exponential assumption on service-time give an adequate explanation of system performance if the service-time is really deterministic?

Returning to the properties of the exponential distribution, it is interesting to note that both the mean and the standard deviation of the exponential random variable are equal to $1/\lambda$. The moments of the exponential random variable as well as many other random variables are readily determined via *Laplace transform* techniques. Towards this end, we define the Laplace transform and state one of its key properties as a theorem, leaving its proof to the exercises.

DEFINITION 2.18 **Laplace-Stieltjes transform.** Let $\tilde{x}$ be a nonnegative random variable with distribution $F_{\tilde{x}}(x)$. Then

$$F_{\tilde{x}}^*(s) = E[e^{-s\tilde{x}}] = \int_0^\infty e^{-sx} dF_{\tilde{x}}(x)$$

is called the *Laplace-Stieltjes transform* of $\tilde{x}$ or the Laplace-Stieltjes transform of $F_{\tilde{x}}(x)$. If $F_{\tilde{x}}(x)$ is differentiable, the same expression is called the *Laplace transform* of $dF_{\tilde{x}}(x)/dx$.

THEOREM 2.2 *Let $\tilde{x}$ be a nonnegative random variable with distribution $F_{\tilde{x}}(x)$, and let $F_{\tilde{x}}^*(s)$ the Laplace-Stieltjes transform of $\tilde{x}$. Then,*

$$E[\tilde{x}^n] = (-1)^n \frac{d^n}{ds^n} F_{\tilde{x}}^*(s)\Big|_{s=0}.$$

□

EXERCISE 2.13 Prove Theorem 2.2.

THEOREM 2.3 *Let $\tilde{x}$ and $\tilde{y}$ be nonnegative random variables having Laplace-Stieltjes transforms $F_{\tilde{x}}^*(s)$ and $F_{\tilde{y}}^*(s)$, respectively. Then the Laplace-Stieltjes transform for the random variable $\tilde{z} = \tilde{x} + \tilde{y}$ is given by the product of $F_{\tilde{x}}^*(s)$ and $F_{\tilde{y}}^*(s)$.* □

EXERCISE 2.14 Prove Theorem 2.3.

EXERCISE 2.15 Let $\tilde{x}$ be an exponentially distributed random variable with parameter λ. Find $F_{\tilde{x}}^*(s)$.

EXERCISE 2.16 Let $\tilde{x}$ be an exponentially distributed random variable with parameter λ. Derive expressions for $E[\tilde{x}]$, $E[\tilde{x}^2]$, and $Var(\tilde{x})$. [*Hint*: Use Laplace transforms.]

EXERCISE 2.17 Let $\tilde{x}$ and $\tilde{y}$ be independent exponentially distributed random variables with parameters α and β, respectively.

1. Find the distribution of $\tilde{z} = min\{\tilde{x}, \tilde{y}\}$. [*Hint*: Note that $\tilde{z} = min\{\tilde{x}, \tilde{y}\}$ and $\tilde{z} > z$ means $\tilde{x} > z$ and $\tilde{y} > z$.]
2. Find $P\{\tilde{x} < \tilde{y}\}$.
3. Show that the conditional distribution $F_{\tilde{z}|\tilde{x}<\tilde{y}}(z) = F_{\tilde{z}}(z)$.

EXERCISE 2.18 Suppose Albert and Betsy run a race repeatedly. The time required for Albert to complete the race, $\tilde{a}$, is exponentially distributed with parameter α and the time required for Betsy to complete, $\tilde{b}$, is exponentially distributed with parameter β. Let $\tilde{n}_b$ denote the number of times Betsy wins before Albert wins his first race. Find $P\{\tilde{n}_b = n\}$ for $n \geq 0$.

EXERCISE 2.19 Let $\{\tilde{x}_i, i = 1, 2, \ldots\}$ be a sequence of exponentially distributed random variables and let $\tilde{n}$ be a geometrically distributed random variable with parameter p, independent of $\{\tilde{x}_i, i = 1, 2, \ldots\}$. Let

$$\tilde{y} = \sum_{i=1}^{\tilde{n}} \tilde{x}_i.$$

Show that $\tilde{y}$ has the exponential distribution with parameter $p\alpha$.

Some interesting properties of the exponential random variables are now summarized together with a brief discussion of their implications. The proofs of these properties are deferred to the exercises. Relative to all of the properties, let $\tilde{x}$ and $\tilde{y}$ be independent random variables with parameters α and β, respectively. Then, we have the following properties.

Properties of exponential random variables:

1. The distribution of $\tilde{z} = min\{\tilde{x}, \tilde{y}\}$ is exponential with parameter $\alpha + \beta$.
2. $F_{\tilde{z}|\tilde{x}<\tilde{y}}(z) = F_{\tilde{z}}(z)$.
3. $P\{\tilde{x} < \tilde{y}\} = \alpha/(\alpha + \beta)$.
4. Two numbers are drawn repeatedly from the distributions for $\tilde{x}$ and $\tilde{y}$. Let $\tilde{n}_x$ denote the number of trials required before the number drawn from $F_{\tilde{y}}(x)$ is smaller than that drawn from $F_{\tilde{x}}(x)$ for the first time. Then

$$P\{\tilde{n}_x = n\} = \left(\frac{\alpha}{\alpha + \beta}\right)^n \left(\frac{\beta}{\alpha + \beta}\right) \quad \text{for} \quad n \geq 0.$$

5. Let $\{\tilde{x}_i, i = 1, 2, \cdots\}$ be a sequence of mutually independent exponentially distributed random variables, and let $\tilde{n}$ be a geometrically distributed random variable with parameter p, independent of $\{\tilde{x}_i, i = 1, 2, \cdots\}$. Let

$$\tilde{y} = \sum_{i=1}^{\tilde{n}} \tilde{x}_i.$$

Then $\tilde{y}$ has the exponential distribution with parameter $p\alpha$.

The implication of Property 1 is that if the state of a process changes whenever the first of two events occurs, and if the time to occurrence of the events are drawn independently from exponential distributions, then the time to change of state is exponentially distributed with parameter equal to the sum of the individual rates. Since exponentiality implies memoryless, the times to occurrence of the individual events start over again whenever either event occurs.

Property 2 states that even if one knows which event caused the change of state, the time to occurrence of the state change is still exponentially distributed with parameter equal to the sum of the rates. It is tempting to conclude that if one knows the state change was caused by the event having its interevent time drawn from the distribution $F_{\tilde{x}}(x)$, then the time to state change is exponentially distributed with parameter α, but this is false. These properties will be found to be very useful in studying queueing systems in which all interevent times are exponentially distributed.

EXERCISE 2.20 This exercise is intended to reinforce the meaning of Property 2 of exponential random variables. Let $\tilde{x}$ and $\tilde{y}$ denote the two independent exponential random variables with rates 1 and 2, respectively, and define $\tilde{z} = min\{\tilde{x}, \tilde{y}\}$. Using a spreadsheet (or a computer programming language), generate a sequence of 100 variables for each of the random variables. Denote the ith variate for $\tilde{x}$ and $\tilde{y}$ by x_i and y_i, respectively, and set $\tilde{z} = min\{\tilde{x}, \tilde{y}\}$ for $i = 1, 2, \ldots, 100$.

Let n denote the number of values of i such that $x_i < y_i$, let i_j denote the jth such value and define $w_j = z_i$, for $j = 1, 2, \ldots, n$. Compute the sample averages for the variates; that is compute $\bar{x} = (1/100)\sum_{i=1}^{100} x_i$, $\bar{y} = (1/100)\sum_{i=1}^{100} y_i$, $\bar{z} = (1/100)\sum_{i=1}^{100} z_i$, and $\bar{w} = (1/100)\sum_{j=1}^{100} w_j$. Compare the results. Is $\bar{w}$ closer to $\bar{x}$ or $\bar{z}$?

Now give an intuitive explanation for the statement, "It is tempting to conclude that if one knows the state change was caused by the event having its interevent time drawn from the distribution $F_{\tilde{x}}(x)$, then the time to state change is exponentially distributed with parameter α, but this is false."

Property 3 states that the probability that the state change was caused by completion of an $\tilde{x}$ event is simply the rate for $\tilde{x}$, α, divided by the sum of the rates, $\alpha + \beta$. Property 4 states if that the number of state transitions due to $\tilde{x}$ completions before the first $\tilde{y}$ completion is geometrically distributed, the parameter being the rate for $\tilde{x}$ divided by the sum of the rates. By symmetry, the number of state transitions due to $\tilde{y}$ completions before the first $\tilde{x}$ completion is geometrically distributed, the parameter being the rate for $\tilde{y}$ divided by the sum of the rates.

The implication of Property 5 is that a geometric sum of exponential random variables is exponential. For example, if a message contains a geometric number of packets having independent and identically distributed exponential transmission times, then the total transmission time of the message is exponential.

Since the types of operations with exponential distributions described above yield exponential distributions, the results are easily extended to the case of n, rather than 2, exponential random variables. This leads to a great deal of simplification in analyzing queueing systems in which all underlying distributions are exponential.

2.4 Poisson Process

The characterization of arrival processes for many queueing systems as Poisson has a solid physical basis, as was first discovered by A. K. Erlang during the 1910's. The Poisson assumption can reduce the analytical complexity of a problem and lead to easily obtained and useful results, but the same assumption may also render the analysis useless. As seen in the examples presented in Chapter 1, while the Poisson characterization is often appropriate, there are many cases in which the Poisson assumption is simply not justifiable, and the distinction between the two cases is not necessarily obvious. Thus an understanding of Poisson processes is enormously important in queueing analysis. Toward this goal, we will present three definitions of the Poisson process, each of which presents a different, but equivalent view.

The Poisson process is perhaps the most important and well known member of a special class of stochastic processes called a *counting process*. Before proceeding to our discussion of the Poisson process, we introduce counting processes and some of their more important properties.

DEFINITION 2.19 **Counting process.** A stochastic process $\{\tilde{n}(t), t \geq 0\}$ is said to be a counting process (CP) if $\tilde{n}(t)$ expresses the number of events that have occurred by time t. Thus,

1. $\tilde{n}(t)$ is integer valued,

2. $\tilde{n}(t)$ is nonnegative,

3. $\tilde{n}(t)$ is nondecreasing, and

4. for $s < t, \tilde{n}(t) - \tilde{n}(s)$ is the number of events that occur in the interval $(s, t]$.

From the above definition, it is clear that a counting process is a process of counting that evolves over time. Simply put, counting processes count the occurrence of events; one can imagine the counting process saying, "1, 2, 3, ..."

Counting processes are characterized by the relationships between events that occur in nonoverlapping intervals of time called increments. In particular, it is of interest to know how the occurrence of events in one interval of time affects the probability of occurrence of events in another, nonoverlapping interval of time. Counting processes are characterized on the basis of whether or not they satisfy the conditions of the following definitions.

DEFINITION 2.20 **Independent increments.** If the numbers of events occurring in disjoint time intervals are independent, then the counting process is said to have independent increments.

DEFINITION 2.21 **Stationary increments.** If the distribution of the number of events that occur in a time interval depends only upon the length of the interval - that is, if $P\{\tilde{n}(t+s) - \tilde{n}(t) = n\}$ is independent of t - then the counting process is said to have stationary increments.

EXERCISE 2.21 Define counting processes which you think have the following properties:

1. independent but not stationary increments,
2. stationary but not independent increments,
3. neither stationary nor independent increments, and
4. both stationary and independent increments.

What would you think would be the properties of the process which counts the number of passengers which arrive to an airport by June 30 of a given year if time zero is defined to be midnight, December 31 of the previous year?

We are now ready to consider our first definition of the Poisson process, which is given in terms of the Poisson distribution.

DEFINITION 2.22 **Poisson process (1).** The counting process $\{\tilde{n}(t), t \geq 0\}$ is said to be a Poisson process with rate λ, $\lambda > 0$, if

1. $\tilde{n}(0) = 0$,

2. $\{\tilde{n}(t), t > 0\}$ has independent increments, and

3. the number of events which occur in any interval of length t is Poisson distributed with parameter λt; that is

$$P\{\tilde{n}(t+s) - \tilde{n}(s) = n\} = \frac{(\lambda t)^n e^{-\lambda t}}{n!} \qquad \text{for all s.}$$

EXERCISE 2.22 Show that $E[\tilde{n}(t+s) - \tilde{n}(s)] = \lambda t$ if $\{\tilde{n}(t), t > 0\}$ is a Poisson process with rate λ.

It is important to note that Property 3 of Definition 2.22 implies that the process has stationary increments; that is, the number of events that occur in an interval of length t is independent of the time at which the observation period begins. Also, note that it is not enough to verify that the distribution of the number of events in a fixed-length interval is Poisson distributed; the number of events counted in all nonoverlapping fixed-length intervals of every length must also be independent.

It is easy to define a process that is not itself Poisson but that results in a Poisson number of arrivals in a fixed-length interval. As an extreme example, suppose arrivals occur in groups every hour on the hour and the group sizes are drawn independently from a Poisson distribution. Then, if measurements of the number of arrivals that occur over intervals having a length of one hour are taken, then the number of arrivals over the measurement period will follow the Poisson distribution. In addition, the number of arrivals in nonoverlapping periods will be independent. The process is also stationary. But, the arrival process is obviously not Poisson; the problem is that if the measurements were taken over intervals of a different fixed-length, say 15 minutes, then the number of arrivals would not follow the Poisson distribution.

We could construct other examples, but suffice it to say at this point that there are many processes wearing Poisson clothing that are not Poisson. Thus, extreme care must be taken in order to avoid making Poisson assumptions inappropriately. More will be said on this topic in Chapter 3; for now, we return to our alternate definitions of the Poisson process.

In order to state the second definition of the Poisson process, we need the notion of a special class of functions, $o(h)$. Functions belonging to this class diminish to zero "faster than linear functions" as their arguments are decreased. The second definition of the Poisson process basically says that over very short intervals, the probability of the occurrence of a single event is proportional to the length of the interval, and the probability of the occurrence of two or more events over the same interval is $o(h)$. Again, increments are stationary and independent. These ideas are now stated more formally.

DEFINITION 2.23 **o(h)** (*"little-oh-of-h"*). A function f is said to be $o(h)$ if f has the property

$$\lim_{h \to 0} \frac{f(h)}{h} = 0.$$

It is easy to show that sums and products of $o(h)$ functions are also $o(h)$. In addition, $o(h)$ functions themselves must tend to 0 as h tends to 0. Exercises are provided below to allow the reader to develop these and other properties of $o(h)$ functions and to gain a better understanding of the concept.

EXERCISE 2.23 For each of the following functions, determine whether the function is $o(h)$ or not. Your determination should be in the form of a formal proof.

1. $f(t) = t$,
2. $f(t) = t^2$,
3. $f(t) = t^{\frac{1}{2}}$,
4. $f(t) = e^{-at}$ for $a, t > 0$
5. $f(t) = te^{-at}$ for $a, t > 0$

EXERCISE 2.24 Suppose that $f(t)$ and $g(t)$ are both $o(h)$. Determine whether each of the following functions is $o(h)$.

1. $s(t) = f(t) + g(t)$
2. $d(t) = f(t) - g(t)$
3. $p(t) = f(t)g(t)$
4. $q(t) = f(t)/g(t)$
5. $i(t) = \int_0^t f(x)\,dx$

We are now ready to introduce our second definition of the Poisson process.

DEFINITION 2.24 **Poisson process (2).** The counting process $\{\tilde{n}(t), t > 0\}$ is said to be a Poisson process with rate λ, $\lambda > 0$, if

1. $\tilde{n}(0) = 0$,
2. $\{\tilde{n}(t), t > 0\}$ has stationary and independent increments,
3. $P\{\tilde{n}(h) = 1\} = \lambda h + o(h)$, and
4. $P\{\tilde{n}(h) \geq 2\} = o(h)$.

EXERCISE 2.25 Show that Definition 1 of the Poisson process implies Definition 2 of the Poisson process.

EXERCISE 2.26 Show that Definition 2 of the Poisson process implies Definition 1 of the Poisson process. [*Hint*: After satisfying the first two properties of Definition 1, establish that $P_0(t) = \exp\{-\lambda t\}$ where $P_n(t) = P\{\tilde{n}(t) = n\}$ and then prove the validity of Property 3 of Definition 1 by induction.]

The Poisson process can also be characterized by its interarrival time which is defined as follows.

DEFINITION 2.25 **Sequence of interarrival times**. Let $\tilde{t}_1$ be the time of the first event from a counting process, and $\tilde{t}_n$ be the time between the $(n-1)$st event and the nth event. Then $\{\tilde{t}_1, \tilde{t}_2, \cdots\}$ is called the sequence of interarrival times. Note that $\tilde{t}_1 > t \Rightarrow \tilde{n}(t) = 0$.

EXERCISE 2.27 Show that the sequence of interarrival times for a Poisson process with rate λ forms a set of mutually independent, identically distributed exponential random variables with parameter λ.

We now turn to the third definition of the Poisson process. From Property 3 of the first definition of the Poisson process, it is easy to see that

$$P\{\tilde{n}(t+s) - \tilde{n}(s) = 0\} = e^{-\lambda t} \qquad \text{for all s.}$$

Thus,

$$P\{\tilde{n}(t) = 0\} = e^{-\lambda t}.$$

Now, the event that there are no events from the process by time t is the same as the event that the first event from the process occurs after time t. That is,

$$P\{\tilde{\tau}_1 => t\} = e^{-\lambda t}.$$

Since $\tilde{t}_1 = \tilde{\tau}_1$, we see that the first interarrival times from a Poisson process is exponentially distributed. Now because the second interarrival time begins at the end of the first interarrival time, and the process has stationary and independent increments, the distribution of $\tilde{t}_2$ is the same as the distribution of $\tilde{t}_1$, and in addition, these random variables are independent. Repeated use of these arguments will reveal that the Poisson process yields a sequence of independent, identically distributed exponential interarrival times.

DEFINITION 2.26 **Poisson process (3)**. Let $\tilde{t}_1, \tilde{t}_2, \cdots$ be *iid* exponential random variables with mean $1/\lambda$. Consider a counting process in which the nth event occurs at time $\tilde{s}_n = \sum_{i=1}^{n} \tilde{t}_i$; then such a counting process is a Poisson process and

$$\tilde{n}(t) = \max\{n : \tilde{s}_n \le t\}.$$

We have argued above that Definition 1 of the Poisson process implies Definition 3. It is left to the exercises to show that Definitions 1 and 2 of the Poisson process stated above are equivalent and that Definition 3 implies Definition 1. Thus all the definitions of the Poisson process given above are equivalent.

EXERCISE 2.28 Show that

$$\frac{d}{dt}P\{\tilde{s}_n \le t\} = \frac{\lambda(\lambda t)^{n-1}e^{-\lambda t}}{(n-1)!}.$$

[*Hint*: Start by noting $\tilde{s}_n \le t \iff \tilde{n}(t) \ge n$].

EXERCISE 2.29 Show that Definition 3 of the Poisson process implies Definition 1 of the Poisson process.

The following additional properties of the Poisson process, stated without proof, are useful in studying queueing systems. They should be part of the working vocabulary of every queueing theorist.

Properties of Poisson processes:

1. Let $\{\tilde{n}_1(t), t \ge 0\}$ and $\{\tilde{n}_2(t), t \ge 0\}$ be independent Poisson processes with rates α and β, respectively. Define $\tilde{n}(t) = \tilde{n}_1(t) + \tilde{n}_2(t)$. Then $\{\tilde{n}(t), t \ge 0\}$ is a Poisson process with rate $\alpha + \beta$.

2. Events occur according to a Poisson process with rate λ. Suppose each event, independent of anything else, is recorded with probability p. Let $\tilde{n}_1(t)$ be the number of events recorded by time t and $\tilde{n}_2(t)$ be the number of events not recorded by time t. Then the processes $\{\tilde{n}_1(t), t \ge 0\}$ and $\{\tilde{n}_2(t), t \ge 0\}$ are independent Poisson processes with rates $p\lambda$ and $(1-p)\lambda$, respectively.

The implications of the above properties of Poisson processes are now discussed briefly. Suppose there are two independent arrival streams of customers converging on a service center. Property 1 says that if the arrival processes of the individual streams are Poisson, then so is the combined stream. This is a direct result of the facts that interarrival times from Poisson processes are exponentially distributed and that the minimum of two independent exponential random variables is also exponential.

The second property covers the following situation. Suppose potential customers arrive to a business establishment according to a Poisson process. Each customer upon approaching the establishment tosses a coin. If "heads" results, the potential customer enters the store, else the potential customer departs without entering. Property 2 says that the customers who actually enter the store do so according to a Poisson process, the process counting the potential customers who choose not to enter is a Poisson process, and furthermore (surprisingly), the two processes are independent.

The above properties, which also apply to more than two streams or choices, are extremely useful in the analysis of networks of exponential queues and in justifying simplified analysis of system bottlenecks. The first of these aspects will be explored in the next chapter. The reader is encouraged to complete the exercises to gain a mastery of these properties.

EXERCISE 2.30 Let $\tilde{n}_1$ and $\tilde{n}_2$ be independent Poisson random variables with rates α and β, respectively. Define $\tilde{n} = \tilde{n}_1 + \tilde{n}_2$. Show that $\tilde{n}$ has the Poisson distribution with rate $\alpha + \beta$. Using this result, prove Property 1 of the Poisson process.

EXERCISE 2.31 Suppose an urn contains $\tilde{n}$ balls, where $\tilde{n}$ is a Poisson random variable with parameter λ. Suppose the balls are either red or green, the proportion of red balls being p. Show that the distribution of the number of red balls, $\tilde{n}_r$, in the urn is Poisson with parameter $p\lambda$, the distribution of green balls, $\tilde{n}_g$ is Poisson with parameter $(1-p)\lambda$, and that $\tilde{n}_r$ and $\tilde{n}_g$ are independent random variables. Use this result to prove Property 2 of the Poisson process. [*Hint*: Condition on the total number of balls in the urn and use the fact that the number of successes in a sequence of n repeated Bernoulli trials has the binomial distribution with parameters n and p.]

EXERCISE 2.32 Events occur at a Poisson rate λ. Suppose all odd numbered events and no even numbered events are recorded. Let $\tilde{n}_1(t)$ be the number of events recorded by time t and $\tilde{n}_2(t)$ be the number of events not recorded by time t. Do the processes $\{\tilde{n}_1(t), t \geq 0\}$ and $\{\tilde{n}_2(t), t \geq 0\}$ each have independent increments? Do they have stationary increments? Are they Poisson processes?

2.5 Markov Chains

In Section 1.2.2, we discussed multiplexing of traffic at the output of an IP switch. The queue length, as seen at the end of a time slot, was described as a discrete valued, discrete parameter stochastic process, and its evolution was described by the dynamical equations

$$\tilde{q}_{k+1} = (\tilde{q}_k - 1)^+ + \tilde{v}_{k+1}.$$

We examined queue length behavior under two different classes of arrival processes. In the first case, $\{\tilde{v}_k, k = 1, 2, \ldots\}$ was assumed to be a sequence of independent, identically distributed binomial random variables with parameters N, and p. A little thought reveals that if we knew the queue length at the end of slot k, then we would be able to determine the distribution of the queue length at the end of slot $k+1$. For example, if $\tilde{q}_k = 2$, then $\tilde{q}_{k+1}$ will be one plus the number of new arrivals to occur over time slot $k+1$. Therefore, $\tilde{q}_{k+1}$ cannot be zero nor can $\tilde{q}_{k+1}$ be more than $N+1$ because no more than N arrivals can occur. In fact, $P\{\tilde{q}_{k+1} = j\} = P\{\tilde{v}_{n+1} = j - 1\}$ for $j = 1, 2, \ldots, N + 1$.

Therefore, for the case where $\{\tilde{v}_k, k = 1, 2, \ldots\}$ was assumed to be a sequence of independent, identically distributed random variables, knowledge of the queue length at the end of time slot k provides complete information about the state of the queueing system, and, from that information alone, we can determine the future evolution of the queueing system. A stochastic process of the type just describe is called a *discrete valued, discrete parameter Markov chain*, the definition of which we now formally state.

DEFINITION 2.27 **Discrete valued, discrete parameter Markov chain.** A stochastic process $\{\tilde{x}_k, k = 0, 1, \ldots\}$ is said to be a *discrete valued, discrete parameter Markov chain* (on the nonnegative integers) if for all integers $k \geq 0$, and all nonnegative integers i, j,

$$P\{\tilde{x}_{k+1} = j \mid \tilde{x}_k = i, \tilde{x}_{k-1} = i_{k-1}, \ldots, \tilde{x}_0 = i_0\} = P\{\tilde{x}_{k+1} = j \mid \tilde{x}_k = i\}.$$

DEFINITION 2.28 **State space.** The set of all possible values of $\tilde{x}_k, k = 0, 1, \ldots$ is called the *state space* of the Markov chain.

Basically, the state space of a discrete valued, discrete parameter Markov chain is the union of the support sets of the random variables $\tilde{x}_k, k = 0, 1, \ldots$.

DEFINITION 2.29 *One-step transition probability matrix.* The probability $p_{ij} = P\{\tilde{x}_{k+1} = j \mid \tilde{x}_k = i\}$ is called the *one-step transition probability from state i to state j*, and the matrix $\mathcal{P} = [p_{ij}]$ is called *the one-step transition probability matrix.* The matrix $\mathcal{P}$ is always square, and its dimension is the same as the cardinality of the state space of the Markov chain, which may be either countably finite or countably infinite.

EXERCISE 2.33 Determine the *one-step transition probability matrix* for the Markov chain $\{q_k, k = 0, 1, \ldots\}$ of Section 1.2.2 for the case where $\{\tilde{v}_k, k = 1, 2, \ldots\}$ is assumed to be a sequence of independent, identically distributed binomial random variables with parameters N and p.

Our interest in discrete parameter Markov chains is generally confined to those having a nonnegative integer-valued state space and the nonnegative integers as their parameter space. We loosely refer to such Markov chains as discrete parameter Markov chains (DPMCs).

In general, the state space for a DPMC may be either countably finite or countably infinite. In either case, if $\{\tilde{x}_k, k = 0, 1, \ldots\}$ is a DPMC, then we define β_k to be the vector of the probability masses of $\tilde{x}_k$; that is, $\tilde{x}_k$ is a discrete random variable whose possible values are the nonnegative integers, and we define

$$\beta_k = [P\{\tilde{x}_k = 0\} \quad P\{\tilde{x}_k = 1\} \quad \ldots]. \tag{2.16}$$

From the definition of β_k and $\mathcal{P}$, we then have

$$\beta_{k+1} = \beta_k \mathcal{P}. \tag{2.17}$$

From (2.17), we then have

$$\beta_{k+1} = \beta_k \mathcal{P} = \beta_{k-1}\mathcal{P}\mathcal{P} = \beta_{k-2}\mathcal{P}\mathcal{P}\mathcal{P} = \ldots = \beta_0 \mathcal{P}^k.$$

Therefore, if $\lim_{k\to\infty} \mathcal{P}^k$ exists, then $\lim_{k\to\infty} \beta_{k+1}$ also exists, and

$$\lim_{k\to\infty} \beta_{k+1} = \beta_\infty = \beta_0 \mathcal{P}^\infty.$$

If, in addition, β_∞ is independent of β_0, then we define

$$\pi = \beta_\infty$$

so that

$$\pi = \beta_0 \mathcal{P}^\infty. \tag{2.18}$$

From (2.18), it is easy to prove that if π is independent of β_0, then the rows of $\mathcal{P}^\infty$ must be identical.

EXAMPLE 2.6 Suppose $\{\tilde{x}_k, k = 0, 1, \ldots\}$ is a Markov chain such that

$$\mathcal{P} = \begin{bmatrix} 0 & 1 \\ 1 & 0 \end{bmatrix}.$$

Then, $\mathcal{P}^k = \mathcal{P}$ if k is odd and $\mathcal{P}^k = I$ if k is even. Thus, $\beta_k = \beta_0$ if k is odd, and $\beta_k = [\,P\{\tilde{x}_0 = 1\} \quad P\{\tilde{x}_0 = 1\}\,]$ if k is even. Suppose, $\beta_0 = [\,0.5 \quad 0.5\,]$. Then, $\beta_k = [\,0.5 \quad 0.5\,]$ for all k so $\lim_{k\to\infty} \beta_k = [\,0.5 \quad 0.5\,]$. But, suppose, $\beta_0 = [\,0.7 \quad 0.3\,]$. Then, $\beta_k = [\,0.7 \quad 0.3\,]$ for k even and $\beta_k = [\,0.7 \quad 0.3\,]$ for k odd, so $\lim_{k\to\infty} \beta_k$ does not exist.

The fundamental reason that $\lim_{k\to\infty} \beta_k$ does not exist is that this Markov chain is *periodic*, and, in general, periodic Markov chains do not have limiting distributions.

EXAMPLE 2.7 Suppose $\{\tilde{x}_k, k = 0, 1, \ldots\}$ is a Markov chain such that

$$\mathcal{P} = \begin{bmatrix} 0.3 & 0.7 & 0 & 0 \\ 0.7 & 0.3 & 0 & 0 \\ 0 & 0 & 0.6 & 0.4 \\ 0 & 0 & 0.5 & 0.5 \end{bmatrix}.$$

Then,

$$\lim_{k\to\infty} \mathcal{P}^k = \begin{bmatrix} \frac{1}{2} & \frac{1}{2} & 0 & 0 \\ \frac{1}{2} & \frac{1}{2} & 0 & 0 \\ 0 & 0 & \frac{5}{9} & \frac{4}{9} \\ 0 & 0 & \frac{5}{9} & \frac{4}{9} \end{bmatrix}.$$

Therefore, $\lim_{k\to\infty} \beta_k$ exists, and $\lim_{k\to\infty} \beta_k = \beta_0 \mathcal{P}^\infty$. Now, suppose, $\beta_0 = [\,p_1 \quad p_2 \quad 0 \quad 0\,]$ for any nonnegative p_1, p_2 such that $p_1 + p_2 = 1$. Then, $\lim_{k\to\infty} \beta_k = [\,\frac{1}{2} \quad \frac{1}{2} \quad 0 \quad 0\,]$. Next, suppose $\beta_0 = [\,0 \quad 0 \quad p_3 \quad p_4\,]$ for any nonnegative p_3, p_4 such that $p_4 + p_4 = 1$. Then, $\lim_{k\to\infty} \beta_k = [\,0 \quad 0 \quad \frac{5}{9} \quad \frac{4}{9}\,]$. In general, if $\beta_0 = [\,p_1 \quad p_2 \quad p_3 \quad p_4\,]$ for any nonnegative p_i such that $p_1 + p_2 + p_3 + p_4 = 1$ then

$$\lim_{k\to\infty} \beta_k = [\,(p_1 + p_2)\tfrac{1}{2} \quad (p_1 + p_2)\tfrac{1}{2} \quad (p_3 + p_4)\tfrac{5}{9} \quad (p_3 + p_4)\tfrac{4}{9}\,].$$

From all of this, it is quite clear that $\lim_{k\to\infty} \beta_k$ exists, but its value is not independent of β_0.

The fundamental reason that $\lim_{k\to\infty} \beta_k$ is not independent of β_0 is that this Markov chain has multiple *classes*. That is, there are two separate sets of states. If the system ever enters state 0 or state 1, the system will never again enter state 3 or state 4. Similarly if the system ever enters either state 3 or state 4, the system will never again enter state 1 or state 2. In general, periodic Markov chains do not have limiting distributions.

In solving queueing problem of arbitrary difficulty, Markov chains of almost any form may be encountered. However, in the current text, we are interested in elementary queueing problems, and our interest lies in those Markov chains having only a single class for which $\lim_{k\to\infty}$ exists and is independent of β_0. If a DPMC has only one class, has only a finite number of states, and is aperiodic, then for that DPMC it is always true $\lim_{k\to\infty}$ exists and is independent of β_0. If a DPMC has an infinite number of states, is aperiodic, and has but one class, then for that DPMC the expected time between successive visits to the same state must be finite in order that $\lim_{k\to\infty}$ exists and is independent of β_0. A DPMC having these properties is said to be *ergodic*. For an indepth treatment of DPMCs, the interested reader is referred to Ross [2003].

EXERCISE 2.34 Staring with (2.18), show that the rows of $\mathcal{P}^\infty$ must be identical. [Hint: First calculate $\mathcal{P}^\infty$ under the assumption $\beta_0 = [\,1 \quad 0 \quad 0 \quad \ldots\,]$. Next, calculate $\mathcal{P}^\infty$ under the assumption $\beta_0 = [\,0 \quad 1 \quad 0 \quad \ldots\,]$. Continue along these lines.]

Alternatively, if $\lim_{k\to\infty} \beta_k$ exists, then by taking the limit on both sides of (2.17), we find $\beta_\infty = \beta_\infty \mathcal{P}$. In addition, for every k, β_k is a vector of probability masses, therefore, the elements of β_k sum to unity. Define $\mathbf{e}$ to be the unit (column) vector conforming to β_k; that is,$\mathbf{e} = [\,1 \quad 1 \quad 1 \quad \ldots\,]$. Then, $\beta_k \mathbf{e} = 1$ for every k, and $\beta_\infty \mathbf{e} = 1$. If, in addition, $\lim_{k\to\infty} \beta_k$ is independent of β_0, then $\pi = \lim_{k\to\infty} \beta_k$, and π satisfies

$$\pi = \pi\mathcal{P} \quad \text{and} \quad \pi\mathbf{e} = 1. \tag{2.19}$$

DEFINITION 2.30 **Stationary vector of a DPMC.** If $\{\tilde{x}_k, k = 0, 1, \ldots\}$ is a DPMC such that its limiting distribution exists and is independent of its initial state probability vector, then the solution to (2.19) is called the *stationary vector of* $\{\tilde{x}_k, k = 0, 1, \ldots\}$.

Two different approaches are commonly used to determine the stationary vector for a DPMC. The first iterates the equation $\beta_{k+1} = \beta_k \mathcal{P}$. An arbitrary vector is chosen for β_0. Then β_1 is computed as $\beta_0 \mathcal{P}$, β_2 is computed as $\beta_1 \mathcal{P}$ and so forth. A stopping criteria is established to determine when enough iterations have taken place so that the result approximates β_∞. At that point, β_∞ is assigned to π.

The second approach starts with (2.19). Then, the matrix equation $\pi(I - \mathcal{P}) = 0$ is formed. Next, any column of $(I - \mathcal{P})$ is replaced by $\mathbf{e}$ and the corresponding element of the zero vector on the right hand side is replaced by 1. The resulting equation is then solved to determine π.

EXERCISE 2.35 Suppose $\{\tilde{x}_k, k = 0, 1, \ldots\}$ is a Markov chain such that

$$\mathcal{P} = \begin{bmatrix} 0.6 & 0.4 \\ 0.5 & 0.5 \end{bmatrix}.$$

Determine the stationary vector of $\{\tilde{x}_k, k = 0, 1, \ldots\}$.

We now discuss the appropriate interpretation of π. The value of the ith element of π, π_i, reveals the long term proportion of all transitions that are into state i. That is, for each k, there is a state transition. At $k = 0$, we start counting the transitions. Suppose that of the K transitions of $\{\tilde{x}_k, k = 0, 1, \ldots\}$ that take place over $k = 1, 2, \ldots, K$, there are n_{Ki} transitions into state i. Then,

$$\pi_i = \lim_{K \to \infty} \frac{n_{Ki}}{K}, i = 0, 1, \ldots. \tag{2.20}$$

We note that in some cases it is possible to solve the equations specified in (2.19) even if $\lim_{k \to \infty} \beta_k$ does not exist. In such cases, the interpretation of the result is the same as that of π, but the implications are different. For example, if

$$\mathcal{P} = \begin{bmatrix} 0 & 1 \\ 1 & 0 \end{bmatrix},$$

then $\lim_{k \to \infty} \beta_k$ does not exist. Nonetheless, if we form the equations $\phi \mathcal{P} = \phi$ and $\phi \mathbf{e} = 1$ and solve, we will find $\phi = [0.5 \quad 0.5]$. Because transitions from state 0 are always into state 1 and vice versa, it is clear that one half of the transitions are into state 0 and one half are into state 1. But, $\phi \neq \lim_{k \to \infty} \beta_k$ because that limit does not exist as shown in Example 2.6.

In the second arrival process considered in Section 1.2.2, packet arrivals on each input line to the switch follow an on-off process. In each time slot, if the arrival process on a given input line is in the on state, the system returns to the on state in the next time slot with probability p_{11} or goes to the off state with probability p_{10}. Similarly, if the arrival process on a given input line is in the off state, the system returns to the off state in the next time slot with probability p_{00} or goes to the on state with probability p_{01}. Since future evolution is based solely on the present state, we see that this particular on-off process is a Markov chain. Specifically, suppose the on state is designated by 1 and the off state by 0, and let $\{\tilde{v}_k^i, k = 0, 1, \ldots\}$ denote the number of arrivals from line i during time slot k. Then, $\{\tilde{v}_k^i, k = 0, 1, \ldots\}$ is a DPMC with state space $\{0, 1\}$, and its one-step state transition matrix is

$$\mathcal{P} = \begin{bmatrix} p_{00} & p_{01} \\ p_{10} & p_{11} \end{bmatrix}.$$

It also turns out that $\{\tilde{v}_k, k = 0, 1, \ldots\}$, where

$$\tilde{v}_k = \sum_{i=1}^{N} \tilde{v}_k^i$$

is the total number of arrivals from the N lines, is a DPMC chain with state space $\{0, 1, \ldots, N\}$.

EXERCISE 2.36 Develop the one-step state probability transition matrix for the special case $\{\tilde{v}_k, k = 0, 1, \ldots\}$ for the special case of $N = 5$.

EXERCISE 2.37 For the example discussed in Section 1.2.2 in the case where arrivals occur according to an on off process, determine whether or not $\{\tilde{q}_k, k = 0, 1, \ldots\}$ is a DPMC. Defend your conclusion mathematically; that is, show that $\{\tilde{q}_k, k = 0, 1, \ldots\}$ either does or does not satisfy the definition of a Markov chain.

In the example of Section 1.2.2, the k transition of $\{q_k, k = 0, 1, \ldots\}$ occurs at the end of the kth time slot. Thus, there is a one-to-one relationship between clock time and the time of the transition times; that is, the elements of the parameter set represent discrete time. However, in general, for a DPMC, there is no direct relationship between clock time and the parameter set.

As an example of a case where the parameter set does not represent real time, consider a digital voice system that is regulated by a token bucket system as analyzed in Blefari-Melazi et. al. [2003]. Voice is typically modeled as an on-off process, alternating between periods of talk and silence, each of which is modeled as a sequence of independent, identically distributed random variables. Voice symbols are generated only during talk periods. A token bucket system has a token bucket of capacity, σ, into which tokens flow at rate ϕ tokens per second. Voice is sampled at some rate, r, and each voice sample is converted to an 8-bit symbol. Each voice symbol arrives to the token bucket, where the system determines whether or not a token is available. If no token is available, the voice symbol is dropped. If a token is available, then a token is removed from the leaky bucket, and the voice symbol may be transmitted into the communication system.

Define $\tilde{n}_k$ to be the number of tokens present at the beginning of the k talk period. The distribution of $\tilde{n}_k$ is an important factor in the quality of service provided by the system. It is not difficult to establish that $\{\tilde{n}_k, k = 0, 1, \ldots\}$ is a DPMC on the state space $\{0, 1, \ldots, \sigma\}$. Also, since the lengths of the talk and silence periods are random variables, it is also clear that the parameter set does not have a one-to-one relationship with clock time. In fact, as a practical matter, the clock time is irrelevant. What is important is the number of tokens available to service the talk spurt, whatever time it may start.

In the case of the voice system just discussed, the time between transitions of the DPMC $\{\tilde{n}_k, k = 0, 1, \ldots\}$ is the sum of the length of a talk period plus the length of a silent period. Denote the length of the kth talk and silent periods by $\tilde{t}_k$ and $\tilde{s}_k$, respectively. Then, the length of time between the kth and the $(k+1)$st transition is $\tilde{t}_k + \tilde{s}_k$.

In many cases of practical interest, the times between successive transitions of a DPMC are exponentially distributed; such a Markov chain is referred to as a continuous-time Markov chain (CTMC). Although a CTMC can be and is often analyzed by first embedding a DPMC at points of transition, it is common to analyze a CTMC directly as a continuous-time process. A formal definition of the CTMC is now presented.

DEFINITION 2.31 **Continuous-time Markov chain.** A stochastic process $\{\tilde{x}(t), t \geq 0\}$ is said to be a continuous-time Markov chain on the nonnegative integers (Ross [1983]) if for all $s, t \geq 0$, and all nonnegative integers $i, j, x(u)$ for $0 \leq u \leq s$,

$$P\{\tilde{x}(t+s) = j \mid \tilde{x}(s) = i, \tilde{x}(u) = x(u)\} = P\{\tilde{x}(t+s) = j \mid \tilde{x}(s) = i\}.$$

The quantity $P\{\tilde{x}(t+s) = j \mid \tilde{x}(s) = i\}$ is called the transition probability from state i to state j over time $(s, s+t]$.

From the three definitions of the Poisson process given above, it can be seen that the Poisson process is a time-homogeneous, continuous-time Markov chain on the nonnegative integers.

DEFINITION 2.32 **Time-homogeneous CTMC (Ross[1989]).** A CTMC is said to be a time-homogeneous CTMC on the nonnegative integers if $\{\tilde{x}(t), t \geq 0\}$ is a CTMC and for all $s, t \geq 0$,

$$P\{\tilde{x}(t+s) = j \mid \tilde{x}(s) = i\} = P\{\tilde{x}(t) = j \mid \tilde{x}(0) = i\} \quad \text{for all } 0 \leq s \leq t.$$

That is, for a time homogeneous CTMC the transition probability from state i to state j over time $(s, s+t]$ is independent of s for all i, j.

DEFINITION 2.33 **Transition probability matrix for CMTC**. For a time-homogeneous CTMC $\{\tilde{x}(t), t \geq 0\}$, , define

$$p_{ij}(t) = P\{\tilde{x}(t+s) = j \mid \tilde{x}(s) = i\}.$$

Then, the matrix $\mathcal{P}(t) = [p_{ij}(t)]$ is called the transition probability matrix over $(s, s+t]$ (Ross[1989]).

DEFINITION 2.34 **Infinitesimal generator for a CTMC (Cohen[1969])**. For a time-homogeneous CTMC $\{\tilde{x}(t), t \geq 0\}$, the matrix $P(t)$ satisfies the following (possibly infinite dimensional) matrix differential equation:

$$\frac{d}{dt}\mathcal{P}(t) = \mathcal{P}(t)\mathcal{Q},$$

with $P(0) = I$. The (possibly infinite dimensional) matrix $\mathcal{Q}$ is called the *infinitesimal generator*, or simply the *generator*, for the CTMC $\{\tilde{x}(t), t \geq 0\}$.

EXERCISE 2.38 Suppose $\{\tilde{x}(t), t \geq 0\}$ is a time-homogeneous CTMC having infinitesimal generator $\mathcal{Q}$ defined as follows:

$$\mathcal{Q}_{ij} = \begin{cases} -\lambda, & \text{if } j = i, \\ \lambda, & \text{if } j = i+1, \\ 0, & \text{otherwise.} \end{cases}$$

Show that $\{\tilde{x}_1(t), t \geq 0\}$ is a Poisson process. [*Hint*: Simply solve the infinite matrix differential equation term by term starting with $\mathcal{P}_{00}(t)$ and completing each column in turn.]

For a CTMC $\{\tilde{n}(t), t \geq 0\}$ define $P_i(t) = P\{\tilde{n}(t) = i\}$ for $i = 0, 1, \ldots$. Then, define the vector of probability masses as $P(t) = [\,P_i(t) \quad P_1(t) \quad \ldots]$. Then,

$$P(t+s) = P(s)\mathcal{P}(t) \quad \text{for all } 0 \leq s \leq t, t \geq 0. \tag{2.21}$$

As in the case of the DPMC, it may be that $\lim_{t\to\infty} P(t)$ exists and is independent of $P(0)$. In that case, we find that

$$P(\infty) = \lim_{t\to\infty} P(t) = P(0)\mathcal{P}(\infty),$$

and we refer to $P(\infty)$ as the equilibrium probability vector for the CTMC. Alternatively, we may obtain the following expression from Definition 2.34:

$$\frac{d}{dt}P_0\mathcal{P}(t) = P_0\mathcal{P}(t)\mathcal{Q}.$$

Because $P(t) = P(0)\mathcal{P}(t)$, we have

$$\frac{d}{dt}P(t) = P(t)\mathcal{Q}.$$

Because $P(t)$ is a vector of probability masses, we have $P(t)\mathrm{e} = 1$ for all t. In the limit, $\frac{d}{dt}P(t) = 0$. Therefore, $P(\infty)$ satisfies the following conditions:

$$P(\infty)\mathcal{Q} = 0 \quad \text{and} \quad P(\infty)\mathrm{e} = 1. \tag{2.22}$$

From (2.22), we may solve for the equilibrium probability vector $P(\infty)$.

We now discuss the appropriate interpretation of $P(\infty)$. The value of the ith element of $P(\infty)$, $P_i(\infty)$, reveals the long term proportion of all time that the system spends in state i. That is, at each time t, the system is in some state. Let $\tau_i(t) = 1$ if the system is in state i at time t and $\tau_i(t) = 0$ otherwise. Then, over the interval $(0, T]$, the total amount of time the system spends in state i over the interval $(0, T]$ is given by the integral of $\tau_i(t)$ over the interval $(0, T]$. Thus, the long-term proportion of time spend in state i is given by

$$P_i(\infty) = \lim_{T\to\infty} \frac{1}{T}\int_0^T \tau_i(t)dt.$$

We then say that $P(\infty)$ is a *time-averaged probability*.

Defining relationships among time- and frequency-averaged probabilities of interrelated stochastic process is an important aspect of solving complex queueing problems. A description of and solution to a queueing problem whose solution requires definition and solution of many subproblems involving Markov chains of many kinds is presented in Daigle and Magalhães [2003]. The problem addressed in that reference relates to the design of a cellular system in which transmission of data requires both queueing and prior reservation.

It is sometimes convenient to analyze CTMC by first embedding a DPMC at points of transition, then solving for the stationary vector of the DPMC, and then calculating the equilibrium probability vector for the CTMC by weighting the stationary vector of embedded DPMC. In some cases, an example of which may be found in Abboud and Daigle [1997], such an approach can drastically reduce the computational complexity of the problem at hand. The transformations necessary are discussed here and used liberally throughout the text when needed.

In order to properly define the transition probabilities for the embedded DPMC, it is necessary only to use the infinitesimal generator and the properties of the exponential distribution presented earlier in this chapter. Specifically, $\mathcal{Q}_{ii}$ reveals the total rate at which the system departs from state i. Since we know the time between transitions of a CTMC is exponential, we know the time spend during each visit to state i is exponentially distributed with parameter $-\mathcal{Q}_{ii}$. Now, this exponentially distributed time spent in state i is the minimum of a number of exponentially distributed random variables. Specifically, if nothing else were going on in the system, the system would remain in state i before transitioning to state j for an exponentially distributed period of time with parameter $\mathcal{Q}_{ij}$, $j \neq i$. Therefore,

$$-\mathcal{Q}_{ii} = \sum_{j \neq i} \mathcal{Q}_{ij}.$$

The probability that the system transitions from state i to state j is then just the probability that the exponential time required to reach state j expires before the exponential time required to reach any other state, which is simply the ratio of $\mathcal{Q}_{ij}$ to $-\mathcal{Q}_{ii}$, as stated in Property 3 of exponential random variables.

Let $\{\tilde{x}(t), t \geq 0\}$ denote the CTMC of interest, $\tilde{t}_k$ denote the time of the kth transition of $\{\tilde{x}(t), t \geq 0\}$, $\tilde{x}_k = \tilde{x}(\tilde{t}_k^+)$ denote the state of the system just after the kth transition, and $\{\tilde{x}_k, k = 0, 1, \ldots\}$ denote the DPMC embedded at points of transition of the CTMC. Then

$$P\{x_{k+1} = j \mid x_k = i\} = \frac{\mathcal{Q}_{ij}}{-\mathcal{Q}_{ii}}.$$

Proceeding in this manner, we can construct $\mathcal{P}$ for the embedded DPMC, and then we can solve for the stationary vector π by solving $\pi\mathcal{P} = \pi$ and $\pi \mathrm{e} = 1$ simultaneously .

Having determined π, we can now do a proper weighting of the elements of π to obtain P_∞. We reason as follows. Let $\tilde{s}_i$ denote the time spent during a visit to state i. Then, as stated in Property 2 of exponential distributions, the time spent in state i is independent of the state into which the next transition occurs. Therefore, $E[\tilde{s}_i] = -1/\mathcal{Q}_{ii}$. Now, suppose that a total of $N(T)$ transitions occur up to time T, and of those, $N_i(T)$ are into state i. Further, denote the time spent in state i on the ℓ_ith visit by $s_{i\ell_i}$ and the total time spent in state i up to time T by $T_i(T)$. Then, for large T, for which the quantity $N_i(T)$ is also large,

$$T_i(T) \approx \sum_{\ell=1}^{N_i(T)} s_{i\ell} = \frac{N_i(T)}{N_i(T)} \sum_{\ell=1}^{N_i(T)} s_{i\ell} = N_i(T) \left[\frac{1}{N_i(T)} \sum_{\ell_i=1}^{N_i(T)} s_{i\ell_i} \right].$$

But, T itself is the sum of the times spent in all states up to time T. Therefore,

$$T \approx \sum_j N_j(T) \left[\frac{1}{N_j(T)} \sum_{\ell_j=1}^{N_j(T)} s_{j\ell_j} \right].$$

The proportion of time the system spends in state i up to time T is given by the $T_i(T)/T$. If we divide both the expression for $T_i(T)$ and T by $N(T)$ and take the limit as $T \to \infty$, we find

$$\lim_{T \to \infty} \frac{N_j(T)}{N(T)} = \pi_j.$$

$$\lim_{T\to\infty} \frac{1}{N_j} \sum_{\ell_j=1}^{N_j(T)} s_{j\ell_j} = E[\tilde{s}_j],$$

and the limit of the proportion of time spent in state i as $\mathcal{P}_i(\infty)$. Therefore we have

$$P_i(\infty) = \frac{\pi_i E[\tilde{s}_i]}{\sum_j \pi_j E[\tilde{s}_j]}. \tag{2.23}$$

This result is also presented in Wolff [1989], pp. 215-216.

If the times between transitions are not exponentially distributed, but instead have general distributions, the parent process of the DPMC embedded at points of transitions is called a semi-Markov process. It is interesting to note that the conversion from frequency to time averages follows along basically the same lines as in the case of the CTMC; that is, conversion is made according to (2.23). It is also straightforward to develop formulae for conversion from frequency based probabilities to time averaged probabilities in more general settings (see Daigle and Magalhães [2003] for an application).

It is also sometimes necessary to compute the stationary vector of the DPMC embedded the points of transition of a CTMC from the equilibrium probabilities of the CTMC. The appropriate transformation is obtained by simply taking the ratio of the total number of transitions into state j to the total number of transitions into all states. In order to determine the total numbers of transitions into each state, we consider a large interval of time, $(0, T]$. Whenever the system is in state i, there are transitions into state ℓ at rate $\mathcal{Q}_{i\ell}$. Thus, the total number of transitions into state ℓ is approximately $P_i(\infty)\mathcal{Q}_{i\ell}T$. To obtain the total number of transitions into state ℓ, we then simply sum over all i. Then to obtain the total number of transitions of the system, we sum over all ℓ. We then form the ratio of the total number of transitions into state j to the total number of transitions of the system over $(0, T]$ and take the limit as $T \to \infty$. The result is

$$\pi_j = \frac{\sum_i P_i(\infty)\mathcal{Q}_{ij}}{\sum_\ell \sum_i P_i(\infty)\mathcal{Q}_{i\ell}}. \tag{2.24}$$

EXERCISE 2.39 Let $\{\tilde{x}(t), t \geq 0\}$ be a CTMC such that

$$\mathcal{Q} = \begin{bmatrix} -2.00 & 1.25 & 0.75 \\ 0.50 & -1.25 & 0.75 \\ 1.00 & 2.00 & -3.00 \end{bmatrix}.$$

1. Solve for $P(\infty)$ directly by solving $P(\infty)\mathcal{Q} = 0$, $P(\infty)\mathrm{e} = 1$.
2. Solve for π for the DPMC embedded at points of state transition using (2.24).
3. Find $\mathcal{P}$ for the DPMC embedded at points of state transition.
4. Show that the value of π found in part 2 of this problem satisfies $\pi = \pi\mathcal{P}$ for the $\mathcal{P}$ found in part 3 of this problem.

Chapter 3

ELEMENTARY CONTINUOUS-TIME MARKOV CHAIN-BASED QUEUEING MODELS

In this chapter, we explore the analysis of several queueing models that are characterized as discrete-valued, continuous-time Markov chains (CTMCs). That is, the queueing systems examined in this chapter will have a countable state space, and the dwell times in each state will be drawn from exponential distributions whose parameters are possibly state-dependent.

The most elementary queueing systems in this class are characterized by one-dimensional birth and death models. The stochastic behavior of these systems at a particular point in time is completely described by a single number, which we shall think of as the "occupancy" of the system. The dwell times for each state are drawn from exponential distributions independently, but, in general, the parameter of the exponential distribution depends upon the current state of the system.

We begin by examining the well known M/M/1 queueing system, which has Poisson arrivals and exponentially distributed service times. For this model, we will consider both time-dependent and equilibrium behavior, with primary emphasis on the latter. In particular, we shall consider both the time-dependent and equilibrium occupancy distributions, the stochastic equilibrium sojourn and waiting time distributions, and the stochastic equilibrium distribution of the length of the busy period. Several related processes, including the departure process, are introduced, and these are used to obtain equilibrium occupancy distributions for simple networks of queues.

After discussing the M/M/1 system, we briefly discuss formulation of the dynamical equations for more general birth-death models in Section 3.2. The time-dependent behavior of finite-state general birth-death models is discussed in Section 3.3. A reasonably complete derivation based upon classical methods is presented herein, and the rate of convergence of the system to stochastic equilibrium is briefly discussed. Additionally, the notion of *randomization*, or equivalently *uniformization*, is introduced. The basic idea is to study a finite-state, continuous-time Markov chain by embedding a finite-state, discrete-time Markov chain whose intertransition times are independent, identically distributed, exponential random variables. Randomization is described in general terms, and an example that illustrates its application is provided.

Section 3.4 presents the balance equation approach to formulating equilibrium state probability equations for birth-death processes and other more general processes. Elementary traffic engineering models are introduced and blocking probabilities for these systems are discussed.

The probability generating function technique for solving balance equations is introduced in Section 3.5. We conclude the chapter with a set of supplementary exercises.

3.1 M/M/1 Queueing System

This section comprises three subsections. In Section 3.1.1, we consider the time-dependent occupancy distribution. We then derive the stochastic equilibrium occupancy, sojourn, and waiting time distributions, together with their means. Along the way, we introduce various related processes, including the occupancy processes as viewed by departing and arriving customers, respectively, which are needed to obtain these results. We also discuss the departure process and its role in obtaining occupancy distributions for simple feedforward networks of queues. In Section 3.1.3, we discuss the dynamics of busy-period processes and derive an expression for the expected length of the busy period in stochastic equilibrium. We also discuss other characteristics of the busy period and briefly discuss the role of busy-period analysis in examining more complicated systems.

3.1.1 Time-Dependent M/M/1 Occupancy Distribution

As mentioned in the introductory section, the M/M/1 queueing system has Poisson arrivals and exponentially distributed service. Due to the memoryless property of both the Poisson process and the exponential distribution, the dynamics of the process that counts the total number of arrivals to and departures from the system over very short periods of time are exactly the same as those of the Poisson process. If there are customers in the system, then the rate for this process is the sum of the arrival and service rates. If there are no customers in service, the rate for the process is simply the arrival rate. Let $\tilde{n}(t)$ denote the system occupancy -the total number of customers in the system, including the one in service, if any- at time t. To simplify notation, let

$$P_n(t) \triangleq P\{\tilde{n}(t) = n\}, \qquad n \geq 0.$$

Clearly, the stochastic process $\{\tilde{n}(t), t \geq 0\}$ is a continuous-time Markov chain, and for $n > 0$, we find

$$\begin{aligned} P\{\tilde{n}(t+h) = n\} &= P\{\tilde{n}(t) = n, \\ &\qquad 0 \text{ arrivals or departures in } (t, t+h]\} \\ &+ P\{\tilde{n}(t) = n-1, \\ &\qquad 1 \text{ arrival and no departures in } (t, t+h]\} \\ &+ P\{\tilde{n}(t) = n+1, \\ &\qquad 1 \text{ departure and no arrivals in} (t, t+h]\} \\ &+ o(h). \end{aligned}$$

Let λ and μ denote the arrival and service rates, respectively. Then, by applying Definition 2 of the Poisson process, we find

$$\begin{aligned} P\{\tilde{n}(t+h) = n\} &= P_n(t)[1 - \lambda h + o(h)][1 - \mu h + o(h)] \\ &+ P_{n-1}(t)[\lambda h + o(h)][1 - \mu h + o(h)] \\ &+ P_{n+1}(t)[\mu h + o(h)][1 - \lambda h + o(h)] + o(h) \\ &= P_n(t)[1 - (\lambda + \mu)h + o(h)] \\ &+ P_{n-1}(t)[\lambda h + o(h)] \\ &+ P_{n+1}(t)[\mu h + o(h)] + o(h). \end{aligned}$$

Upon rearranging the terms of the previous equation, we find

$$\begin{aligned} P_n(t+h) - P_n(t) &= -(\lambda + \mu)hP_n(t) + \lambda h P_{n-1}(t) \\ &\quad + \mu h P_{n+1}(t) + o(h). \end{aligned} \tag{3.1}$$

Finally, division of both sides of (3.1) by h, taking limits, and applying the definition of $o(h)$ leads to the following dynamical equation relating the state probabilities to each other:

$$P_n'(t) = -(\lambda + \mu)P_n(t) + \lambda P_{n-1}(t) + \mu P_{n+1}(t) \qquad \text{for } n > 0. \tag{3.2}$$

Similarly, we find for $n = 0$,

$$P_0'(t) = -\lambda P_0(t) + \mu P_1(t). \tag{3.3}$$

The solution to the system (3.2) and (3.3) depends upon the initial number of customers, i, in the system. Then, from L. Takács [1962], pp. 23-26, we find

$$\begin{aligned} P_n(t; i) &= e^{-(\lambda+\mu)t}[(\lambda/\mu)^{(n-i)/2} I_{n-i}(2\sqrt{\lambda\mu}\, t) \\ &\quad + (\lambda/\mu)^{(n-i+1)/2} I_{n+i+1}(2\sqrt{\lambda\mu}\, t) \\ &\quad + (1 - \lambda/\mu)(\lambda/\mu)^n \sum_{\nu=i+n+2}^{\infty} (\lambda/\mu)^{-\nu/2} I_\nu(2\sqrt{\lambda\mu}\, t)], \end{aligned} \tag{3.4}$$

where $I_\nu(x)$ for $\nu = 0, \pm 1, \pm 2, \ldots$ is the modified Bessel function of order ν. For $\nu \geq 0$,

$$I_\nu(x) = \sum_{j=0}^{\infty} \frac{(x/2)^{\nu+2j}}{(j+\nu)!j!}$$

and

$$I_{-\nu}(x) = I_\nu(x). \tag{3.5}$$

For an application of (3.4) to a flow control problem in a computer communication network, see Stern [1979].

Evaluation of (3.4) would appear to be a formidable task. First of all, the results are given in the form of an infinite series of modified Bessel functions. Secondly, each of the modified Bessel functions is itself expressed as an infinite series. Fortunately, as indicated in many references, (3.4) and (3.5) are not the most efficient starting point for evaluating the time-dependent state probabilities. The most efficient starting point for numerical work appears to be an integral equation expression. For a discussion of numerical methods for computing these probabilities and other time-dependent quantities of interest, the reader is referred to two excellent papers: Abate and Whitt [1988] and Abate and Whitt [1989]. For a treatment of the time-dependent behavior of a more complicated version of the M/M/1 system, the reader is referred to Daigle and Magalhães [1989] and the references therein.

3.1.2 Stochastic Equilibrium M/M/1 Distributions

In the previous section, we obtained the time-dependent probability distribution for the system occupancy. In most cases of practical interest, the time-dependent probability distribution converges to a unique solution as time increases beyond bound. This solution is called the *stochastic equilibrium* solution, stochastic equilibrium meaning that the distribution is no longer changing as a function of time.

We note in passing that equilibrium is never actually reached, except in the sense of a limit, unless the initial distribution is chosen as the equilibrium distribution. On the other hand, for most applications, an understanding of the stochastic equilibrium behavior of the system is sufficient. In that case, we can solve (3.2) and (3.3) for the equilibrium probabilities and use those results to derive the stochastic equilibrium sojourn time and waiting time distributions.

Let $\tilde{n}$ denote the queue occupancy at an arbitrary point in time after the system has reached stochastic equilibrium. We define $P_n = P\{\tilde{n} = n\}$, or equivalently, $P_n = \lim_{t\to\infty} P_n(t)$. We then expect that $P_n'(t) \to 0$ (although this is not absolutely necessary from a mathematical point of view) and (3.3) and (3.2) become (3.5) and (3.6), respectively. That is,

$$\mu P_1 = \lambda P_0, \tag{3.6}$$

and

$$\mu P_{n+1} = (\lambda + \mu)P_n - \lambda P_{n-1}. \tag{3.7}$$

Upon substitution of (3.7) into (3.6) with $n = 1$, we find

$$P_2 = \frac{\lambda}{\mu} P_1,$$

and solving (3.6) for P_1 yields $P_1 = (\lambda/\mu)P_0$. Thus, we find

$$P_2 = \left(\frac{\lambda}{\mu}\right)^2 P_0.$$

Repeating this procedure leads to the general expression

$$P_n = \left(\frac{\lambda}{\mu}\right)^n P_0 \quad \text{for} \quad n \geq 0.$$

Because the probabilities sum to unity, we find that if $\lambda/\mu < 1$,

$$P_0 = 1 - \frac{\lambda}{\mu}.$$

Thus, in general, the stochastic equilibrium occupancy probabilities are given by

$$P_n = \left(1 - \frac{\lambda}{\mu}\right)\left(\frac{\lambda}{\mu}\right)^n \quad \text{for} \quad n \geq 0. \tag{3.8}$$

Note that $E[\,\text{number in system at time}\ t\,] \geq \lambda t - \mu t = (\lambda - \mu)t$ since $E[$ number of arrivals by time $t\,] = \lambda t$ and $E[$ number of service completions by time $t\,] \leq \mu t$. Thus, if $\lambda > \mu$, then $\lim_{t\to\infty} E[\,\text{number in system at time}\ t\,]$ grows beyond bound. So, in order to have an equilibrium solution, we cannot have $\lambda > \mu$; that is, the arrival rate cannot exceed the service rate. In fact, for there to be an equilibrium solution, we actually need $\lambda < \mu$. To see why this is true intuitively, we can draw an analogy between the system occupancy and the position of a *random walker* on the nonnegative integers.

A random walker steps either to the left or right according to the following rules. If the walker is at position zero, one step to the right is taken with probability one. If the walker is not at position zero then before taking a step, a coin is flipped. If the result is "heads", the walker steps one step to the right, else one step to the left is taken. It is easy to see that if the probability of "heads" exceeds one-half, then the walker tends to drift to the right. The longer the experiment continues, the further to the right we would expect the walker to be; no stochastic equilibrium distribution would be reached. On the other hand, if the probability of "heads" is less than one-half, then the walker tends to drift to the left. It would be possible for the walker to roam any distance to the right through a series of "heads" outcomes, but the positive tendency to move to the left would tend to return the walker to position zero occasionally. Thus one would expect a stochastic equilibrium solution to exist.

More formally, the position of the walker, measured in steps to the right from zero, is the state of an irreducible discrete-time Markov chain having a countable number of states. From the theory of Markov chains (see, for example, Wolff [1989]), it is well known that the states are *positive recurrent* if $P\{\text{heads}\} < 0.5$, *null recurrent* if $P\{\text{heads}\} = 0.5$, and *transient* if $P\{\text{heads}\} > 0.5$. An equilibrium solution exists if and only if all states are positive recurrent.

EXERCISE 3.1 Carefully pursue the analogy between the random walk and the occupancy of the M/M/1 queueing system. Determine the probability of an increase in the queue length, and show that this probability is less than 0.5 if and only if $\lambda < \mu$.

In the case of single-server queueing systems without state-dependent arrival and service rates, the quantity λ/μ is called the *traffic intensity*, and it is usually designated by ρ; that is, $\rho \equiv \lambda/\mu$. Since $\tilde{n}$ denotes the number of customers in the system at an arbitrary point in time after the system has reached stochastic equilibrium, and $P_n \equiv P\{\tilde{n} = n\}$, we have

$$P_n = \rho^n (1 - \rho). \tag{3.9}$$

The stability condition for the queueing system is then stated as $\rho < 1$.

From (3.9), we may find the probability that the total number in the system exceeds n. In particular,

$$P\{\tilde{n} > n\} = \sum_{j=n+1}^{\infty} \rho^j (1 - \rho) = \rho^{n+1}. \tag{3.10}$$

Graphs of the quantity $P\{\tilde{n} > n\}$, which is called the survivor function or complementary distribution for the number of customers in the system, are shown in Figure 3.1 for several values of traffic intensity. From these graphs we see that as traffic intensity nears unity, relatively small changes in traffic intensity result in large changes in the probability that the occupancy exceeds a given value. For example, at $\rho = 0.9$, $P\{\tilde{n} > 40\} \approx 0.01$, but at $\rho = 0.95$, $P\{\tilde{n} > 40\} > 0.1$. These probabilities are not to be confused with blocking probabilities, which are discussed in a later section.

We now turn to the computation of averages for the number in the system and the time spent in the system. In order to compute average values, we make use of the following theorems, the proofs of which we leave to the exercises.

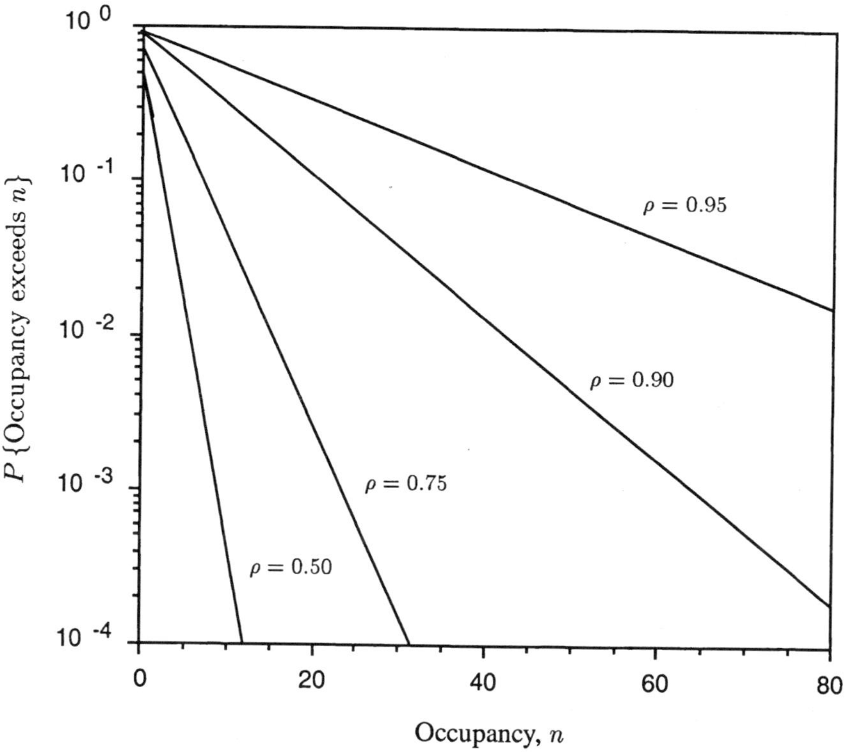

Figure 3.1. Survivor function for system occupancy for several values of ρ.

THEOREM 3.1 *Let $\tilde{x}$ be a nonnegative integer-valued random variable. Then*

$$E[\tilde{x}] = \sum_{n=0}^{\infty} P\{\tilde{x} > n\}.$$

□

EXERCISE 3.2 Prove Theorem 3.1 and its continuous analog

$$E[\tilde{x}] = \int_0^{\infty} P\{\tilde{x} > x\}dx.$$

THEOREM 3.2 *Let $\tilde{x}$ and $\tilde{y}$ be any two nonnegative random variables. Then*

$$E[\min \tilde{x}, \tilde{y}] < \min E[\tilde{x}], E[\tilde{y}].$$

□

EXERCISE 3.3 Prove Theorem 3.2.

From Theorem 3.1, we find

$$E[\tilde{n}] = \sum_{n=0}^{\infty} P\{\tilde{n} > n\}.$$

Substitution of (3.10) into this equation yields

$$E[\tilde{n}] = \frac{\rho}{1-\rho}. \tag{3.11}$$

Then, if we assume ergodicity[1] and let $N(t)$ denote the number of customers in the system at time t for a typical sample path,

$$\begin{aligned} E[\tilde{n}] &\triangleq \lim_{t\to\infty} \frac{\int_0^t n(s)ds}{t} \\ &= \frac{\rho}{1-\rho}. \end{aligned}$$

That is, for a particular system under study, $E[\tilde{n}]$ is the expected number of customers in the system when averaged over time.

Figure 3.2 shows a graph of the mean occupancy as a function of traffic intensity. Again we see the effect of increasing occupancy due to increasing traffic intensity. As $\rho \to 1$ and the system nears instability, the mean occupancy grows without bound, as expected.

Another quantity of interest is the sojourn time, the total time customers spend in the system including both waiting time and service time. Following our notation of Chapter 2, let $\tilde{s}$ denote the stochastic equilibrium value for this quantity with $F_{\tilde{s}}$ being its distribution. The set of events $\{\tilde{n} = n, n = 0, 1 \ldots\}$ partitions the sample space, so we have

$$E[\tilde{s}] = \sum_{n=0}^{\infty} E[\tilde{s} \mid \tilde{n} = n] P\{\tilde{n} = n\}. \tag{3.12}$$

Now, the sojourn time is measured from the time an *arbitrary arriving customer* enters the system, but $\tilde{n}$ represents the view of an *arbitrary observer.* The following exercise illustrates that these points of view are not necessarily the same.

[1]Ergodicity is a very technical concept, but basically it implies that time averages are equal to ensemble averages. That is, if we collect statistics at a single point in time from a large number of systems that are operating in stochastic equilibrium, then those measurements will be statistically the same as measurements taken from a single system over a long period of time.

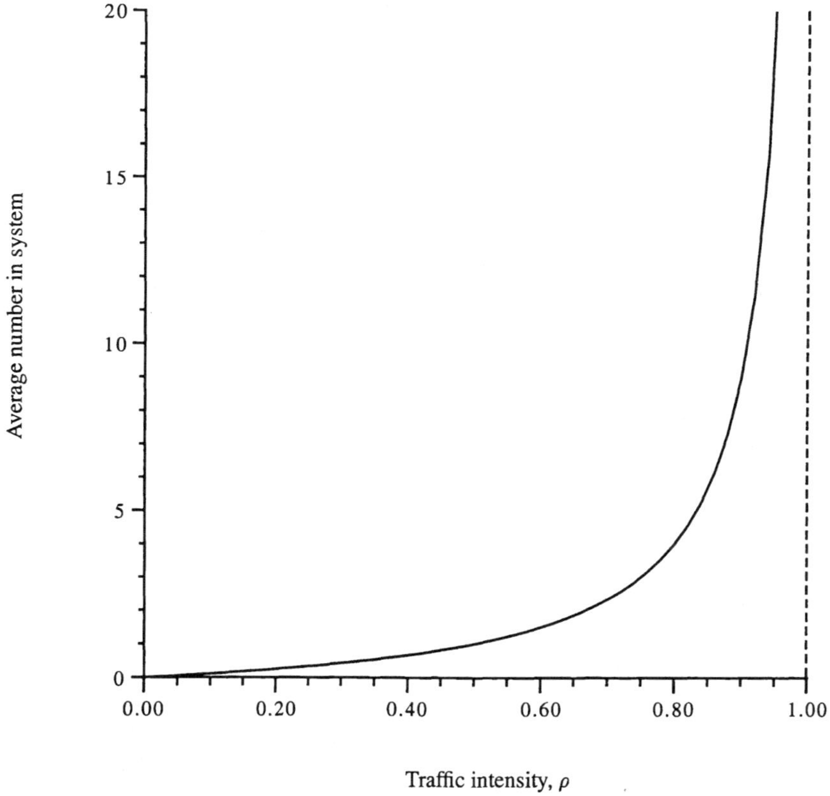

Figure 3.2. Schematic diagram of a single-server queueing system.

EXERCISE 3.4 Suppose customers arrive to a system at the end of every even-numbered second and each customer requires exactly one second of service. Compute the stochastic equilibrium occupancy distribution, that is, the time-averaged distribution of the number of customers found in the system. Compute the occupancy distribution as seen by arriving customers. Compare the two distributions. Are they the same?

The following exercise shows that for the special case of the M/M/1 queueing system, the limiting distribution of the number of customers seen by a departing customer is equal to the limiting distribution of the number of customers found in the system by arriving customers, and these distributions are both equal to the stochastic equilibrium distribution. Thus, for the M/M/1 queueing system, the arrivals see the stochastic equilibrium occupancy distribution as given in (3.8). However, before presenting the exercise, we introduce some definitions and notation.

Define $\tilde{n}_d(k)$, for $k = 1, 2, \ldots$, to be the number of customers left in the system by the kth departing customer, and define

$$\pi_{dj} = \lim_{k \to \infty} P\{\tilde{n}_d(k) = j\} \quad \text{for} \quad j = 0, 1, \ldots.$$

Then, the random process $\{\tilde{n}_d(k), k = 1, 2, \ldots\}$ is a discrete parameter Markov chain defined on the nonnegative integers and is called an embedded Markov chain. In particular, the process $\{\tilde{n}_d(k), k = 1, 2, \ldots\}$ is called the *occupancy process embedded at points immediately following customer departure*. For $\lambda < \mu$, the stationary probability vector, $\pi_d = [\pi_{d0} \quad \pi_{d1} \quad \cdots]$, exists and satisfies the system

$$\pi_d = \pi_d \mathcal{P}_d \quad \text{with} \quad \pi_d \mathbf{e} = 1,$$

where $\mathcal{P}_d$ is the one-step transition probability matrix (Ross[1989]) for the embedded Markov chain $\{\tilde{n}_d(k), k = 1, 2, \ldots\}$ and $\mathbf{e}$ is the column vector in which each element is unity. For $i, j = 0, 1, \ldots$, the probability

$$\mathcal{P}_{d_{ij}} = P\{\tilde{n}_d(k+1) = j | \tilde{n}_d(k) = i\}$$

is called the one-step transition probability from state i to state j. We note that $P\{\tilde{n}_d(k+1) = j | \tilde{n}_d(k) = i\}$ is simply the probability of having exactly $j-(i-1)^+$ arrivals during the $(k+1)$th service time, where $(c)^+ = \max\{0, c\}$. Given the properties of the exponential distribution, we can readily determine the transition probabilities. For example, for $j = 0, 1, \ldots$,

$$\mathcal{P}_{d_{0j}} = \mathcal{P}_{d_{1j}} = \left(\frac{\lambda}{\lambda + \mu}\right)^j \left(\frac{\mu}{\lambda + \mu}\right).$$

Similarly, define $\tilde{n}_a(k)$, for $k = 1, 2, \ldots$, to be the number of customers found in the system by the kth arriving customer and

$$\pi_{aj} = \lim_{k \to \infty} P\{\tilde{n}_a(k) = j\} \quad \text{for} \quad j = 0, 1, \ldots.$$

Then the random process $\{\tilde{n}_a(k), k = 1, 2, \ldots\}$ is called the *occupancy process embedded at points immediately prior to customer arrival*. Again, for $\lambda < \mu$, the vector $\pi_a = [\pi_{a0} \quad \pi_{a1} \quad \cdots]$ exists and satisfies the system

$$\pi_a = \pi_a \mathcal{P}_a \quad \text{with} \quad \pi_a \mathbf{e} = 1,$$

where $\mathcal{P}_a$ is the one-step transition probability matrix for the embedded Markov chain $\{\tilde{n}_a(k), k = 1, 2, \ldots\}$. In this case, $P\{\tilde{n}_a(k+1) = j | \tilde{n}_a(k) = i\}$ is simply the probability of having exactly $i + 1 - j$ service completions during the $(k+1)$th interarrival time for $j = 0, 1, \ldots, i+1$; for $j > i+1$, this probability is equal to zero.

EXERCISE 3.5 For the ordinary M/M/1 queueing system, determine the limiting distribution of the system occupancy

1. as seen by departing customers, [*Hint*: Form the system of equations $\pi_d = \pi_d \mathcal{P}_d$, and then solve the system as was done to obtain $P\{\tilde{n} = n\}$.]

2. as seen by arriving customers, and [*Hint*: First form the system of equations $\pi_a = \pi_a \mathcal{P}_a$, and then try the solution $\pi_a = \pi_d$.]

3. at instants of time at which the occupancy changes. That is, embed a Markov chain at the instants at which the occupancy changes, defining the state to be the number of customers in the system immediately following the state change. Define $\pi = [\pi_0 \quad \pi_1 \quad \cdots]$ to be the stationary probability vector and P to be the one-step transition probability matrix for this embedded Markov chain. Determine π, and then compute the stochastic equilibrium distribution for the process $\{\tilde{n}(t), t \geq 0\}$ according to the following well known result from the theory of Markov chains as discussed in Chapter 2:

$$P_i = \frac{\pi_i E[\tilde{s}_i]}{\sum_{i=0}^{\infty} \pi_i E[\tilde{s}_i]},$$

where $\tilde{s}_i$ denotes the time the systems spends in state i on each visit.

Observe that the results of parts 1, 2, and 3 are identical, and that these are all equal to the stochastic equilibrium occupancy probabilities determined previously.

The results of the above exercise have several implications. First, the stationary departure and arrival distributions are equal. That is, for any n, the proportion of departing customers who leave n customers in the system must equal the proportion of arriving customers who find n customers in the system. A little thought will reveal that this must be the case for systems in which arrivals and departures occur one by one. Suppose, for example, an arriving customer finds n customers in the system. This represents a change in system occupancy from n to $n+1$. If there is ever to be another transition in system occupancy from n to $n+1$, then there must be a transition from $n+1$ to n in the interim. This means that the actual number of departures who find n in the system can never differ by more than the number of arrivals that find n in the system. In the limit as time goes to infinity, the two proportions must then be equal. For a formal proof, see Cooper [1981].

The second implication is that, in the case of the M/M/1 queue, the stationary arrival and stochastic equilibrium distributions are equal. This is a special case of the well known result in queueing theory: Poisson arrivals see time averages (PASTA) (see Wolff [1970,1982]), where time averages imply stochastic equilibrium distributions for ergodic systems. The PASTA property and a more general property, arrivals see time averages (ASTA) and its implications, are discussed in detail in Melamed and Whitt [1990] and the references therein. For completeness, the reader is also referred to Green and Melamed [1990] and Wolff [1990] for discussions of Anti-PASTA, all arrivals do not see time averages. These articles are of only peripheral interest to our current discussion, except for the fact that the equivalence between the stochastic equilibrium behavior of the system and the behavior of the system as viewed by an arbitrary arrival is highly dependent on the nature of the arrival process and is, in general, not a system property.

Returning to our discussion of sojourn times, we find that because the service times are exponentially distributed with parameter μ, $E[\tilde{s} \mid \tilde{n} = n] = (n+1)/\mu$. Upon substituting this expression and (3.8) into (3.12), we find that

$$E[\tilde{s}] = \frac{1}{\mu - \lambda} = \frac{1/\mu}{1-\rho}. \tag{3.13}$$

From (3.13) we see that the mean sojourn time displays the same kind of exponential increase as does the mean system occupancy as $\rho \to 1$.

Now, from (3.11) and (3.13), we see that

$$\frac{E[\tilde{n}]}{E[\tilde{s}]} = \frac{\rho}{1-\rho} \Big/ \left(\frac{1/\mu}{1-\rho} \right) = \lambda,$$

or, equivalently,

$$E[\tilde{n}] = \lambda E[\tilde{s}]. \tag{3.14}$$

The relationship (3.14) is usually written $L = \lambda W$ and is called *Little's result* (Little [1961]). Although we obtained this relationship for the M/M/1 queueing system, it is also true for most other complex queueing systems. The more general statement of Little's result is now stated as a theorem.

THEOREM 3.3 **Little's result.** *The expected number of customers in the system is equal to the product of the arrival rate of customers entering the system and the expected amount of time customers spend in the system.* □

The system need not be an entire service system; for example, the system can be defined as the server only or the waiting line only. In a network of queues, the system may include the entire network or all the servers of the network. For the purposes of this theorem, the arrival rate is defined as the average number of entities that arrive to the system per unit of time.

The primary constraint for the applicability of Little's result is that the notion of a time average for the quantities of interest must make sense in the system under consideration. This is always true if the stochastic processes of interest have a stochastic equilibrium distribution. Of course, in order to obtain correct results, a great deal of care must be taken to assure that L, λ, and W are all defined properly for exactly the same system.

As an example, let $\tilde{n}_q$ denote the number of customers in the queue (that is, the number of customers in the system not including the one in service, if any), $\tilde{w}$ denote the waiting time of the customers in the queue, and λ_q denote the arrival rate of the customers to the queue. Then, from Little's result, $E[\tilde{n}_q] = \lambda_q E[\tilde{w}]$.

By using Little's result, we can derive the mean waiting time in the system in a very straightforward and intuitive manner. It is left as an exercise to show that the probability that the server is busy is given by the quantity ρ. Now, a customer who has just arrived to the queue has to wait an average of $1/\mu$ for each customer in the queue and $1/\mu$ for the customer in service, if any. Thus

$$E[\tilde{w}] = \frac{1}{\mu}E[\tilde{n}_q] + \frac{1}{\mu}\rho.$$

But $E[\tilde{n}_q] = \lambda E[\tilde{w}]$, so

$$E[\tilde{w}] = \frac{1}{\mu}\lambda E[\tilde{w}] + \frac{1}{\mu}\rho.$$

Solving for $E[\tilde{w}]$, we find

$$E[\tilde{w}] = \frac{1}{\mu}\frac{\rho}{1-\rho}. \tag{3.15}$$

EXERCISE 3.6 Using Little's result, show that the probability that the server is busy at an arbitrary point in time is equal to the quantity (λ/μ).

We now provide a proof of Little's result, which is not altogether rigorous, but which captures the basic elements of a rigorous proof. For more rigorous proofs, the reader is referred to the references following this proof.

Proof of Little's result. Customers accumulate system time linearly while they are in the system. Let $N(t)$ denote the number of customers in the system at time t for a typical sample path; the total amount of time in the system accumulated by all customers up to time τ is given by

$$\int_0^\tau N(t)\,dt.$$

Also, let $T_i(\tau)$ denote the amount of time the ith customer spends in the system up to time τ, and let $M(\tau)$ denote the total number of customers who have arrived to the system by time τ. Then the total time spent in the system up to time τ by all customers is given by

$$\sum_{i=0}^{M(\tau)} T_i(\tau).$$

Thus, for any given sample path, it is always true that

$$\sum_{i=0}^{M(\tau)} T_i(\tau) = \int_0^\tau N(t)\,dt. \tag{3.16}$$

Now, so long as $\tau > 0$, we can divide both sides of (3.16) by τ. Also, so long as $M(\tau) > 0$, we may multiply the numerator and denominator of (3.16) by $M(\tau)$. It thus follows that

$$\frac{M(\tau)}{\tau}\frac{1}{M(\tau)}\sum_{i=0}^{M(\tau)} T_i(\tau) = \frac{1}{\tau}\int_0^\tau N(t)\,dt. \tag{3.17}$$

If the system has a stochastic equilibrium, then it is clear that

$$\lim_{\tau\to\infty}\frac{M(\tau)}{\tau},$$

$$\lim_{\tau\to\infty}\frac{1}{M(\tau)}\sum_{i=0}^{M(\tau)} T_i(\tau),$$

and

$$\lim_{\tau\to\infty}\frac{1}{\tau}\int_0^\tau N(t)\,dt$$

all exist individually. The first limit expression above defines λ, the second defines $E[\tilde{s}]$, and the third defines $E[\tilde{n}]$. Thus, Little's result follows by taking limits as $\tau \to \infty$ on both sides of (3.17). □

Note that there is no assumption here about the form of the interarrival-time distribution or the service-time distribution. A somewhat different (heuristic) proof due to Paul Burke is given in Cooper [1972, 1981], and more formal proofs are given in Little [1961], Jewell [1967], and Stidham [1974].

We now turn to the derivation of the equilibrium system sojourn time distribution $P\{\tilde{s} \le x\}$, which we shall denote by $F_{\tilde{s}}(x)$. Consider the sojourn time of an arbitrary customer, the *tagged customer*, who arrives to the system at an arbitrary point in time, t_0, after the system has reached stochastic equilibrium. Now, the arrivals to the system, being Poisson, see the system in stochastic equilibrium. Hence the probability that the tagged customer finds n customers in the system is $P\{\tilde{n} = n\}$. Also the service times are independently drawn from a memoryless distribution, and therefore the sojourn time of the tagged customer, given the tagged customer finds n customers in the system, can be expressed as the sum of $n+1$ independent service times. In particular,

$$\tilde{s} \mid \{\tilde{n} = n\} = \sum_{i=1}^{n+1} \tilde{x}_i,$$

where $\tilde{x}_i$ denotes the service time of the ith customer to receive service after time t_0, with the $(n+1)$th service being that of the tagged customer. [2] Thus we have

$$\begin{aligned} E\left[e^{-s\tilde{s}}\right] &= \sum_{n=0}^{\infty} E\left[e^{-s\tilde{s}} \mid \tilde{n} = n\right] P\{\tilde{n} = n\} \\ &= \sum_{n=0}^{\infty} E\left[e^{-s\sum_{i=1}^{n+1}\tilde{x}_i}\right] P\{\tilde{n} = n\} \\ &= \sum_{n=0}^{\infty} E^{n+1}\left[e^{-s\tilde{x}_i}\right](1-\rho)\rho^n \\ &= (1-\rho)E\left[e^{-s\tilde{x}_i}\right]\frac{1}{1-\rho E\left[e^{-s\tilde{x}_i}\right]}, \end{aligned} \tag{3.18}$$

where the equality between the second and third steps results from the fact that $e^{-s\tilde{x}_i}$ and $e^{-s\tilde{x}_j}$ for $i \neq j$ are independent random variables, and the expectation of the product of independent random variables is the product of the expectations of the individual random variables.

We showed earlier that if $\tilde{x}$ is an exponentially distributed random variable with parameter α, then

$$E\left[e^{-s\tilde{x}}\right] = \frac{\alpha}{s+\alpha}.$$

Because the service times are drawn from exponential distributions with parameter μ, we find $E\left[e^{-s\tilde{x}_i}\right] = \mu/(s+\mu)$ so that

$$\begin{aligned} E\left[e^{-s\tilde{s}}\right] &= \frac{(1-\rho)\mu}{s+\mu}\frac{1}{1-[\rho\mu/(s+\mu)]} \\ &= \frac{(1-\rho)\mu}{s+(1-\rho)\mu}. \end{aligned} \tag{3.19}$$

[2] In general, the notation $\tilde{z} = \{\tilde{x} \mid E\}$, where $\tilde{x}$ and $\tilde{z}$ are random variables and E is an event, means that $P\{\tilde{z} \leq z\} = P\{\tilde{x} \leq z \mid E\}$.

Thus we find that $\tilde{s}$ has the exponential distribution with parameter $(1-\rho)\mu$, and

$$F_{\tilde{s}}(x) = 1 - e^{-\mu(1-\rho)x}, \qquad \text{for } x \geq 0. \tag{3.20}$$

By following similar arguments, we can determine the distribution of the waiting time to be

$$F_{\tilde{w}}(x) = 1 - \rho e^{-\mu(1-\rho)x}, \qquad \text{for } x \geq 0. \tag{3.21}$$

This derivation is left as an exercise.

EXERCISE 3.7 Let $\tilde{w}$ and $\tilde{s}$ denote the length of time an arbitrary customer spends in the queue and in the system, respectively, in stochastic equilibrium. Let $F_{\tilde{s}}(x) \equiv P\{\tilde{s} \leq x\}$ and $F_{\tilde{w}}(x) \equiv P\{\tilde{w} \leq x\}$. Show that

$$F_{\tilde{s}}(x) = 1 - e^{-\mu(1-\rho)x}, \qquad \text{for } x \geq 0,$$

and

$$F_{\tilde{w}}(x) = 1 - \rho e^{-\mu(1-\rho)x}, \qquad \text{for } x \geq 0,$$

without resorting to the use of Laplace-Stieltjes transform techniques.

Another important stochastic process associated with the M/M/1 queueing system is its departure process. The characteristics of this process are now briefly addressed. For a much more detailed treatment, the reader is referred to Disney and Kiessler [1987]. This process is also discussed in many other books on probabilistic modeling including Ross [1990] and Bertsakis and Gallagher [1987]. We shall see that the departure process from the M/M/1 system in stochastic equilibrium is Poisson with the same parameter as the arrival process. After presenting a definition and the main result, we provide a brief discussion of the implications.

DEFINITION 3.1 **Departure process.** Let $\tilde{d}_i$ denote the time between the ith and the $(i+1)$th departures from a queueing system. Then $\tilde{d}_i$ is called the ith interdeparture time for the system. The process $\{\tilde{d}_i, i = 0, 1, \cdots\}$ is called the departure process. A typical interdeparture time will be denoted by $\tilde{d}$ and the distribution of $\tilde{d}$ will be denoted by $F_{\tilde{d}}$.

THEOREM 3.4 **Burke's Theorem** *(Burke [1956]). The sequence of interdeparture times for the M/M/1 system in stochastic equilibrium is a sequence of independent, identically distributed exponential random variables with parameter identical to that of the arrival process; that is, the departure process from the M/M/1 queueing system having arrival rate λ is a Poisson process with parameter λ.* □

EXERCISE 3.8 **M/M/1 Departure Process.** Show that the distribution of an arbitrary interdeparture time for the M/M/1 system in stochastic equilibrium is exponential with the same parameter as the interarrival-time distribution. Argue that the interdeparture times are independent so that the departure process for this system is Poisson with the same rate as the arrival process (Burke [1956]). [*Hint*: Use the fact that the Poisson arrival sees the system in stochastic equilibrium. Then condition on whether or not the ith departing customer leaves the system empty.]

Proof of Burke's theorem can be accomplished very simply by using the concept of *reversibility* (see Asmussen [2003], pp. 56-58 for a brief introduction). We now briefly sketch the main ideas. Consider a general stochastic process $\{\tilde{x}(t), t \geq 0\}$ for which a stochastic equilibrium distribution exists. To assure that the system is operating in stochastic equilibrium, assume the distribution of $\tilde{x}(0)$ is the same as the stochastic equilibrium distribution so that the time derivatives of the occupancy probabilities are all equal to zero. Now observe the probability structure of the process at a very large point in time, say t_0. If the probability structure of the process looking forward in time from t_0 is identical to the probability structure of the process looking backward in time from t_0, then the process is said to be time-reversible.

Ross [1989], pp. 277-78 provides a simple proof that all *birth-death processes* (which are defined in Section 3.2) are time-reversible. The occupancy process for the M/M/m system is a special case of a birth-death process, and is therefore time-reversible. This means that for the M/M/m system, the instants at which the occupancy increases when looking backwards in time have exactly the same probability structure as the instants at which the occupancy increases when looking forward in time. Now, the instants at which the occupancy increases when looking backwards in time are exactly the instants of customer departure. Because the instants at which the occupancy increases when looking forward in time are the instants of arrivals from a Poisson process, we see that the departure process is also Poisson with the same parameter as the arrival process.

Remark. It is interesting to note that when a queueing process is reversible, then the Markov chain embedded just after points of departure is the reverse process for the Markov chain embedded just prior to points of arrival. The stationary probabilities of the two embedded chains are then equal as has been shown in the specific case of the M/M/1 system. The interested reader is referred to Ross [1989], pp. 173-184 for an elementary treatment of time-reversibility of Markov chains and to Disney and Kiessler [1987], p. 99 for a proof of the result given in this remark.

Since all birth-death processes are time reversible, we see that Burke's theorem applies not only to single-server queueing systems but also to the $M/M/m$

and M/M/∞ systems as well. The implications of this theorem in analyzing the occupancy process for systems having Poisson arrivals and exponential service are significant. For example, we showed in Chapter 2 that sums of Poisson processes are Poisson, and randomly split Poisson processes form two independent Poisson streams. Because the departure processes are also Poisson, complex systems of exponential servers can be analyzed by first determining the average arrival rates to each of the queues, and then analyzing the individual queues independently. The results of the independent analyses are then combined to analyze the system as a whole.

We now provide a simple example, leaving to a later section a more general treatment of networks of queues.

EXAMPLE 3.1 Consider the system of Figure 3.3. Exogenous arrivals (that is, from outside the system) occur according to a Poisson process at rate λ to an exponential server having service rate μ. Following service, each customer decides with probability p, independently of everything, whether or not to enter the second service system, which has exponential service with rate α. Customers who decide not to enter the second service system proceed immediately to the third system, which has service rate β. There, they join the waiting line along with customers departing the second service system. We wish to determine the joint equilibrium state occupancy distribution for the three queues.

Solution: Because the departure process from the first queue is Poisson with rate λ, arrivals to the second queue are Poisson with rate $p\lambda$. The departure process from the second queue is therefore Poisson with rate $p\lambda$, and this processes is independent of the process due to customers who decide not to enter the second system. The stream of customers entering the third service system is the result of combining independent Poisson streams with rates $p\lambda$ and $(1-p)\lambda$, and is therefore Poisson with rate λ.

Stochastic equilibrium exists if $\lambda < \min\{\mu, \alpha/p, \beta\}$, and, in that case,

$$P\{\tilde{n}_1 = n_1\} = (1 - \lambda/\mu)(\lambda/\mu)^{n_1},$$

$$P\{\tilde{n}_2 = n_2\} = (1 - p\lambda/\alpha)(p\lambda/\alpha)^{n_2},$$

and

$$P\{\tilde{n}_3 = n_3\} = (1 - \lambda/\beta)(\lambda/\beta)^{n_3},$$

where $\tilde{n}_i$ denotes the occupancy at queue i. The joint queue length distribution is then the product of the individual occupancy distributions.

It is worth pausing at this point to reflect on the implications of Burke's theorem. While Burke's theorem does state that the interdeparture times are a sequence of iid exponential random variables, the theorem does not say that the departure process is independent of the state of the occupancy process.

In Example 3.1, the fact that the joint probability mass function for the occupancies of the three servers is given by the product of the marginal mass probabilities means that the server occupancies are independent random variables. On the other hand, the waiting times at the servers are not independent because the interdeparture times from a given server are not independent of the occupancy of that server, and the waiting time at the server is dependent upon the occupancy at that node. Finally, sojourn times of customers at different nodes are not independent.

The result is that joint occupancies, the waiting-time distribution at individual servers, and average network delays may be computed via elementary analysis, but higher moments of network delay are more difficult to obtain. Thus we must exercise extreme care in drawing deep conclusions from elementary analysis of this form.

The following exercise emphasizes that a knowledge of the ergodic occupancy distribution for even a simple queueing system is insufficient information from which to compute the waiting-time distribution. The interested reader is referred to Disney and Kiessler [1987] for a more thorough discussion. We note that the aggregate arrival process to the queue defined in this exercise is not a Poisson process (see Disney, McNickle, and Simon [1980] and Disney and Kiessler [1987], pp. 124-125).

EXERCISE 3.9 M/M/1 with Instantaneous Feedback. A queueing system has exogenous Poisson arrivals with rate λ and exponential service with rate μ. At the instant of service completion, each potentially departing customer rejoins the service queue, independent of system state, with probability p.

1. Determine the distribution of the total amount of service time the server renders to an arbitrary customer.

2. Compute the distribution of the number of customers in the system in stochastic equilibrium. How does your solution compare to that of the M/M/1 queueing system? What explains this behavior? [*Hint*: Consider the remaining service time required for each customer in the queue. Suppose customers that required additional increments of service returned immediately to service rather than joining the tail of the queue. What would be the effect on the queue occupancy?]

3. Argue that the departure process from the system is a Poisson process with rate λ.

4. Compute the average sojourn time for this system and comment on computation of the distribution of the sojourn time.

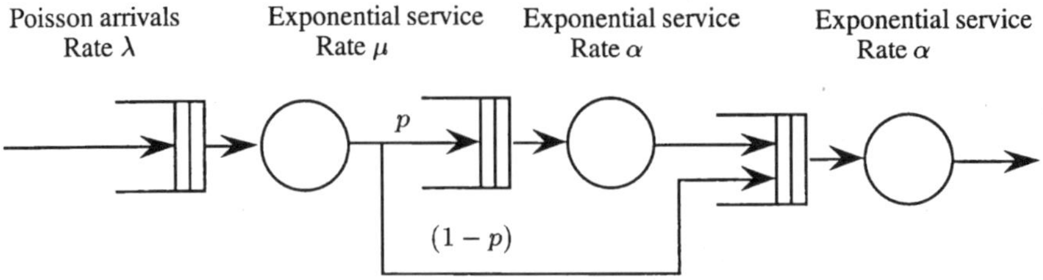

Figure 3.3. Schematic diagram of a simple network of queues.

3.1.3 Busy Period for M/M/1 Queueing System

Recall that $\tilde{n}(t)$ is defined to be the number of customers in the system at time t. The system is said to be idle at time t if $\tilde{n}(t) = 0$ and busy at time t if $\tilde{n}(t) > 0$. A busy period begins at any instant in time at which the value of $\tilde{n}(t)$ increases from zero to one and ends at the first instant in time, following entry into a busy period, at which the value of $\tilde{n}(t)$ again reaches zero. An idle period begins when a given busy period ends and ends when the next busy period begins. From the perspective of the server, the M/M/1 queueing system alternates between two distinct types of periods: *idle periods* and *busy periods*, as illustrated in Figure 3.4. These types are descriptive; the busy periods are periods during which the server is *busy* servicing customers, and the idle periods are those during which the server is not servicing customers. For the ordinary M/M/1 queueing system, the server is never idle when there is at least one customer in the system.

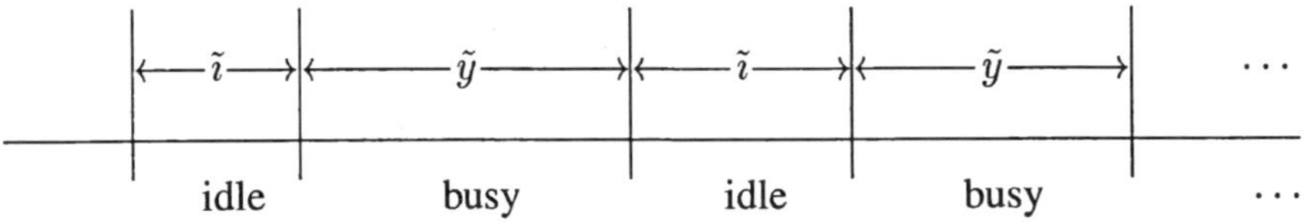

Figure 3.4. Sequence of busy and idle periods.

Because of the memoryless property of both the exponential distribution and the Poisson process, the length of an idle period is the same as the length of time between two successive arrivals from a Poisson process with parameter λ. The length of a busy period, on the other hand is dependent upon both the arrival and service processes. The busy period begins upon the arrival of its first customer, say customer 1, whom we shall denote by C_1. During the service time of C_1, the length of which we shall denote by $\tilde{x}_1$, K_1 additional customers arrive. If $K_1 > 0$, we call the K_1 customers *second-generation customers* and denote them by $C_{11}, C_{12}, \cdots, C_{1K_1}$. The service times of these customers follow that of C_1 in their order of arrival. During the service time of C_{11}, additional customers may arrive; they are denoted by $C_{111}, C_{112}, \cdots, C_{11K_{11}}$.

Additional arrivals that occur during the service times of the ith second-generation customer are denoted by $C_{1i1}, C_{1i2}, \cdots, C_{1iK_{1i}}$, and the collection of all these customers constitute the third generation. Service for third-generation customers follows completion of service of second-generation customers. Arrivals occurring while the nth-generation customers are receiving service are classed $(n+1)$th-generation customers, and their service begins following completion of nth-generation servicing. The service and arrival processes continue until there are no longer any remaining customers, and at that point in time the system returns to an idle period. Thus, the length of a busy period is the total amount of time required to service all of the customers of all of the generations of the first customer of the busy period. Consequently, we can think of the busy period as being *generated* by its first customer. Alternatively, we can view the server as having to work until all of the first customer's descendents die out.

We shall denote the length of a generic busy period by $\tilde{y}$, the length of a generic idle period by $\tilde{\imath}$, and the number of customers served during a generic busy period by $\tilde{h}$. The service time of the ith customer served in a generic busy period will be denoted by $\tilde{x}_i$. The length of the busy period is then the sum of the service times, or

$$\tilde{y} = \sum_{i=1}^{\tilde{h}} \tilde{x}_i. \tag{3.22}$$

The diagram of Figure 3.5 illustrates servicing during the busy period.

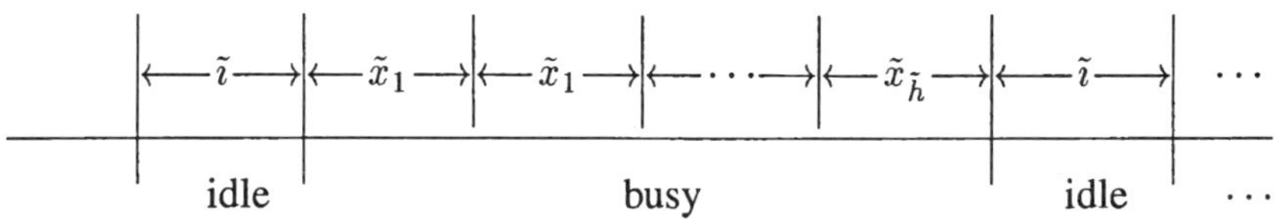

Figure 3.5. Sequence of service times during a generic busy period.

The distribution of the length of a busy period is of interest in its own right, but an understanding of the behavior of busy-period processes is also extremely helpful in understanding waiting time and queue length behavior in both ordinary and priority queueing systems. An alternate and instructive way to view the busy-period process is to separate the busy period into two parts: the part occurring before the first customer arrival after the busy period has started, and the part occurring after the first customer arrival after the busy period has started, if such an arrival occurs.

Let $\tilde{t}_1$ denote the length of the first interarrival time after the busy period has begun, and let $\mathcal{D}$ denote the event that the first customer completes service before the first arrival after the busy period begins; that is, let $\mathcal{D}$ denote the event that $\tilde{x}_1 < \tilde{t}_1$. Further, let $\tilde{z}_1 = \min\{\tilde{x}_1, \tilde{t}_1\}$.

We have shown previously that if $\tilde{x}_1$ and $\tilde{t}_1$ are exponentially distributed random variables with parameters μ and λ, respectively, then $\tilde{z}_1$ is an exponentially distributed random variable with parameter $\mu + \lambda$. We also showed in an earlier exercise that the random variables $\tilde{z}_1|\{\tilde{x}_1 < \tilde{t}_1\}$ and $\tilde{z}_1|\{\tilde{x}_1 > \tilde{t}_1\}$ are also exponentially distributed random variables with parameter $\mu + \lambda$; that is; the distribution of $\tilde{z}_1$ is independent of whether $\tilde{x}_1 < \tilde{t}_1$ or $\tilde{x}_1 > \tilde{t}_1$. Additionally, if $\tilde{x}_1 < \tilde{t}_1$, then the busy period ends after the initial interval so that

$$\tilde{y}|\{\tilde{x}_1 < \tilde{t}_1\} = \tilde{z}_1. \tag{3.23}$$

On the other hand, if $\tilde{x}_1 > \tilde{t}_1$, then a period of length $\tilde{z}_1$ will have expired, but due to the memoryless property of the exponential distribution, the remaining service time of the first customer will be the same as it was initially. Thus, for all practical purposes, the service time starts over. The remaining time in the busy period is therefore equivalent to the length of a busy period in which there are initially two customers present rather than one. We denote the length of such a period by $\tilde{y}_2$. Thus, we find that

$$\tilde{y}|\{\tilde{x}_1 > \tilde{t}_1\} = \tilde{z}_1 + \tilde{y}_2. \tag{3.24}$$

Now, the length of a busy period is independent of the order in which the customers of the busy period are served. The length of the busy period is simply the sum of the lengths of the service times of the customers that are served as shown in (3.22) and Figure 3.5. A little thought will reveal that the length of the busy period generated by two customers is simply the sum of the lengths of the *sub-busy periods* generated by the first and second customers, respectively. That is,

$$\tilde{y}_2 = \tilde{y}_{21} + \tilde{y}_{22} \tag{3.25}$$

where $\tilde{y}_{21}$ and $\tilde{y}_{22}$ denote the lengths of the sub-busy periods generated by the first and second customers, respectively. Additionally, $\tilde{y}_{21}$ and $\tilde{y}_{22}$ are independent, and their distributions are the same as that of an ordinary busy period, $\tilde{y}$. Figure 3.6 illustrates the decomposition of the busy period from this point of view.

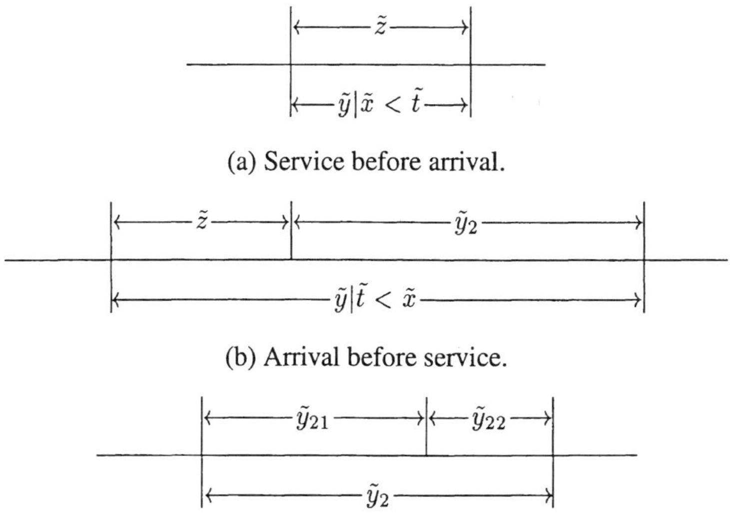

(a) Service before arrival.

(b) Arrival before service.

(c) Decomposition of remaining busy period.

Figure 3.6. Busy period decompositions depending upon interarrival versus service times.

We now turn our attention to the determination of $E[\tilde{y}]$. We define $\mathcal{A}$ as the complement of $\mathcal{D}$. Then, upon conditioning on the occurrence or nonoccurrence of $\mathcal{D}$, we find

$$E[\tilde{y}] = E[\tilde{y}|\mathcal{D}]P\{\mathcal{D}\} + E[\tilde{y}|\mathcal{A}]P\{\mathcal{A}\}. \tag{3.26}$$

Upon substituting (3.23), (3.24), and (3.25) into (3.26), we find that

$$E[\tilde{y}] = E[\tilde{z}_1]P\{\mathcal{D}\} + E[\tilde{z}_1 + \tilde{y}_{21} + \tilde{y}_{22}]P\{\mathcal{A}\}. \tag{3.27}$$

Using the fact that $\tilde{y}_{21}$ and $\tilde{y}_{22}$ each have the same distribution as $\tilde{y}$ in (3.27) leads to

$$\begin{aligned} E[\tilde{y}] &= E[\tilde{z}_1] + 2E[\tilde{y}]P\{\mathcal{A}\} \\ &= \frac{E[\tilde{z}_1]}{1 - 2P\{\mathcal{A}\}}. \end{aligned} \tag{3.28}$$

We showed earlier that $P\{\mathcal{A}\} = \lambda/(\mu + \lambda)$ and $E[\tilde{z}_1] = 1/(\mu + \lambda)$. Substituting these values into (3.28) leads to

$$E[\tilde{y}] = \frac{1}{\mu} \frac{1}{1 - \lambda/\mu}, \tag{3.29}$$

or equivalently,

$$E[\tilde{y}] = \frac{1/\mu}{1-\rho}. \tag{3.30}$$

The techniques leading to (3.27) are extremely useful in busy-period analysis, and they can be applied to determine $E[\tilde{h}]$ and $E\left[e^{-s\tilde{y}}\right]$. The arguments are also useful in studying the behavior of other queueing disciplines, such as last-come-first-serve (LCFS). Examination of these aspects of busy-period analysis is left to the exercises.

EXERCISE 3.10 For the M/M/1 queueing system,

1. find $E[\tilde{h}]$, the expected number of customers served in busy period, and

2. find $E[e^{-s\tilde{y}}]$, the Laplace-Stieltjes transform of the distribution of the length of a busy period. Show that $(d/dy)F_{\tilde{y}}(y) = 1/(y\sqrt{\rho})e^{-(\lambda+\mu)y}I_1(2y\sqrt{\lambda\mu})$. A Laplace transform pair,

$$\frac{\sqrt{s+2a}-\sqrt{s}}{\sqrt{s+2a}+\sqrt{s}} \Longleftrightarrow \frac{1}{t}e^{-at}I_1(at),$$

taken from *Mathematical Tables from the Handbook of Physics and Chemistry*, will be useful in accomplishing this exercise.

We have previously stated that

$$\tilde{y} = \sum_{i=1}^{\tilde{h}} \tilde{x}_i.$$

Given the formula for the expected length of the busy period, one can readily determine the expected number of customers served during a busy period through the application of *Wald's equation* (Ross[1989]) which states that the expected value of the sum of a random number, $\tilde{n}$, of identically distributed random variables, $\tilde{x}_i, 0 \le i \le \tilde{n}$, is given by the product of the expected values of $\tilde{n}$ and $\tilde{x}_i$ provided that $\tilde{n}$ is a *stopping time* for the sequence of random variables $\{\tilde{x}_i, i = 1, 2, \ldots\}$ For $\tilde{n}$ to be a stopping time for the sequence $\{\tilde{x}_i, i = 1, 2, \ldots\}$, it is sufficient to show that $\tilde{n}$ is independent of $\tilde{x}_{\tilde{n}+1}$[3].

[3]It is interesting to note that the last service time of a busy period is stochastically shorter than the other service times because the last service time contains no arrivals with probability one. However, the $\tilde{x}_i$ are still drawn independently from a common distribution in exactly the same way as a gambler's winnings on the ith game. The gambler always loses on the last game, but the winnings on the ith game are drawn before the game is played. Similarly, the ith service time is drawn from the common distribution before it is decided whether or not it is the last service time of the busy period.

EXERCISE 3.11 For the M/M/1 queueing system, argue that $\tilde{h}$ is a stopping time for the sequence $\{\tilde{x}_i, i = 1, 2, \ldots\}$ illustrated in Figure 3.5. Find $E[\tilde{h}]$ by using the results given above for $E[\tilde{y}]$ in combination with Wald's equation.

EXERCISE 3.12 For the M/M/1 queueing system, argue that $E[\tilde{s}]$, the expected amount of time a customer spends in the system, and the expected length of a busy period are equal. [*Hint*: Consider the expected waiting time of an arbitrary customer in the M/M/1 queueing system under a nonpreemptive LCFS and then use Little's result.]

EXERCISE 3.13 Let $\tilde{s}_{\mathrm{LCFS}}$ denote the total amount of time an arbitrary customer spends in the M/M/1 queueing system under a nonpreemptive discipline. Determine the Laplace-Stieltjes transform for the distribution of $\tilde{s}_{\mathrm{LCFS}}$.

EXERCISE 3.14 Determine the Laplace-Stieltjes transform for the length of the busy period for the M/M/2 queueing system, the system having Poisson arrivals, exponential service, two parallel servers, and an infinite waiting room capacity. [*Hint*: Condition on whether or not an arrival occurs prior to the completion of the first service of a busy period. Then note that there is a very close relationship between the time required to reduce the occupancy from two customers to one customer in the M/M/2 and the length of the busy period in the ordinary M/M/1 system.]

EXERCISE 3.15 We have shown that the number of arrivals from a Poisson process with parameter λ, that occur during an exponentially distributed service time with parameter μ, is geometrically distributed with parameter $\mu/(\mu + \lambda)$; that is, the probability of n arrivals during a service time is given by $[\lambda/(\lambda + \mu)]^n[\mu/(\lambda + \mu)]$. Determine the mean length of the busy period by conditioning on the number of arrivals that occur during the first service time of the busy period. For example, let $\tilde{n}_1$ denote the number of arrivals that occur during the first service time, and start your solution with the statement

$$E[\tilde{y}] = \sum_{n=0}^{\infty} E[\tilde{y}|\tilde{n}_1 = n]P\{\tilde{n}_1 = n\}.$$

[*Hint*: The arrivals segment the service period into a sequence of intervals.]

3.2 Dynamical Equations for General Birth-Death Process

A variation to the M/M/1 queueing system is a system with exponentially distributed interarrival times and service times, but having state-dependent arrival and service rates. The arrival rate when there are n customers in the system is λ_n, and the service rate when there are n customers in the system is μ_n. The occupancy for such a system is modeled by a general *birth-death* or *birth-and-death process*.

Examples of queueing systems for which the occupancy can be modeled by a birth-death process are numerous. For example, the $M/M/s$ queueing system is the system having s servers, Poisson arrivals at rate λ, and exponential service at rate μ. In this system, the arrival rate λ is independent of the current occupancy, but the service rate is $n\mu$ if the occupancy is less than s and $s\mu$ if the occupancy equals or exceeds s; that is, $\lambda_i = \lambda$ for $i \geq 0$, but

$$\begin{cases} \mu_n = n\mu, & \text{for } n < s; \\ \mu_n = s\mu, & \text{for } n \geq s. \end{cases}$$

This model is useful in modeling a circuit switching system for a system in which a large population of users share a relatively small number of lines and the customers are allowed to join a queue while waiting for a line to become available. A variation of this system, the *Erlang loss system* is considered in a later section.

Another example is the M/M/1/K queueing system. There is a finite population, K, of customers, each operating in a constant *think-wait-service* cycle. The length of time the customer remains in the *think* state is drawn from an exponential distribution with rate λ, independent of everything; a customer may generate a request for service only while in the think state. Upon departure from the think state, the customer joins the queue to await service. Upon reaching the head of the queue, the customer receives service, the length of which is drawn from an exponential distribution with rate μ. For this model, we find $\mu_n = \mu$, independent of occupancy, but $\lambda_n = (K - n)\lambda$ for $0 \leq n \leq K$.

The dynamical equations for the general birth-death process are the same as those for the M/M/1 queueing system except that the arrival and service rates are replaced by state-dependent arrival and service rates. The resulting dynamical equations, the development of which are left as an exercise, are as follows:

$$P_n'(t) = \begin{cases} -(\lambda_n + \mu_n)P_n(t) + \lambda_{n-1}P_{n-1}(t) & \\ +\mu_{n+1}P_{n+1}(t) & \text{for } n > 0; \\ \lambda_0 P_0(t) + \mu_1 P_1(t) & \text{for } n = 0. \end{cases} \tag{3.31}$$

We shall consider special cases of birth-death processes when we study the balance-equation approach to solving elementary queueing systems.

EXERCISE 3.16 Suppose that the arrival and service time distributions are memoryless, but that their rates depend upon the number of customers in the system. Let the arrival rate when there are k customers in the system be λ_k, and let the service rate when there are k customers in the system be μ_k. Show that the dynamical equations are as follows:

$$P_n'(t) = \begin{cases} -(\lambda_n + \mu_n)P_n(t) + \lambda_{n-1}P_{n-1}(t) & \\ +\mu_{n+1}P_{n+1}(t), & \text{for } n > 0; \\ \lambda_0 P_0(t) + \mu_1 P_1(t) & \text{for } n = 0. \end{cases}$$

3.3 Time-Dependent State Probabilities for Finite-State Systems

In this section, we discuss approaches for obtaining the time-dependent probabilities for the special case in which the queueing system can be modeled as a continuous-time, finite-state, Markov chain. Our discussion focuses on the finite-state birth-death process, but extensions to the more general case are obvious.

Two methods of analysis are discussed: classical eigensystem analysis, and *randomization.* The latter is also often referred to in the literature as *uniformization* for reasons stated at the end of this section. Following Grassman [1990], we adopt the name *Jensen's method*, which Grassman argues is more appropriate.

We limit the maximum queue occupancy to K. For this special case, we find that $\lambda_n = 0$ for $n \geq K$ and μ_n has an arbitrary value for $n > K$. Under these conditions, the system (3.31) leads to the following system of (K+1) linear differential equations:

$$\frac{d}{dt}P(t) = P(t)\mathcal{Q} \tag{3.32}$$

where $P(t)$ is the row vector of state probabilities,

$$P(t) = [\, P_0(t) \quad P_1(t) \quad \cdots \quad P_K(t)\,],$$

and

$$\mathcal{Q} = \begin{bmatrix} -\lambda_0 & \lambda_0 & 0 & \cdots & 0 & 0 & 0 \\ \mu_1 & -(\lambda_1 + \mu_1) & \lambda_1 & \cdots & 0 & 0 & 0 \\ \vdots & \vdots & \vdots & \cdots & \vdots & \vdots & \vdots \\ 0 & 0 & 0 & \cdots & \mu_{K-1} & -(\lambda_{K-1} + \mu_{K-1}) & \lambda_{K-1} \\ 0 & 0 & 0 & \cdots & 0 & \mu_K & -\mu_K \end{bmatrix}$$

is the infinitesimal generator matrix for the (finite) Markov chain $\{\tilde{n}(t), t \geq 0\}$ (Cohen, [1969]).

It is well known and easily shown that the above equation has the general solution

$$P(t) = P(0)e^{Qt}, \tag{3.33}$$

where $P(0)$ denotes the vector of initial state probabilities. Thus, at least in principle, we may easily determine the time-dependent state probabilities for particular values of t.

Remark. The form (3.33) of the solution to the vector first order differential equation (3.32) has inspired the development of numerous ways to evaluate the required matrix exponential. A summary of the most prominent of these is given in Moler and van Loan [1978]. Matrix exponentiation is not, however, necessarily the best way to solve for the time-dependent solution to (3.32). In fact, it may be faster, computationally, to simply solve (3.32) directly using a standard ordinary differential equations solution package. The reader is referred to Giffin [1978] for a pedagogical presentation of this subject matter. Grassman [1990] provides a perspective on computational complexity issues and on the pros and cons of the various computational approaches. The expression (3.33) is, nonetheless, very useful in discussing the behavior of the solution.

3.3.1 Classical Approach

Observation of (3.32) reveals that $\mathcal{Q}$ is a tridiagonal matrix, and the off-diagonal terms have the same sign. Thus, the matrix $\mathcal{Q}$ is similar to the symmetric matrix $\hat{\mathcal{Q}}$ in which the diagonal terms are the same as those of $\mathcal{Q}$, and the off-diagonal elements are given by

$$\hat{q}_{i,i+1} = \sqrt{q_{i,i+1}q_{i+1,i}} \quad \text{for} \quad i = 0, 1, \ldots, K-1. \tag{3.34}$$

That is,

$$\hat{\mathcal{Q}} = R^{-1}\mathcal{Q}R, \tag{3.35}$$

where

$$\begin{aligned} R &= \operatorname{diag}(1, \sqrt{\frac{q_{1,0}}{q_{0,1}}}, \sqrt{\frac{q_{2,1}}{q_{1,2}}}, \ldots, \\ &\quad \sqrt{\frac{q_{1,0}}{q_{0,1}}\frac{q_{2,1}}{q_{1,2}} \cdots \frac{q_{K,K-1}}{q_{K-1,K}}}). \end{aligned} \tag{3.36}$$

It is readily verified that the matrix $\hat{\mathcal{Q}}$ is negative semidefinite (Noble and Daniel [1977]), so the eigenvalues of $\hat{\mathcal{Q}}$, and therefore of $\mathcal{Q}$, are nonpositive. In addition, the columns of $\mathcal{Q}$ sum to a null column vector, so one of the eigenvalues of $\mathcal{Q}$ is equal to zero. This means that $\lim_{t\to\infty} P(t)$ exists, and the maximum negative eigenvalue of $\mathcal{Q}$ (the one closest to zero) determines the rate at which $P(t)$ converges to its limiting value. The inverse of this maximum negative eigenvalue is sometimes referred to as the *relaxation time* of the system (Keilson [1979]).

If the eigenvalues of $\mathcal{Q}$ are distinct, then $\mathcal{Q}$ is similar to a diagonal matrix with the eigenvalues as the diagonal elements. That is, we may write diag $(\sigma_0, \sigma_1, \ldots, \sigma_K) = \mathcal{M}^{-1}\,\mathcal{Q}\,\mathcal{M}$, or equivalently,

$$\mathcal{Q} = \mathcal{M} \text{ diag } (\sigma_0, \sigma_1, \ldots, \sigma_K)\, \mathcal{M}^{-1}, \tag{3.37}$$

where $\mathcal{M}$ is a nonsingular matrix spanning the $(K+1)$-dimensional space, σ_i denotes the ith eigenvalue of $\mathcal{Q}$, and $0 = \sigma_0 > \sigma_1 > \cdots > \sigma_K$. Indeed, the ith column of $\mathcal{M}$ is (proportional to) the eigenvector corresponding to σ_i. Thus, we can rewrite (3.33) as

$$P(t) = P(0)\, \mathcal{M} \text{ diag } \left(e^{\sigma_0 t}, e^{\sigma_1 t}, \ldots, e^{\sigma_K t}\right) \mathcal{M}^{-1}. \tag{3.38}$$

Because the eigenvalues are all nonpositive and we have labeled them in decreasing order, we find that σ_1 determines the rate at which $P(t)$ converges to its equilibrium value P.

For example, suppose $K = 1$, $\lambda_0 = \lambda$, and $\mu_1 = \mu$. Then we have

$$\frac{d}{dt}[\, P_0(t) \quad P_1(t)\,] = [\, P_0(t) \quad P_1(t)\,]\begin{bmatrix} -\lambda & \lambda \\ \mu & -\mu \end{bmatrix}. \tag{3.39}$$

The eigenvalues of $\mathcal{Q}$ are found to be 0 and $-(\lambda+\mu)$ and their corresponding eigenvectors are proportional to $[\,1 \quad 1\,]^T$ and $[\,-\lambda \quad \mu\,]^T$, respectively. Thus we find

$$\begin{aligned} P(t) &= [\, P_0(0) \quad P_1(0)\,]\begin{bmatrix} 1 & -\lambda \\ 1 & \mu \end{bmatrix}\begin{bmatrix} 1 & 0 \\ 0 & e^{-(\lambda+\mu)t} \end{bmatrix} \\ &\quad \begin{bmatrix} \mu/\mu+\lambda & \lambda/\mu+\lambda \\ -1/\mu+\lambda & 1/\mu+\lambda \end{bmatrix}. \end{aligned} \tag{3.40}$$

The time-dependent state probabilities can be computed from (3.40).

In case the equilibrium probabilities are needed, we find

$$\begin{aligned} \lim_{t\to\infty} P(t) &= [\, P_0(0) \quad P_1(0)\,]\begin{bmatrix} 1 & -\lambda \\ 1 & \mu \end{bmatrix}\begin{bmatrix} 1 & 0 \\ 0 & 0 \end{bmatrix} \\ &\quad \begin{bmatrix} \mu/\mu+\lambda & \lambda/\mu+\lambda \\ -1/\mu+\lambda & 1/\mu+\lambda \end{bmatrix} \\ &= [\,\mu/\lambda+\mu \quad \lambda/\lambda+\mu\,], \end{aligned} \tag{3.41}$$

which is as expected from direct evaluation of the equilibrium probabilities.

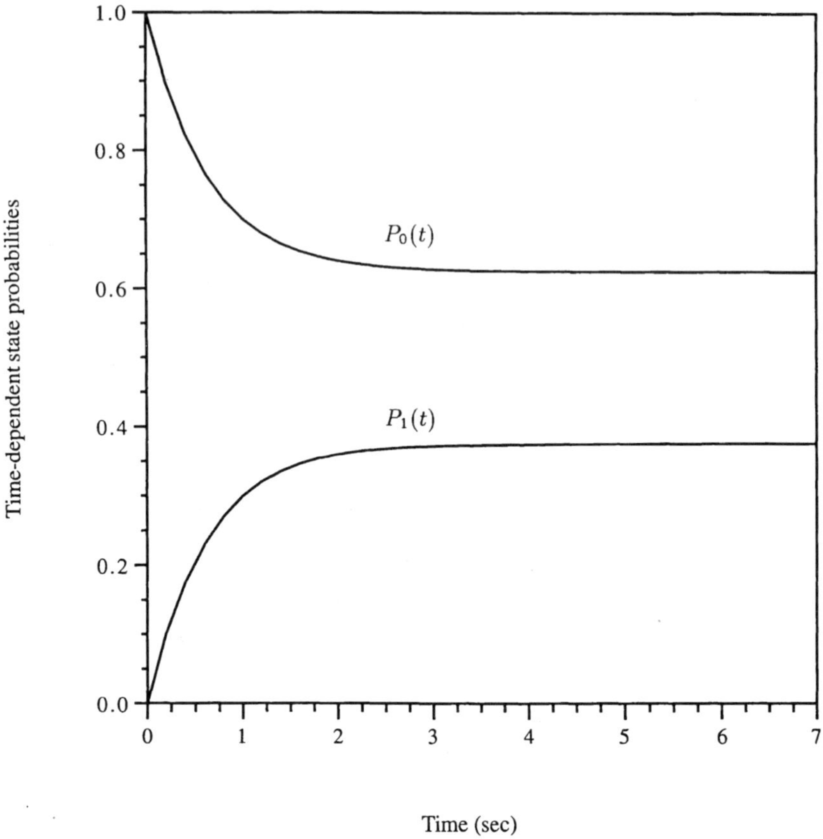

Figure 3.7. Time-dependent state probabilities corresponding to Example 3.1.

EXAMPLE 3.2 Suppose $\mu = 1$, $\lambda = 0.6$, and $P(0) = [1 \quad 0]$. Then, (3.40) and (3.41) reduce to

$$P(t) = [0.675 + 0.375e^{-1.6t} \quad 0.375(1 - e^{-1.6t})],$$

and

$$\lim_{t\to\infty} p(t) = [0.625 \quad 0.375].$$

Figure 3.7 shows graphs of $P_0(t)$ and $P_1(t)$ as a function of time. Note that the limiting values of $P_0(t)$ and $P_1(t)$ are reached to a very high degree of accuracy by the time $t = 4$, which is between six and seven times the quantity $1/(\mu + \lambda)$.

In the above case, we note that the equilibrium probabilities are proportional to the left eigenvector of $\mathcal{Q}$ corresponding to the eigenvalue $\sigma_0 = 0$. To see that this is always the case, we consider

$$\lim_{t \to \infty} P'(t) = 0 = P\,\mathcal{Q}.$$

That is,

$$P\,\mathcal{Q} = 0.$$

By definition, if M_0 is a left eigenvector of $\mathcal{Q}$ corresponding to the eigenvalue σ_0, then

$$M_0 \mathcal{Q} = \sigma_0 M_0.$$

But with $\sigma_0 = 0$,

$$M_0 \mathcal{Q} = 0.$$

Thus, P is proportional to M_0. The implication is that the equilibrium probabilities can always be determined by normalizing the left eigenvector of $\mathcal{Q}$ corresponding to the eigenvalue zero. Thus we find

$$P = \frac{1}{M_0 \mathbf{e}} M_0, \tag{3.42}$$

where $\mathbf{e}$ is the column vector in which each element is unity.[4]

It is sometimes desirable to obtain the left eigenvectors of $\mathcal{Q}$ via hand calculation. With regard to this possibility, we state the following theorems.

THEOREM 3.5 *Let $\mathcal{Q}$ be a $(K+1)$-dimensional square matrix whose distinct eigenvalues and their corresponding left eigenvectors are $\sigma_0, \sigma_1, \ldots, \sigma_K$ and $M_0, M_1, \ldots, M_K$, respectively. Then M_i is proportional to the rows of the adjoint of the matrix $(\sigma_i I - \mathcal{Q})$. That is,*

$$\mathrm{adj}\,(\sigma_i I - \mathcal{Q}) = \begin{bmatrix} c_0 M_i \\ c_1 M_i \\ \vdots \\ c_K M_i \end{bmatrix},$$

where c_0, c_1, …,c_K are nonzero constants. □

THEOREM 3.6 *Let $\mathcal{Q}$ be a $(K+1)$-dimensional square matrix having distinct eigenvalues $\sigma_0, \sigma_1, \ldots, \sigma_K$. Then the rows of* $\mathrm{adj}\,(\sigma_i I - \mathcal{Q})$ *are proportional to each other, and the columns of* $\mathrm{adj}\,(\sigma_i I - \mathcal{Q})$ *are proportional to each other.* □

[4]We will use this definition for $\mathbf{e}$ in the remainder of the text.

The proofs of these theorems are left as exercises.

EXERCISE 3.17 Prove Theorem 3.5.

EXERCISE 3.18 Prove Theorem 3.6.

EXERCISE 3.19 Let $K = 1$. Use Definition 2 of the Poisson process to write an equation of the form

$$\frac{d}{dt}[\,P_0(t) \quad P_1(t)\,] = [\,P_0(t) \quad P_1(t)\,]\,Q.$$

Show that the eigenvalues of the matrix Q are real and nonnegative. Solve the equation for $P_0(t), P_1(t)$ and show that they converge to the solution given in Example 3.2 regardless of the values $P_0(0), P_1(0)$. [*Hint*: First, do a similarity transformation on the matrix Q, which converts the matrix to a symmetric matrix $\hat{Q}$. Then show that the matrix $\hat{Q}$ is negative semi-definite.]

3.3.2 Jensen's Method

An alternative method of computing the time-dependent probabilities can be formulated via the introduction of some additional state transitions into the dynamics of the system in such a way as to *uniformize* the amount of time the system spends in each state. That is, we introduce self transitions into each state so that the amount of time spent in each state, on each visit, is exponentially distributed with identical parameter, say ν. This will allow us to study the system as though it were a discrete-time Markov chain with the transition epochs occurring according to a Poisson process with parameter ν. The latter is referred to as *randomization of time.*

Mathematically, we proceed as follows. First, we rewrite (3.33) as

$$P(t) = P(0)\; e^{\{-\nu I + \nu I + Qt\}}. \tag{3.43}$$

Then, because νIt commutes with $\{\nu I + Q\}t$, then so do $e^{\nu It}$ and $e^{\{\nu I + Q\}}t$. Thus the right hand side of (3.43) can be rewritten as the product of two matrices:

$$P(t) = P(0)\; e^{-\nu It} e^{\{\nu I + Q\}t}. \tag{3.44}$$

But, since $e^{\nu It} = e^{\nu t} I$, we find that

$$P(t) = P(0)e^{-\nu t} e^{\nu\{I + (1/\nu)Q\}t}. \tag{3.45}$$

Expanding the matrix exponential on the right hand side of (3.45) in a Maclaurin series, we obtain

$$P(t) = P(0)e^{-\nu t} \sum_{n=0}^{\infty} \frac{1}{n!} \left(\nu\{I + \frac{1}{\nu}Q\}t\right)^n. \tag{3.46}$$

Upon regrouping the terms of (3.46), we find

$$P(t) = P(0) \sum_{n=0}^{\infty} \frac{(\nu t)^n}{n!} e^{-\nu t} \left[I + \frac{1}{\nu} \mathcal{Q} \right]^n . \tag{3.47}$$

In terms of our former description, we can view (3.47) as describing the dynamics of a discrete-time Markov chain having state transition probability matrix $[I + (1/\nu)\mathcal{Q}]$ and whose transition epochs are generated according to a Poisson process with rate ν. That is, the probability of n transitions in a period of length t is given by $(\nu t)^n e^{-\nu t}/n!$, the n-step transition matrix is $[I + (1/\nu)\mathcal{Q}]^n$, and the initial state probabilities are given by $P(0)$.

For the above interpretation to be valid, we must have ν at least as large as the magnitude of the maximal term on the diagonal of $\mathcal{Q}$ because the diagonal terms of the matrix $[I + (1/\nu)\mathcal{Q}]$ must be nonnegative. These terms are simply $1 + q_{ii}/\nu$, where q_{ij} represents the (i, j)th term of the $\mathcal{Q}$ matrix. The term q_{ii} represents the (exponential) rate at which the system departs state i, whenever it is in state i, while the term $1 + q_{ii}/\nu$ represents the probability that the system will return immediately to state i upon its departure. The terms q_{ij}/ν represent the probability of entering state j given a departure from state i, and q_{ij} denotes the rate at which the system enters state j from state i.

To illustrate what is happening here, consider the M/M/1 queueing system with a maximum occupancy of 1, as before. Then, from (3.39), we find

$$\mathcal{Q} = \begin{bmatrix} \lambda & -\lambda \\ -\mu & \mu \end{bmatrix}, \tag{3.48}$$

and

$$\left(I + \frac{1}{\nu} \mathcal{Q} \right) = \begin{bmatrix} 1 - \lambda/\nu & \lambda/\nu \\ \mu/\nu & 1 - \mu/\nu \end{bmatrix}. \tag{3.49}$$

Assuming $\lambda < \mu$, which is not required in this case, let us choose $\nu = \mu$. Then we find

$$\left(I + \frac{1}{\nu} \mathcal{Q} \right) = \begin{bmatrix} 1 - \rho & \rho \\ 1 & 0 \end{bmatrix}. \tag{3.50}$$

Figure 3.8 illustrates the randomization process. The original state diagram for the M/M/1 system with finite waiting room of capacity 1 is shown in Figure 3.8(a). In Figure 3.8(b), additional self transitions have been added to each state such that the total departure rate from each state is ν. In Figure 3.8(c), time is scaled so that the mean occupancy time in each state on each visit is unity. Finally, in Figure 3.8(d), ν is chosen to be μ so that the resulting diagram corresponds to the above example.

Thus, in the randomized system, the system always returns to state 0 whenever it leaves state 1, just as it does in the real system, but, unlike in the real system, the randomized system also returns immediately to state 0 whenever it

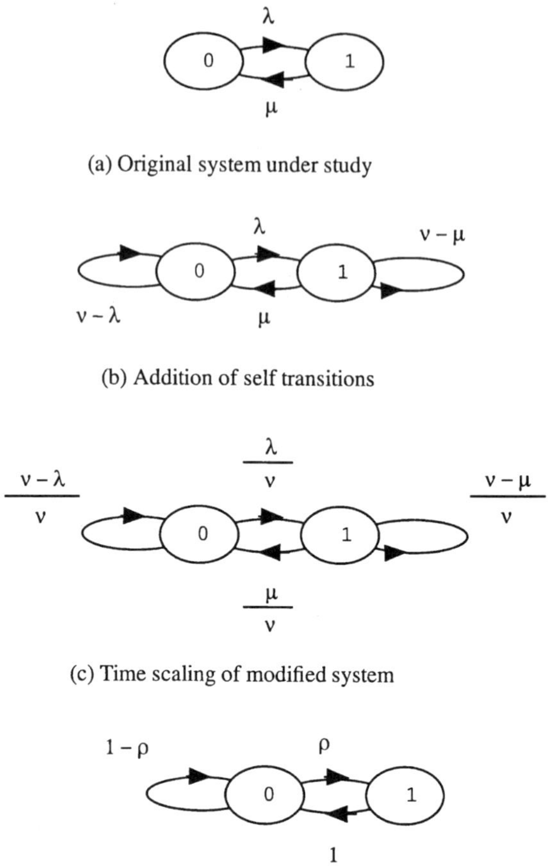

(a) Original system under study

(b) Addition of self transitions

(c) Time scaling of modified system

(d) Result after choosing $\nu = \mu$

Figure 3.8. Steps involved in randomization.

departs state 0 with probability $1 - \rho$. The system therefore returns to state 0 a geometric number of times, with the probability of departure equal to ρ, before entering state 1.

With our choice of $\nu = \mu$, the rate at which the system departs state 0 is μ, so that the sojourn time in state 0 is exponential with rate μ. Thus, the total amount of time spent in state 0 before returning to state 1 is the geometric sum of exponentials at rate μ, where the parameter of the geometric random variable is ρ. We have shown that the latter quantity of time is exponentially distributed with rate $\rho\mu = \lambda$, where we have used the fact that the geometric sum of exponentially distributed random variables is exponentially distributed.

Thus, in the randomized system, the sojourn time on each visit to each state is exponential, rate μ, but the total amount of time that the system spends in a state before it enters a different state is the same as that of the original system.

EXERCISE 3.20 For the specific example given here show that the equilibrium probabilities for the embedded Markov chain are the same as those for the continuous-time Markov chain.

Randomization apparently originated with Jensen [1953], but seems to have been independently developed by Keilson and Wishart [1964]. It has been described in several books including Keilson [1979] and Ross [1989]. The technique has been applied to the study of numerous systems in areas ranging from software reliability (Sumita and Shantikumar [1986]) to local area networks (Beuerman and Coyle [1987]). Grassman [1990], who provides an historical perspective on randomization, has advocated that this concept should be referred to as *Jensen's method.*

In this section, we presented the basics of Jensen's method and illustrated its use in the context of the finite capacity M/M/1 queueing system. Note that uniformization techniques can be applied to obtain state-dependent probability distributions for any finite-state continuous-time Markov chain. Readers seriously interested in using Jensen's method are urged to study Grassman [1990], where serious issues such as computational complexity and difficulty of use are addressed in depth.

EXERCISE 3.21 Show that the equilibrium probabilities for the embedded Markov chain underlying the continuous-time Markov chain are equal to the equilibrium probabilities for the continuous-time Markov chain.

EXERCISE 3.22 For the special case of the finite capacity M/M/1 queueing system with $K = 2$, $\lambda_0 = \lambda_1 = 0.8$, and $\mu_1 = \mu_2 = 1$, determine the time-dependent state probabilities by first solving the differential equation (3.32) directly and then using uniformization for $t = 0.0, 0.2, 0.4, \ldots, 1.8, 2.0$ with $P_0(0) = 1$, plotting the results for $P_0(t)$, $P_1(t)$, and $P_2(t)$. Compare the quality of the results and the relative difficulty of obtaining the numbers.

3.4 Balance Equation Approach for Systems In Equilibrium

Suppose all interarrival and service time distributions are exponential. Then from any point in time, the amount of time until the state changes is exponentially distributed. Previously, we wrote differential equations for $P_n(t)$ and then let $P_n'(t) \to 0$. Instead, we could write the equations directly.

In equilibrium, the rate of entry into a state must equal the rate of departure from the same state; that is, the entrance and departure rates must *balance.*

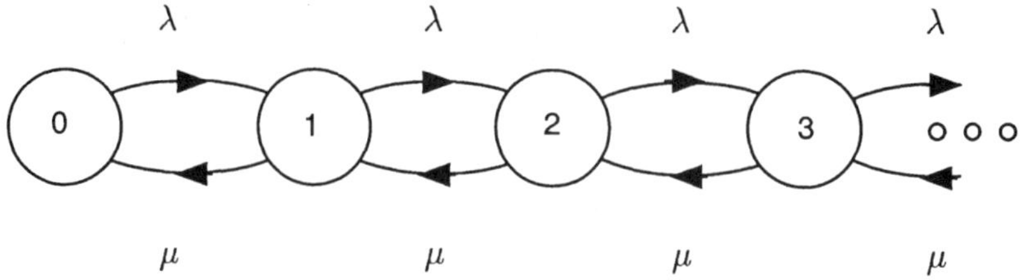

Figure 3.9. State diagram for M/M/1 System.

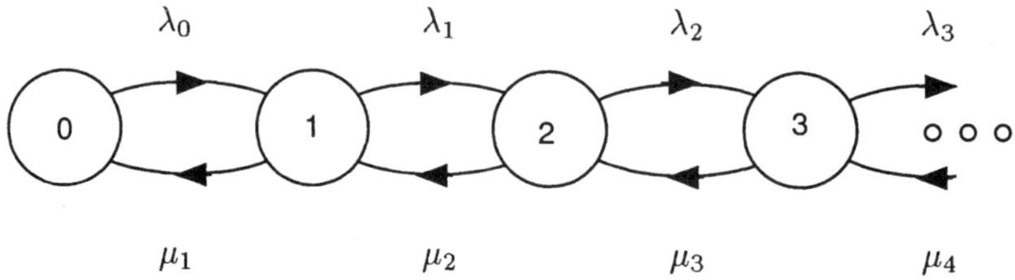

Figure 3.10. State diagram for general birth-death process.

For example, Figure 3.9 shows a state diagram for the M/M/1 system, and the following table expresses the concept of balance.

state	rate leaves		rate enters	
0	λP_0	=	μP_1	(3.51)
1	$(\lambda + \mu)P_1$	=	$\lambda P_0 + \mu P_2$	(3.52)
2	$(\lambda + \mu)P_2$	=	$\lambda P_1 + \mu P_3$	(3.53)
	...		...	
n	$(\lambda + \mu)P_n$	=	$\lambda P_{n-1} + \mu P_{n+1}$	(3.54)

In the above, (3.51), (3.52), (3.53), and (3.54) are called "balance equations." More generally, we have a similar notion of balance in the case of state-dependent arrival and service rates, or equivalently, for general birth-death processes. That is, we might have

$$\lambda_n = \text{arrival rate when } n \text{ are in system}$$

and

$$\mu_n = \text{service rate when } n \text{ are in system.}$$

Figure 3.10 shows the state diagram for the general birth-death process, and the following table expresses the concept of balance.

state	rate leaves		rate enters
0	$\lambda_0 P_0$	=	$\mu_1 P_1$
1	$(\lambda_1 + \mu_1) P_1$	=	$\lambda_0 P_0 + \mu_2 P_2$
2	$(\lambda_2 + \mu_2) P_2$	=	$\lambda_1 P_1 + \mu_3 P_3$
		$\cdots$	$\cdots$
n	$(\lambda_n + \mu_n) P_n$	=	$\lambda_{n-1} P_{n-1} + \mu_{n+1} P_{n+1}$

In that case, we find that

$$\begin{aligned} P_1 &= \frac{\lambda_0}{\mu_1} P_0, \\ P_2 &= \frac{\lambda_1}{\mu_2} P_1 = \frac{\lambda_1}{\mu_2}\frac{\lambda_0}{\mu_1} P_0, \\ &\vdots \\ P_n &= \left[\frac{\prod_{i=0}^{n-1} \lambda_i}{\prod_{i=1}^{n} \mu_i} \right] P_0, \end{aligned} \tag{3.55}$$

with

$$\sum_{i=0}^{\infty} P_i = 1. \tag{3.56}$$

Then

$$1 = P_0 + \sum_{n=1}^{\infty} \left[\frac{\prod_{i=0}^{n-1} \lambda_i}{\prod_{i=1}^{n} \mu_i} \right] P_0$$

or

$$P_0 = \frac{1}{1 + \sum_{n=1}^{\infty} \left[\left(\prod_{i=0}^{n-1} \lambda_i \right) / \left(\prod_{i=1}^{n} \mu_i \right) \right]}. \tag{3.57}$$

So, for an equilibrium solution to exist, we must have

$$\sum_{i=1}^{\infty}\left[\left(\prod_{i=0}^{n-1}\lambda_i\right)/\left(\prod_{i=1}^{n}\mu_i\right)\right] < \infty.$$

Otherwise, $P_0 = 0 \Rightarrow P_1 = 0 \Rightarrow P_2 = 0$, and so on.

EXAMPLE 3.3 Suppose that we have $\mu_i = \mu$ for all i, and $\lambda_i = \lambda$ for $0 \leq i \leq K$, and $\lambda_i = 0$ for all $i > K$. That is, we have an M/M/1 queueing system with finite waiting room of size K including the customer in service. Arrivals that occur while the system is in state K are not allowed to enter the system; that is, they are blocked. Then, (3.57) becomes

$$\begin{aligned} P_0 &= \frac{1}{1+\sum\limits_{n=1}^{K}[\lambda^n/\mu^n]} \\ &= \frac{1-\lambda/\mu}{1-(\lambda/\mu)^{K+1}}. \end{aligned} \tag{3.58}$$

Using this result in (3.55) leads to

$$P_n = \left[\frac{1-\lambda/\mu}{1-(\lambda/\mu)^{K+1}}\right](\lambda/\mu)^n. \tag{3.59}$$

When the waiting room's capacity is finite, customers attempting to enter the queue may be blocked, and it is of interest to specify the blocking probability. The blocking probability is defined as the proportion of the customers seeking admission to the queueing system who are denied. We can readily compute the blocking probability from the state probabilities.

Assuming a finite waiting room's of capacity K, the average number of customers seeking admission to the system over a long period of time of length τ, once the system has reached stochastic equilibrium, is given by $\sum_{n=0}^{K}\lambda_n P_n \tau$. Note that λ_K does not play a role in determining the equilibrium probabilities since customers arriving while the system is in state K are blocked. On the other hand, the average number of customers blocked under the same condition is simply $\lambda_K P_K \tau$. Thus the probability that an arbitrary customer is blocked, which we shall denote by $P_B(K)$, is simply

$$P_B(K) = \frac{\lambda_K P_K}{\sum_{n=0}^{K}\lambda_n P_n}. \tag{3.60}$$

In the case of the finite-capacity M/M/1 system, the right-hand side of (3.60) reduces to P_K.

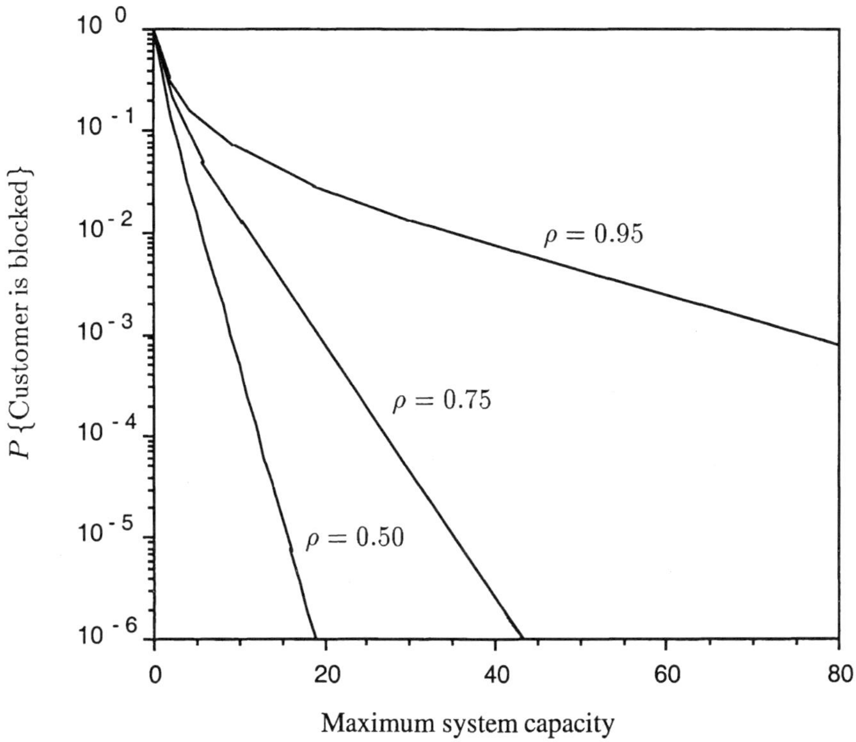

Figure 3.11. State diagram for general birth-death process.

Figure 3.11 shows a graph of $P_B(K)$ as a function of K. From this figure, we can readily compare $P_B(K)$ to $P\{\tilde{n} > K\}$ as obtained for the ordinary M/M/1 system. For example, at $\rho = 0.95$ and $K = 77$, $P_B(K) = 0.001$ while $P\{\tilde{n} > 77\} = (0.95)^{77} = 0.0193$; that is, the probability of exceeding the given occupancy level in the ordinary M/M/1 system is over 19 times as large as the probability of blocking for the capacity-limited system.

EXERCISE 3.23 For the special case of the finite-capacity M/M/1 system, show that for $K = 1, 2, \ldots$,

$$P_B(K) = \frac{\rho P_B(K-1)}{1 + \rho P_B(K-1)},$$

where $P_B(0) = 1$.

EXERCISE 3.24 For the finite-state general birth-death process, show that for $K = 1, 2, \ldots,$

$$P_B(K) = \frac{(\lambda_K/\mu_K)P_B(K-1)}{1 + (\lambda_K/\mu_K)P_B(K-1)},$$

where $P_B(0) = 1$.

An important special case of the birth-death process that finds broad application in traffic engineering is the *Erlang loss system.* This system has Poisson arrivals and s exponential servers [5], each serving at rate μ. Customers who arrive to the system when all servers are busy are *cleared* from the system; that is, they are blocked from entry. Thus an important measure of system performance is the proportion of customers who are lost. Since arrivals are Poisson, the proportion of customers who are lost is simply P_K.

For the Erlang loss system, we find

$$\mu_i = i\mu \quad \text{for} \quad i \leq s.$$

Also, potential arrivals to the system while the system is in state s are blocked, and

$$\lambda_i = \begin{cases} \lambda, & \text{for } i \leq s \\ 0, & \text{otherwise.} \end{cases}$$

Then, from (3.58) and (3.55), we find

$$P_n = \frac{(\lambda/\mu)^n/n!}{\sum_{i=0}^{s} (\lambda/\mu)^i/i!} \quad \text{for} \quad 0 \leq n \leq s. \tag{3.61}$$

Since potential customers arrive to the system according to a (state-independent) Poisson process, the blocking probability is given by P_s. Thus, for the Erlang loss system,

$$P\left\{\text{Customer is Blocked}\right\} = \frac{(\lambda/\mu)^s/s!}{\sum_{i=0}^{s} (\lambda/\mu)^i/i!}. \tag{3.62}$$

It is customary to express the blocking probability in terms of the *offered load*, a, which is defined as the ratio of the total arrival rate to the service rate of a single server; that is, the offered load is defined as

$$a = \lambda/\mu. \tag{3.63}$$

[5]Exponentiality is not required in order that the result hold for this case; that is, in this case, the results are *insensitive* to the form of the service-time distribution. There are many cases in which insensitivity holds in queueing systems; see Kelly [1979].

The blocking probability is then obtained from (3.62) and (3.63) and, in the standard notation of traffic engineering, is found to be

$$B(s,a) = a^s/s!/\sum_{i=0}^{s} a^i/i!. \tag{3.64}$$

This equation is called the *Erlang loss formula.* Another important term is the *carried load,* a', which is defined as the average number of busy servers for the system. It is easy to see that

$$a' = a[1 - B(s,a)]. \tag{3.65}$$

A typical application of the Erlang loss formula is to specify the number of lines needed to satisfy a certain level of blocking. For example, suppose a local division of a company knows the rate at which long distance calls are generated and the average call holding time. Suppose further that the company wants these long-distance calls to be blocked less than 1% of the time. Then, the company can use the Erlang loss formula to determine the minimum number of long distance lines that need to be available, provided that the assumption of Poisson arrivals for the calls is justified. Tables are provided in traffic engineering books (and some queueing books) for this purpose. We include a supplementary exercise which examines the difference between finite-population and infinite-population models of blocking at the end of this chapter. For a more thorough discussion and an historical perspective, the reader is referred to Cooper [1981].

Returning to the balance-equation approach, we note in passing that we can also write the differential equations by inspection by noting that the rate of change in the probabilities is given by the difference between the rate entering the state at time t and the rate departing the state at time t.

EXERCISE 3.25 Let K be arbitrary. Use the balance equation approach to write an equation of the form

$$\frac{d}{dt}P(t) = P(t)Q$$

where $P(t) = [\,P_0(t) \quad P_1(t) \quad \cdots \quad P_K(t)\,]$. Show that the eigenvalues of the matrix Q are real and nonpositive.

The above discussion presents the concept of *detailed* or *global* balance. It is sometimes easier to solve balance equations if they are initially written in terms of boundaries separating sets of states. For example, in Figure 3.12 we can consider everything to the left of the vertical line as one set of states and everything to the right as another set. Then, the rate of flow out of the set of states to the left must equal the rate of flow into the set of states to the right;

the concept underlying this solution technique is called *local balance*. This concept, which has broad application in the analysis of networks of queues, will be mentioned again later in the text.

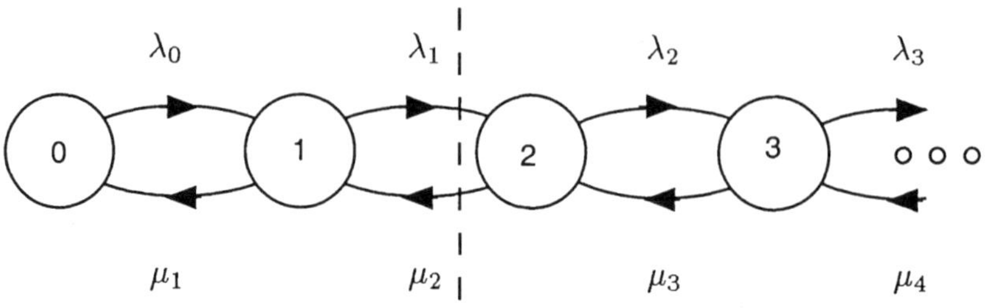

Figure 3.12. State diagram illustrating local balance.

EXERCISE 3.26 Using the concept of local balance, write and solve the balance equations for the general birth-death process shown in Figure 3.12.

3.5 Probability Generating Function Approach to Solving Equilibrium Equations

The solution of balance equations is not always straightforward. In some instances in which the solutions are not obvious, it is helpful to transform the system of equations, solve the transform equations, and then invert the transform to obtain the equilibrium probabilities. A useful transform is the *probability generating function (PGF)*.

DEFINITION 3.2 **Probability generating function.** Let $\tilde{x}$ be a nonnegative, integer-valued random variable. Then $\mathcal{F}_{\tilde{x}}(z) \stackrel{\Delta}{=} E[z^{\tilde{x}}] = \sum_{i=0}^{\infty} z^i P\{\tilde{x} = i\}$ is called the probability generating function for $\tilde{x}$.

A thorough treatment of probability generating functions is presented in Hunter [1983]. Among the properties of the PGF which we shall find useful are the following:

$$E[\tilde{x}(\tilde{x}-1)\cdots(\tilde{x}-n+1)] = \left.\frac{d^n}{dz^n}\mathcal{F}_{\tilde{x}}(z)\right|_{z=1} \tag{3.66}$$

and

$$P\{\tilde{x} = n\} = \left.\frac{1}{n!}\frac{d^n}{dz^n}\mathcal{F}_{\tilde{x}}(z)\right|_{z=0}. \tag{3.67}$$

We note in passing that the latter property follows directly from the uniqueness of the Maclaurin series expansion of the function $\mathcal{F}_{\tilde{x}}(z)$, which is

$$\mathcal{F}_{\tilde{x}}(z) = \sum_{n=0}^{\infty} \left[\frac{1}{n!} \frac{d^n}{dz^n} \mathcal{F}_{\tilde{x}}(z) \bigg|_{z=0} \right] z^n,$$

and its comparison to the definition of the PGF.

EXERCISE 3.27 Prove (3.66).

EXERCISE 3.28 Prove (3.67).

Recall that for M/M/1, we found from detailed balance that

$$\lambda P_0 = \mu P_1, \tag{3.68}$$

$$(\lambda + \mu) P_n = \lambda P_{n-1} + \mu P_{n+1} \qquad \text{for } n \geq 1, \tag{3.69}$$

and from local balance (discussed at the close of Section 3.4) that

$$\lambda P_n = \mu P_{n+1} \quad \text{for all} \quad n. \tag{3.70}$$

In order to illustrate the use of the probability generating function approach to the solution of balance equations, we solve the system of equations (3.70).

For the specific case in which the random variable of interest is the M/M/1 occupancy, we find

$$\begin{aligned} \mathcal{F}_{\tilde{n}}(z) &= \sum_{n=0}^{\infty} z^n P\{\tilde{n} = n\} \\ &= \sum_{n=0}^{\infty} z^n P_n. \end{aligned}$$

After multiplying both sides of (3.70) by z^n, we find

$$\lambda z^n P_n = \mu z^n P_{n+1}.$$

Thus

$$\lambda \sum_{n=0}^{\infty} z^n P_n = \mu \sum_{n=0}^{\infty} z^n P_{n+1}.$$

After applying the definition of $\mathcal{F}_{\tilde{n}}(z)$ to the above equation, we find

$$\begin{aligned} \lambda \mathcal{F}_{\tilde{n}}(z) &= \frac{\mu}{z} \sum_{n=0}^{\infty} z^{n+1} P_{n+1} \\ &= \frac{\mu}{z} [\mathcal{F}_{\tilde{n}}(z) - P_0]. \end{aligned}$$

Thus

$$\begin{aligned}\mathcal{F}_{\tilde{n}}(z) &= \frac{\mu P_0}{\mu - \lambda z} \\ &= \frac{P_0}{1 - \rho z}.\end{aligned}$$

But, from the properties of probability generating functions, $\mathcal{F}_{\tilde{n}}(1) = 1$, so $P_0 = 1 - \rho$. Finally, we obtain

$$\mathcal{F}_{\tilde{n}}(z) = \frac{1-\rho}{1-\rho z}. \tag{3.71}$$

We note

$$\frac{1-\rho}{1-\rho z} = (1-\rho)\sum_{n=0}^{\infty}(\rho z)^n.$$

Thus

$$\mathcal{F}_{\tilde{n}}(z) = \sum_{n=0}^{\infty}[(1-\rho)\rho^n]z^n.$$

But, by definition,

$$\mathcal{F}_{\tilde{n}}(z) = \sum_{n=0}^{\infty} P_n z^n.$$

So, by matching coefficients, we find $P_n = (1-\rho)\rho^n$ as expected.

We will now work with (3.68) and (3.69) to illustrate how to handle slightly more complicated problems. To begin, multiply both sides of (3.68) by z^n to obtain

$$z^n(\lambda+\mu)P_n = \lambda z^n P_{n-1} + \mu z^n P_{n+1}. \tag{3.72}$$

Now sum both sides of (3.72) from $n = 1$ to $n = \infty$ to obtain

$$\sum_{n=1}^{\infty} z^n(\lambda+\mu)P_n = \lambda \sum_{n=1}^{\infty} z^n P_{n-1} + \mu \sum_{n=1}^{\infty} z^n P_{n+1}.$$

After using the definition of $\mathcal{F}_{\tilde{n}}(z)$ in the above equation, we find

$$\begin{aligned}(\lambda+\mu)\left[\sum_{n=0}^{\infty} z^n P_n - P_0\right] &= \lambda z \sum_{n=0}^{\infty} z^n P_n \\ &\quad + \frac{\mu}{z}\left[\sum_{n=0}^{\infty} z^n P_n - zP_1 - P_0\right], \\ (\lambda+\mu)\left[\mathcal{F}_{\tilde{n}}(z) - P_0\right] &= \lambda z \mathcal{F}_{\tilde{n}}(z) + \frac{\mu}{z}\left[\mathcal{F}_{\tilde{n}}(z) - zP_1 - P_0\right].\end{aligned}$$

But, from (3.68), we know that $\lambda P_0 - \mu P_1 = 0$. Substituting this equality into the previous equation and solving for $\mathcal{F}(z)$ we get

$$\mathcal{F}_{\tilde{n}}(z) = \frac{\mu(1-z)P_0}{\lambda z^2 - (\lambda + \mu)z - \mu}.$$

Finally, upon dividing the numerator and denominator of the last equation by $\mu(1-z)$, we obtain the same result as before for the probability generating function. That is,

$$\mathcal{F}_{\tilde{n}}(z) = \frac{P_0}{1 - \rho z}.$$

The remainder of the solution is as before.

| EXERCISE 3.29 Use (3.66) to find $E[\tilde{n}]$ and $E[\tilde{n}^2]$.

3.6 Supplementary Problems

3-1 Messages arrive to a statistical multiplexing system according to a Poisson process having rate λ. Message lengths, denoted by $\tilde{m}$, are specified in octets, groups of 8 bits, and are drawn from an exponential distribution having mean $1/\mu$. Messages are multiplexed onto a single trunk having a transmission capacity of C bits per second according to a FCFS discipline.

(a) Let $\tilde{x}$ denote the time required for transmission of a message over the trunk. Show that $\tilde{x}$ has the exponential distribution with parameter $\mu C/8$.

(b) Let $E[\tilde{m}] = 128$ octets and $C = 56$ kilobits per second (kb/s). Determine $\lambda_{\max}$, the maximum message-carrying capacity of the trunk.

(c) Let $\tilde{n}$ denote the number of messages in the system in stochastic equilibrium. Under the conditions of part (b), determine $P\{\tilde{n} > n\}$ as a function of λ. Determine the maximum value of λ such that $P\{\tilde{n} > 50\} < 10^{-2}$.

(d) For the value of λ determined in part (c), determine the minimum value of s such that $P\{\tilde{s} > s\} < 10^{-2}$, where $\tilde{s}$ is the total amount of time a message spends in the system.

(e) Using the value of λ obtained in part (c), determine the maximum value of K, the system capacity, such that $P_{\mathcal{B}}(K) < 10^{-2}$.

3-2 A finite population, K, of users attached to a statistical multiplexing system operate in a continuous cycle of *think, wait, service*. During the think phase, the length of which is denoted by $\tilde{t}$, the user generates a message. The message then waits in a queue behind any other messages, if any, that may be awaiting transmission. Upon reaching the head of the queue, the user receives service and the corresponding message is transmitted over a communication channel. Message service times, $\tilde{x}$, and think times, $\tilde{t}$, are drawn from exponential distributions with rates μ and λ, respectively. Let the state of the system be defined as the total number of users waiting and in service and be denoted by $\tilde{n}$.

(a) The first passage time from state i to state $i-1$ is the total amount of time the system spends in all states from the time it first enters state i until it makes its first transition to the state $i-1$. Let $\tilde{s}_i$ denote the total cumulative time the system spends in state i during the first passage time from state i to state $i-1$. Determine the distribution of $\tilde{s}_i$.

(b) Determine the distribution of the number of visits from state i to state $i+1$ during the first passage time from state i to $i-1$.

(c) Show that $E[\tilde{y}_K]$, the expected length of a busy period, is given by the following recursion:

$$E[\tilde{y}_K] = \frac{1}{\mu}\left(1 + \lambda(K-1)E\left[\tilde{y}_{K-1}\right]\right) \quad \text{with} \quad E\left[\tilde{y}_0\right] = 0.$$

[*Hint*: Use the distribution found in part (b) in combination with the result of part (a) as part of the proof.]

(d) Let $P_0(K)$ denote the stochastic equilibrium probability that the communication channel is idle. Determine $P_0(K)$ using ordinary birth-death process analysis.

(e) Let $E[\tilde{\imath}_K]$ denote the expected length of the idle period for the communication channel. Verify that $P_0(K)$ is given by the ratio of the expected length of the idle period to the sum of the expected lengths of the idle and busy periods; that is,

$$P_0(K) = \frac{E[\tilde{\imath}_K]}{E[\tilde{\imath}_K] + E[\tilde{y}_K]}$$

which can be determined iteratively by

$$P_0(K) = \frac{1}{1 + [(K\lambda)/\mu]\left\{1 + (K-1)\lambda\, E[\tilde{y}_{K-1}]\right\}}.$$

That is, show that $P_0(K)$ computed by the formula just stated is identical to that obtained in part (d).

3-3 *Traffic engineering with finite population.* Ten students in a certain graduate program share an office that has four telephones. The students are always busy doing one of two activities: *doing queueing homework* (work state) or *using the telephone* (service state); no other activities are allowed - ever. Each student operates continuously as follows: the student is initially in the work state for an exponential, rate β, period of time. The student then attempts to use one of the telephones. If all telephones are busy, then the student is blocked and returns immediately to the work state. If a telephone is available, the student uses the telephone for a length of time drawn from an exponential distribution with rate μ and then returns to the work state.

(a) Define an appropriate state space for this service system.

(b) Draw a state diagram for this system showing all transition rates.

(c) Write the balance equations for the system.

(d) Specify a method of computing the ergodic blocking probability for the system - that is the proportion of attempts to join the service system that will be blocked - in terms of the system parameters and the ergodic state probabilities.

(e) Specify a formula to compute the average call generation rate.

(f) Let $\mu = 1/3$ calls per minute; that is, call holding times have a mean of three minutes. Compute the call blocking probability as a function of β for $\beta \in (0, 30)$.

(g) Compare the results of part (f) to those of the Erlang loss system having 4 servers and total offered traffic equal to that of part (f). That is, for each value of β, there is a total offered traffic rate for the system specified in this problem. Use this total offered traffic to obtain a value of λ, and then obtain the blocking probability that would result in the Erlang loss system, and plot this result on the same graph as the results obtained in (f). Then compare the results.

3-4 A company has six employees who use a leased line to access a database. Each employee has a *think* time which is exponentially distributed with parameter λ. Upon completion of the think time, the employee needs the database and joins a queue along with other employees who may be waiting for the leased line to access the database. Holding times are exponentially distributed with parameter μ. When the number of waiting employees reaches a level 2, use of an auxiliary line is authorized. The time required for the employee to obtain the authorization is exponentially distributed with rate τ. If the authorization is completed when there are less than three employees waiting or if the number of employees waiting drops below two at any time while the extra line is in use, the extra line is immediately disconnected.

(a) Argue that the set $\{0, 1, 2, 3, 3r, 3a, 4r, 4e, 5r, 5a, 6r, 6a\}$, where the numbers indicate the number of employees waiting and in service, the letter *r* indicates that authorization has been requested, and the letter *a* indicates that the auxiliary line is actually available for service, is a suitable state space for this process.

(b) The situation in state 4r is that there are employees waiting and in service and an authorization has been requested. With the process in state 4r at time t_0, list the events that would cause a change in the state of the process.

(c) Compute the probability that each of the possible events listed in part (b) would actually cause the change of state, and specify the new state of the process following the event.

(d) What is the distribution of the amount of time the system spends in state 4r on each visit? Explain.

(e) Draw the state transition rate diagram.

(f) Write the balance equations for the system.

3-5 Messages arrive to a statistical multiplexer at a Poisson rate λ for transmission over a communication line having a capacity of C in octets per second. Message lengths, specified in octets, are exponentially distributed with parameter μ. When the waiting messages reach a level 3, the capacity of the transmission line is increased to C_e by adding a dial-up line. The time required to set up the dial-up line to increase the capacity is exponentially distributed with rate τ. If the connection is completed when there are less than three messages waiting or if the number of messages waiting drops below two at any time while the additional capacity is in use, the extra line is immediately disconnected.

(a) Define a suitable state space for this queueing system.

(b) Draw the state transition-rate diagram.

(c) Organize the state vector for this system according to level, where the level corresponds to the number of messages waiting and in service, and write the vector balance equations for the system.

(d) Determine the infinitesimal generator for the underlying Markov chain for this system and comment on its structure relative to matrix geometric solutions.

3-6 Consider the M/M/2 queueing system, the system having Poisson arrivals, exponential service, 2 parallel servers, and an infinite waiting room capacity.

(a) Determine the expected first passage time from state 2 to state 1. [*Hint*: How does this period of time compare to the length of the busy period for an ordinary M/M/1 queueing system?]

(b) Determine the expected length of the busy period for the ordinary M/M/2 queueing system by conditioning on whether or not an arrival occurs before the first service completion of the busy period and by using the result from part (a).

(c) Define $\tilde{c}$ as the length of time between successive entries into busy periods, that is, as the length of one busy/idle cycle. Determine the probability that the system is idle at an arbitrary point in time by taking the ratio of the expected length of an idle period to the expected length of a cycle.

(d) Determine the total expected amount of time the system spends in state 1 during a busy period. Determine the probability that there is exactly one customer in the system by taking the ratio of the expected amount of time that there is exactly one customer in the system during a busy period to the expected length of a cycle.

(e) Check the results of (c) and (d) using classical birth-death analysis.

(f) Determine the expected sojourn time, $E[\tilde{s}]$, for an arbitrary customer by conditioning on whether an arbitrary customer finds either zero, one, or two or more customers present. Consider the nonpreemptive last-come-first-serve discipline together with Little's result and the fact that the distribution of the number of customers in the system is not affected by order of service.

Chapter 4

ADVANCED CONTINUOUS-TIME MARKOV CHAIN-BASED QUEUEING MODELS

In this chapter, we continue our analysis of queueing models that are characterized as discrete-valued, continuous-time Markov chains (CTMCs).

In Section 4.1, we discuss simple networks of exponential service stations of the feedforward, open, and closed varieties. We discuss the form of the joint state probability mass functions for such systems, which are of the so-called product form type. We discuss in detail a novel technique, due to Gordon [1990], for obtaining the normalizing constant for simple closed queueing networks in closed form.

In Section 4.2, we address the solution of a two-dimensional queueing model in which both the arrival and service rates are determined by the state of a single independent continuous-time Markov chain. This type of two-dimensional Markov chain is called a quasi-birth and death process (QBD), which is a vector version of the scalar birth-death process discussed in Chapter 3. A number of techniques for solving such problems are developed. The first approach uses the probability generating function approach, which was introduced in Chapter 3. We make extensive use of eigenvalue/eigenvector analysis to resolve unknown probabilities. Next, the matrix geometric technique is introduced and used to solve for the state probabilities of the QBD model. Next, a technique based on solving eigensystems for finding the rate matrix of the matrix geometric method, which reveals the entire solution, is discussed. Finally, a generalized state space approach is developed.

In Section 4.3, we introduce distributions of the phase (PH) type by modifying the class of models discussed in Section 4.2, and we provide the equilibrium occupancy distribution for the M/PH/1 system in matrix geometric form. We conclude the chapter with a set of supplementary exercises.

4.1 Networks of Single-Server Exponential Service Stations

In the previous chapter, we showed that the superposition and decomposition of Poisson processes form other Poisson processes. In addition, we presented a theorem that states that the output process from the M/M/1 queueing system is a Poisson process. It is then apparent that a feedforward network of exponential servers with exogenous Poisson arrivals behaves as though it were a collection of independently operating M/M/1 queueing systems provided the service times of the entities are chosen independently at the various servers in the network.

Even in cases when the network has feedback, so that the arrival process to each node is not Poisson, the marginal occupancy distribution at each node can be computed as though the arrival process were Poisson, and the joint occupancy distribution for the system is simply the product of the marginal distributions. This property also carries through to the case of closed networks of exponential servers under a certain broad class of assumptions. In the case of closed networks, however, the solution contains an unknown constant that must be computed by normalizing the joint distribution so that the joint probabilities sum to unity. An interesting aspect of our coverage is that we include a technique, due to Gordon [1990], for specifying the normalizing constant of closed networks of single-server queues in closed form.

Simple networks of exponential queues have been used successfully in a broad variety of modeling environments. Performance evaluation of computing systems is discussed extensively in Lazowska, Zahorjan, Graham and Sevcik [1984], Chandy and Sauer [1981], Trivedi [1982], and Kobayashi [1978]. Kleinrock [1976] and Schwartz [1987] address the application of queueing networks to design problems in computer communications. All of these books provide significant coverage of the theory underlying networks of exponential servers, and provide references for further study.

Single-server networks of this nature will be described in the following sections. Many of the results presented herein may be modified so that they apply to networks of multiserver exponential queues as well as queues having other than exponential service under other service disciplines such as LCFS. The reader interested in applying the methodology to large problems may also wish to consult Schwartz [1987], Gelenbe and Pujolle [1987] and, particularly, the references given there. For an excellent, highly readable, discussion on the merits of applying queueing network methodology to practical problems and a discussion of why usable results can be obtained with minimal effort, the reader is referred to Lazowska, Zahorjan, Graham and Sevcik [1986]. The reader seriously interested in traffic processes in networks of queues is strongly encouraged to consult Disney and Kiessler [1987]. Other significant books of interest in this area include Kelly [1979] and Walrand [1988].

4.1.1 Feedforward Networks of Single Servers (Fixed Routing)

Consider an arbitrarily connected network of N sources and destinations, M exponential servers, and Poisson exogenous arrivals. Assume that routing in the network is fixed; that is, there is a specific path that all customers having a particular source/destination pair must follow. Assume further that once a customer has received service from a particular server, the customer can never return to the same server; that is, the network allows no feedback.

Define $\delta_{jk}(i) = 1$ if units going from source j to destination k traverse server i and $\delta_{jk}(i) = 0$ otherwise; γ_{ij} as the rate units destined for destination j arrive to source i; μ_i as the service rate of server i; and λ_i as the aggregate unit arrival rate to server i. For each of the M servers, we have

$$\lambda_i = \sum_{j=1}^{N} \sum_{k=1}^{N} \gamma_{jk} \delta_{jk}(i).$$

As in the case of the M/M/1 system, the marginal occupancy density for server i is given by

$$P_i(n) = P\{\tilde{n}_i = n_i\} = (1 - \rho_i)\rho_i^{n_i}, \tag{4.1}$$

where $\rho_i = \lambda_i/\mu_i$, and the joint occupancy density is given by

$$\begin{aligned} P\{n_1, n_2, \ldots, n_M\} &= P\{\tilde{n}_1 = n_1, \tilde{n}_2 = n_2, \ldots, \tilde{n}_M = n_M\} \\ &= \prod_{i=1}^{M} (1 - \rho_i)\rho_i^{n_i}. \end{aligned} \tag{4.2}$$

Thus, the expected delay at node i is

$$E[\tilde{s}_i] = \frac{1/\mu_i}{1 - \rho_i}. \tag{4.3}$$

The average network delay for traffic entering node j destined for node k is therefore given by

$$E[\tilde{s}_{jk}] = \sum_{i=1}^{M} \delta_{jk}(i) E[\tilde{s}_i]. \tag{4.4}$$

Note that the logic leading to (4.4) does not apply to any moment of the waiting time distribution other than the first.

We now turn to the computation of the average delay through the network. From Little's result, we know that the average number of customers present at server i is $E[\tilde{n}_i] = \lambda_i E[\tilde{s}_i]$. Thus the expected number of customers in the system is

$$E[\tilde{n}] = \sum_{i=1}^{M} \lambda_i E[\tilde{s}_i] \tag{4.5}$$

or equivalently,

$$E[\tilde{n}] = \sum_{i=1}^{M} \frac{\rho_i}{1-\rho_i}. \tag{4.6}$$

We also know that the total number of customers entering the system per unit time is

$$\gamma = \sum_{k=1}^{N} \sum_{j=1}^{N} \gamma_{jk}. \tag{4.7}$$

Because we know from Little's result that $E[\tilde{n}] = \gamma E[\tilde{s}]$, it follows that the total average time spent in the system is given by

$$E[\tilde{s}] = \frac{1}{\gamma} \sum_{i=1}^{M} \frac{\rho_i}{1-\rho_i}, \tag{4.8}$$

where γ is given by (4.7).

From (4.8), we see that the network delay may be dominated by a single server if the capacities of the servers are chosen arbitrarily. In the design of systems, sometimes capacities are assigned to minimize $E[\tilde{s}]$ for a given traffic pattern; this problem is called the capacity assignment problem (Kleinrock [1976]). An example in which (4.8) was used as a major factor in a network design algorithm is Gavish and Altinkemer [1990].

4.1.2 Arbitrary Interconnections (Random Routing)

We now turn our attention to the analysis of a network with arbitrary random routing among M single exponential servers. That is, a customer enters a particular service station, say station i, obtains service at station i and then with probability r_{ij} proceeds next to station j independent of his past history. Customers depart the system from node i with probability r_{id}; that is,

$$\sum_{j=1}^{M} r_{ij} \leq 1$$

with equality if and only if customers cannot depart the system from node i. Routing among the stations of the network is thus governed by a Markov chain with an $M \times M$ routing matrix $R = (r_{ij})$. We assume exogenous arrivals to server i to be Poisson with parameter γ_i and that the service rate for server i is μ_i for $i = 1, 2, \ldots M$.

The arrival rate at a particular node is the sum of the exogenous arrival rate and the arrival rate due to customers entering from neighboring service stations. Thus, the total arrival rate at node j is given by

$$\lambda_j = \gamma_i + \sum_{i=1}^{M} r_{ij}\lambda_i \qquad \text{for } j = 1, 2, \ldots M, \tag{4.9}$$

or, in matrix form,

$$\lambda = \gamma + \lambda R, \tag{4.10}$$

where $\lambda = [\lambda_1, \lambda_2, \dots, \lambda_M]$ and $\gamma = [\gamma_1, \gamma_2, \dots, \gamma_M]$. Thus we find that

$$\lambda = \gamma [I - R]^{-1}. \tag{4.11}$$

Although the composite arrival processes at the service stations are not Poisson, the marginal occupancy density for server i is given by

$$P_i(n) = P\{\tilde{n}_i = n_i\} = (1 - \rho_i)\rho_i^{n_i}, \tag{4.12}$$

where $\rho_i = \lambda_i / \mu_i$, and the joint occupancy density is given by

$$P\{n_1, n_2, \dots, n_M\} = \prod_{i=1}^{M} (1 - \rho_i)\rho_i^{n_i}. \tag{4.13}$$

Equivalently,

$$P\{n_1, n_2, \dots, n_M\} = \frac{1}{G(M)} \prod_{i=1}^{M} \rho_i^{n_i}, \tag{4.14}$$

where

$$G(M) = \prod_{i=1}^{M} (1 - \rho_i)^{-1}. \tag{4.15}$$

This system is said to have a *product-form* solution, and the above result is called Jackson's theorem (Jackson [1963]).

EXERCISE 4.1 Using Little's result, determine the average time spent in the system for an arbitrary customer when the system is in stochastic equilibrium.

Results similar to the above are available for many networks including those with finite population and state-dependent servers. For an excellent summary of the results, the reader is referred to Chapter 3 of Kobayashi [1978].

4.1.3 Closed Networks of Single Servers (Random Routing)

In a closed network, there is no exogenous traffic arriving to the system nor is there traffic leaving the system. Instead, we view the network as representing a system in which a fixed number of jobs continually circulate. Such networks have a surprising array of applications. For example, they are sometimes used to analyze flow control behavior in communication networks that limit the total number of messages present in the system at any given time.

Closed queueing networks also have product form solutions of the type described above (Gordon and Newall [1967]). That is, the joint occupancy probabilities for the network have the form of a product of marginal probabilities. That is,

$$\begin{aligned} P(n_1, n_2, \ldots, n_M) &= P\{\tilde{n}_1 = n_1, \tilde{n}_2 = n_2, \ldots, \tilde{n}_M = n_M\} \\ &= k \prod_{i=1}^{M} \left(\frac{\lambda_i}{\mu_1}\right)^{n_i}, \end{aligned} \tag{4.16}$$

where k is a normalizing constant. In the case of closed networks, however, k is not determined as simply as it was in the previous two network types. In fact, in closed networks, the total occupancy of the system is limited to N, so that we always have $\sum_{i=1}^{M} n_i = N$. Thus, to emphasize the dependence upon N and M, (4.16) is usually written as:

$$P\{n_1, n_2, \ldots, n_M\} = \frac{1}{g(N,M)} \prod_{i=1}^{M} \left(\frac{\lambda_i}{\mu_i}\right)^{n_i}, \tag{4.17}$$

and $g(N, M)$ is thought of as the normalizing constant.

A peculiarity of closed queueing networks is that the flow balance equation analogous to (4.10) has the form

$$[I - R]\,\lambda = 0, \tag{4.18}$$

so that the vector λ is the eigenvector of the matrix $[I - R]$ corresponding to its zero eigenvalue. Thus the vector of traffic intensities can be determined only to within a multiplicative constant. Obviously, the choice of λ influences the computation of the normalizing constant, but not the occupancy probabilities.

EXERCISE 4.2 Argue that the matrix R is stochastic, and that, therefore, the vector λ is proportional to the equilibrium probabilities of the Markov chain for which R is the one-step transition probability matrix.

If the state space of a closed queueing network is large, the determination of the normalizing constant via brute force would require the addition of $\binom{N+M-1}{N-1}$ scaled probabilities. Numerous algorithms have been developed to avoid summing this large number of terms, the major results being summarized in Kobayashi [1978]. Although very efficient algorithms have been developed, none seem to have resulted in a closed-form expression for $g(N, M)$.

However, Harrison [1985] found a closed-form expression for $g(N, M)$ for the special case of single-server systems under discussion here. Gordon [1990], apparently encouraged by Harrison's work, reformulated the problem in an elegant way and derived Harrison's result, in addition to many other results that will be mentioned below, via a more direct approach.

We now turn to our discussion of Gordon's approach to specifying the normalizing constant for closed queueing networks. Recall that there are always a total of N customers in the system, so

$$\sum_{i=1}^{M} n_i = N, \tag{4.19}$$

where n_i is the number of customers at node i. Define $\mathcal{S}_{N,M}$ to be the set of all admissible states, that is,

$$\mathcal{S}_{N,M} = \left\{ (n_1, n_2, \cdots, n_M) | \sum_{i=1}^{M} n_i = N \right\}. \tag{4.20}$$

We therefore have from the law of total probability that

$$\sum_{(n_1,n_2,\cdots,n_M)\in\mathcal{S}_{N,M}} P\{\tilde{n}_1 = n_1, \tilde{n}_2 = n_2, \ldots, \tilde{n}_M = n_M\} = 1. \tag{4.21}$$

From (4.17) and (4.21), we then have

$$g(N, M) = \sum_{(n_1,n_2,\cdots,n_M)\in\mathcal{S}_{N,M}} \prod_{i=1}^{M} \left(\frac{\lambda_i}{\mu_i}\right)^{n_i}. \tag{4.22}$$

The key to Gordon's success is replacement of the finite sum on the right hand side of (4.22) by an infinite sum. Gordon does this by introducing an appropriate *delta* function into the summation. The delta function, a function of n, is defined as follows:

$$\delta(n - n_0) = \begin{cases} 1, & \text{if } n = n_0, \\ 0, & \text{otherwise,} \end{cases} \tag{4.23}$$

where n_0 is usually referred to as the *location of the delta function.* This function has the following representation as a contour integral on the complex plane:

$$\delta(n - n_0) = \frac{1}{j2\pi} \oint_C \phi^{(n-n_0)} \frac{d\phi}{\phi}, \tag{4.24}$$

where $j = \sqrt{-1}$ and $\oint_C$ indicates the integral around the unit circle, a closed contour, of the complex plane. It is readily verified, by performing the indicated integration using the residue theorem (see Churchill [1960]), that (4.24) and (4.23) are equivalent.

EXERCISE 4.3 Let x denote any integer. Show that

$$\frac{1}{j2\pi} \oint_C \phi^x d\phi = \begin{cases} 1, & \text{for } x = -1 \\ 0, & \text{otherwise.} \end{cases}$$

by direct integration.

From (4.23) and (4.24), we see that

$$\delta\left(\sum_{i=1}^{M} n_i - N\right) = \begin{cases} 1, & \text{if } \sum_{i=1}^{M} n_i = N \\ 0, & \text{otherwise,} \end{cases}$$

and

$$\begin{aligned} \delta\left(\sum_{i=1}^{M} n_i - N\right) &= \frac{1}{j2\pi}\oint_C \phi^{\left(\sum_{i=1}^{M} n_i - N\right)}\frac{d\phi}{\phi} \\ &= \frac{1}{j2\pi}\oint_C \phi^{\left(\sum_{i=1}^{M} n_i\right)}\frac{d\phi}{\phi^{N+1}} \\ &= \frac{1}{j2\pi}\oint_C \prod_{i=1}^{M}\phi^{n_i}\frac{d\phi}{\phi^{N+1}}. \end{aligned} \tag{4.25}$$

Now, if we multiply $\prod_{i=1}^{M}(\lambda_i/\mu_i)^{n_i}$ by $\delta\left(\sum_{i=1}^{M} n_i - N\right)$, then this product will be zero if $(n_1, n_2, \cdots, n_M) \ni \mathcal{S}_{N,M}$, where $\ni$ stands for the relationship *not in*. Therefore, if we perform the above multiplication in (4.22), we find

$$g(N,M) = \sum_{n_1=0}^{\infty}\sum_{n_2=0}^{\infty}\cdots\sum_{n_M=0}^{\infty}\prod_{i=1}^{M}\left(\frac{\lambda_i}{\mu_i}\right)^{n_i}\delta\left(\sum_{i=1}^{M} n_i - N\right),$$

which is alternately represented in contour integral form by

$$\begin{aligned} g(N,M) &= \frac{1}{j2\pi}\oint_C \sum_{n_1=0}^{\infty}\sum_{n_2=0}^{\infty} \\ &\quad \cdots\sum_{n_M=0}^{\infty}\prod_{i=1}^{M}\phi^{n_i}\prod_{i=1}^{M}(\lambda_i/\mu_i)^{n_i}\frac{d\phi}{\phi^{N+1}} \\ &= \frac{1}{j2\pi}\oint_C \sum_{n_1=0}^{\infty}(\rho_1\phi)^{n_1}\sum_{n_2=0}^{\infty}(\rho_2\phi)^{n_2} \\ &\quad \cdots\sum_{n_M=0}^{\infty}(\rho_M\phi)^{n_M}\frac{d\phi}{\phi^{N+1}}, \end{aligned} \tag{4.26}$$

where, as usual, $\rho_i = \lambda_i/\mu_i$. Upon performing the indicated infinite summations, which converge for $|\rho_i\phi| < 1$, we find

$$g(N,M) = \frac{1}{j2\pi}\oint_C g(N,M,\phi)\frac{d\phi}{\phi^{N+1}}, \tag{4.27}$$

where we have defined

$$g(N,M,\phi) = \prod_{i=1}^{M}\frac{1}{1-\rho_i\phi}. \tag{4.28}$$

Now, $g(N, M, \phi)$ has been obtained from a finite product of infinite polynomials of the form $\sum_{j=0}^{\infty} (\rho_i \phi)^j$. Therefore, it is clear that $g(N, M, \phi)$ itself can be written in the form

$$g(N, M, \phi) = \sum_{j=0}^{\infty} g_i \phi^j. \tag{4.29}$$

In fact, the expression $\sum_{j=0}^{\infty} (\rho_i \phi)^j$ is the generating function (Hunter [1983]) for a sequence $\{a(i)\} = \{a_j(i), j = 0, 1, \ldots\}$ in which $a_j(i) = \rho_i{}^j$. Because $g(N, M, \phi)$ is the product of the generating functions for the M sequences $\{a(1)\}$,$\{a(2)\}$,...,$\{a(M)\}$, it follows from the properties of sequences (Hunter [1983]) that the sequence $\{g_i\}$ is just the (m-fold) convolution of the sequences $\{a(1), a(2), \ldots, a(M)\}$.

Upon substitution of (4.29) into (4.27), we find

$$\begin{aligned} g(N, M) &= \frac{1}{j2\pi} \oint_C \sum_{j=0}^{\infty} g_i \phi^j \frac{d\phi}{\phi^{N+1}} \\ &= \frac{1}{j2\pi} \oint_C \sum_{j=0}^{\infty} g_i \phi^{(j-N)} \frac{d\phi}{\phi}. \end{aligned} \tag{4.30}$$

From the residue theorem, it readily follows that

$$g(N, M) = g_N,$$

the coefficient of ϕ^N in the expression for $g(N, M, \phi)$. Now, if $g(N, M, \phi)$ is viewed as the generating function for the convolution of M sequences, (4.30) is not very surprising; this is exactly the (unpleasant) message conveyed by (4.22). The determination of the coefficient via convolution requires the addition of $\binom{N+M-1}{M-1}$ scaled state probabilities.

However, the form of (4.28) suggests that the determination of this coefficient can be carried out in a much more efficient fashion. In particular, $g(N, M, \phi)$ can be rewritten using partial fraction expansions (Hunter [1983]), and once this is done, the coefficient of ϕ^N will be obvious. For example, for the special case in which the ρ_i are distinct, we can rewrite $g(N, M, \phi)$ in the following form:

$$g(N, M, \phi) = \sum_{i=1}^{M} \frac{c_i}{1 - \rho_i \phi}. \tag{4.31}$$

We then find by expanding $1/(1 - \rho_i \phi)$ in geometric series form that the coefficient of ϕ^M for the ith partial fraction is simply $c_i \rho_i^N$. Thus, upon summing the values due to the respective partial fractions, we find

$$g(N, M) = \sum_{i=1}^{M} c_i \rho_i^N. \tag{4.32}$$

Using elementary calculus, we can readily determine that

$$c_i = \frac{\rho_i^{M-1}}{\prod\limits_{\substack{1 \le j \le M \\ j \ne i}} (\rho_i - \rho_j)}. \tag{4.33}$$

Upon substitution of (4.33) into (4.32), we find

$$g(N, M) = \sum_{i=1}^{M} \frac{\rho_i^{N+M-1}}{\prod\limits_{\substack{1 \le j \le M \\ j \ne i}} (\rho_i - \rho_j)}, \tag{4.34}$$

which is the result given by Harrison [1985] and Gordon [1990].

EXERCISE 4.4 Suppose that the expression for $g(N, M, \phi)$ can be written as

$$g(N, M, \phi) = \prod_{i=1}^{r} \frac{1}{(1 - \sigma_i \phi)^{\nu_i}}, \tag{4.35}$$

where $\sum_{i=1}^{r} \nu_i = M$. That is, there are exactly r distinct singular values of $g(N, M, \phi)$- these are called σ_1,σ_2,. . .,σ_r - and the multiplicity of σ_i is ν_i. We may rewrite (4.35) as

$$g(N, M, \phi) = \sum_{i=1}^{r} \sum_{j=1}^{\nu_j} \frac{c_{ij}}{(1 - \sigma_i \phi)^j}. \tag{4.36}$$

Show that

$$\begin{aligned} c_{ij} &= \frac{1}{(\nu_i - j)!} \left(-\frac{1}{\sigma_i} \right)^{(\nu_i - j)} \\ &\quad \frac{d^{(\nu_i - j)}}{d\phi^{(\nu_i - j)}} \left[(1 - \sigma_i \phi)^{\nu_i} g(N, M, \phi) \right] \Big|_{\phi = 1/\sigma_i}. \end{aligned} \tag{4.37}$$

EXERCISE 4.5 Define b_{nN} to be the coefficient of ϕ^N in the expansion of $(1 - \sigma_i \phi)^{-n}$. Show that

$$b_{nN} = \binom{N + n - 1}{N} \sigma_i^N. \tag{4.38}$$

EXAMPLE 4.1 To illustrate the application of Gordon's ideas to a problem not specifically solved in Gordon[1990], we consider a window flow control technique in a communications network as in Figure 3-36 of Schwartz [1987]. Figure 4.1 shows the diagram for the system. We wish to determine the state probabilities for the network, which has 5 queues, a maximum occupancy of four, and a simple routing matrix.

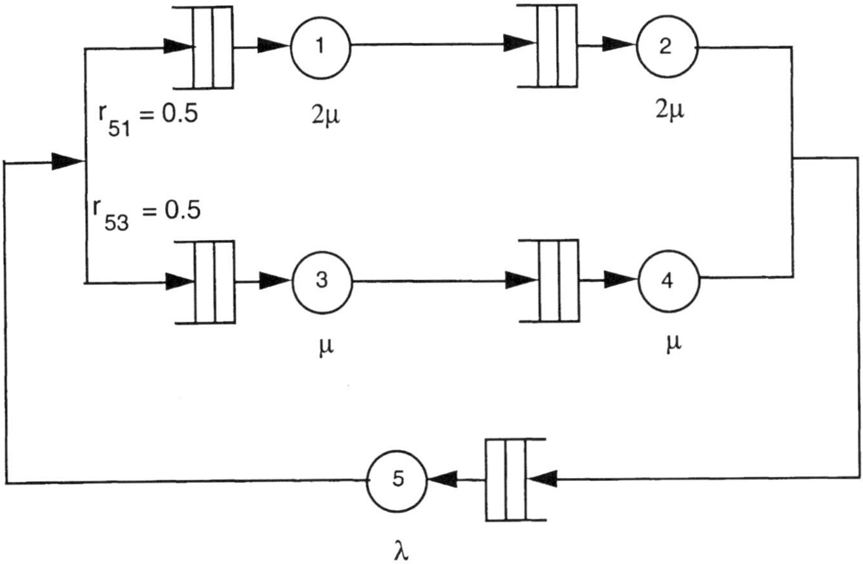

Figure 4.1. Block diagram for window flow control network.

Solution: From the diagram, we readily see that $\lambda_1 = \lambda_2 = \lambda_3 = \lambda_4 = 0.5\lambda_5$. Because $\mu = [2 \quad 2 \quad 1 \quad 1 \quad 1]$, we can choose $\rho_5 = \lambda_5 = \rho$, $\rho_1 = \rho_2 = \rho/4$, and $\rho_3 = \rho_4 = \rho/2$. We thus find from (4.17) that

$$\begin{aligned} P\{n_1, n_2, \cdots, n_5\} &= \frac{1}{g(4,5)} \rho_1^{n_1} \rho_2^{n_2} \rho_3^{n_3} \rho_4^{n_4} \rho_5^{n_5} \\ &= \frac{1}{g(4,5)} \left(\frac{\rho}{4}\right)^{n_1+n_2} \left(\frac{\rho}{2}\right)^{n_3+n_4} \rho^{n_5}, \end{aligned}$$

and from (4.28) that

$$\begin{aligned} g(4,5,\phi) &= \frac{1}{1-\rho_1\phi} \frac{1}{1-\rho_2\phi} \frac{1}{1-\rho_3\phi} \frac{1}{1-\rho_4\phi} \frac{1}{1-\rho_5\phi} \\ &= \frac{1}{[1-(\rho/4)\phi]^2} \frac{1}{[1-(\rho/2)\phi]^2} \frac{1}{1-\rho\phi}. \end{aligned}$$

On the basis of the results of Exercises 3.29 and 3.30, we first find that

$$g(N,5,\phi) = \frac{-16/9}{[1-(\rho/4)\phi]} + \frac{-1/3}{[1-(\rho/4)\phi]^2} + \frac{-4}{[1-(\rho/2)\phi]^2} + \frac{64/9}{(1-\rho\phi)},$$

and then we find

$$g(N,5) = -\frac{16}{9}\left(\frac{\rho}{4}\right)^N - \frac{1}{3}(N+1)\left(\frac{\rho}{4}\right)^N - 4(N+1)\left(\frac{\rho}{2}\right)^N + \frac{64}{9}\rho^N.$$

Now, in order to assure convergence of the infinite summation required to obtain $g(N, M, \phi)$ in closed form, we required that each $|\rho_i \phi|$ be less than unity. Thus the choice of the λ_i, and hence ρ_i, affects only the range of ϕ over which the summation converges. For consistency with Schwartz [1987], we choose $\rho = 4$. We then find that $g(4,5) = 1497$ as given in Table 5-5 of Schwartz [1987]. We then find the joint queue occupancy probabilities to be

$$P\{n_1, n_2, \cdots, n_5\} = \frac{1}{1497} 2^{(n_3+n_4)} 4^{n_5}. \tag{4.39}$$

The reader should verify that there are a total of 70 possible states and that the probabilities obtained sum to unity.

EXERCISE 4.6 Verify that the probabilities as specified by (4.39) sum to unity.

Now that we have specified a procedure to obtain a closed-form expression for $g(N, M)$, it seems natural to ask whether or not it is possible to specify (marginal) node occupancy probabilities and moments of the node occupancy distribution in simple closed forms as well. As we shall see, the answer is "yes." In what follows, we shall first obtain simple expressions for the node occupancy probabilities and then use these results to obtain a simple expression for the expected node occupancy.

Recall from (4.17), with $\rho_i = \lambda_i / \mu_i$, we have

$$P\{n_1, n_2, \cdots, n_M\} = \frac{1}{g(N, M)} \prod_{i=1}^{M} \rho_i^{n_i}. \tag{4.40}$$

To obtain the marginal occupancy probability for node i, we simply sum over all possible joint occupancy probabilities with $\tilde{n}_i = n$. Without loss of generality, we may reorder the nodes so that $i = M$ and consider node M to be arbitrary. Then, because the set of values over which $n_M = n$ is given by the set

$$\mathcal{S}_{N-n,M-1} = \left\{ (n_1, n_2, \cdots, n_{M-1}) | \sum_{i=1}^{M-1} n_i = N - n \right\},$$

we readily find that

$$\begin{aligned} P\{\tilde{n}_M = n\} &= \sum_{(n_1,n_2,\ldots,n_{M-1} \in \mathcal{S}_{N-n,M-1}} \\ &\quad P\{\tilde{n}_1, \ldots, \tilde{n}_{M-1}\} \\ &= \sum_{(n_1,n_2,\ldots,n_{M-1}) \in \mathcal{S}_{N-n,M-1}} \end{aligned}$$

$$
\begin{aligned}
& \frac{1}{g(N,M)} \prod_{i=1}^{M-1} \rho_i^{n_i} {\rho_M}^n \\
= & \frac{\rho_M^n}{g(N,M)} \sum_{(n_1,n_2,\dots,n_{M-1})\in\mathcal{S}_{N-n,M-1}} \\
& \frac{g(N-n,M-1)}{g(N-n,M-1)} \prod_{i=1}^{M-1} \rho_i^{n_i} \\
= & \frac{\rho_M^n}{g(N,M)} g(N-n,M-1). \qquad (4.41)
\end{aligned}
$$

Now, (4.41) is in a reasonably simple form, but it involves terms of the form $g(\cdot, M-1)$, and it would be nicer to have all constants in the form $g(\cdot, M)$ because our normalizing constants are specified in closed forms for each M with N as a variable. From (4.41) and the law of total probability, we find

$$
1 = \sum_{n=0}^{N} P\{\tilde{n}_m = n\} = \frac{1}{g(N,M)} \sum_{n=0}^{N} {\rho_M}^n g\left(N-n, M-1\right),
$$

so that

$$
g(N,M) = \sum_{n=0}^{N} {\rho_M}^n g(N-n, M-1), \qquad (4.42)
$$

where we define $g(0,M) = 1$ for $M \geq 1$, $g(N,0) = 0$ for all $N \geq 0$, and $g(N,M) = 0$ for all $N < 0$. Expanding (4.42), we find for $N, M \geq 1$,

$$
\begin{aligned}
g(N,M) &= g(N,M-1) + \sum_{n=1}^{N} {\rho_M}^n g(N-n, M-1) \\
&= g(N,M-1) + \rho_M \sum_{n=0}^{N-1} {\rho_M}^n g(N-1-n, M-1).
\end{aligned}
$$

But, from (4.42), we recognize the summation of the right-hand side of the previous equation to be $g(N-1,M)$. Thus we have

$$
g(N,M) = g(N,M-1) + \rho_M g(N-1,M), \qquad (4.43)
$$

with $g(0,M) = 1$ for $M \geq 1$ and $g(N,0) = 0$ for all $N \geq 0$, as previously stated.

We note in passing that the recurrence equation (4.43) provides a handy way of generating the normalizing constants recursively for an arbitrary closed network of single-server queues. Kobayashi [1978] presents the same recursion for the special case described here.

The complexity of obtaining $g(N, M)$ for this special case via (4.34) is not substantially different from that of using (4.43). However, the power in Gordon's approach is that it makes it possible to obtain closed-form results for a variety of more complicated systems. In particular, Gordon easily derives closed-form expressions for single-server queues in the following special cases: $\rho_{M-1} = \rho_M$; $\rho_i = \rho$ for all i. Extensions to other special cases of the single-server class of networks are simply a matter of applying partial fraction expansion rules to obtain the coefficient of ϕ^N in (4.34).

In addition, Gordon derives a closed-form expression for the case in which each of the service stations may have a finite number, s_i, of servers, and the fractions ρ_i/s_i are distinct, and he indicates how this method may be extended to the case in which the fractions ρ_i/s_i are not distinct. These closed-form expressions and the extensions to more general cases do not seem to have appeared previously in the literature.

It is interesting to observe that the methods discussed by Kobayashi [1978] in explaining the recursive expressions also depend upon infinite summations and generating functions of exactly the same form as those used by Gordon. However, the relationship of the results to contour integration and the resulting utility of partial fraction expansions in obtaining closed-form results appear to have originated with Gordon.

Returning to our specification of the marginal occupancy probabilities, we note that the specification of N in (4.43) is arbitrary. We therefore can substitute $N - n$ for N, and after rearranging, we have

$$g(N-n, M-1) = g(N-n, M) - \rho_M g(N-n-1, M). \tag{4.44}$$

Upon substitution of (4.44) into (4.41), we obtain

$$P\{\tilde{n}_M = n\} = \frac{{\rho_M}^n}{g(N, M)} [g(N-n, M) - \rho_M g(N-n-1, M)]. \tag{4.45}$$

We have seen earlier that expectations may be computed by summing complementary distributions; for example,

$$E[\tilde{n}] = \sum_{n=0}^{\infty} P\{\tilde{n} > n\}.$$

Clearly, $P\{\tilde{n}_M > N\} = 0$ since N is the population size. Thus, we find from (4.45) that

$$P\{\tilde{n}_M > N-1\} = P\{\tilde{n}_M = N\} = \frac{{\rho_M}^N}{g(N, M)}.$$

By successive substitutions into (4.45), we find

$$P\{\tilde{n}_M > n\} = \begin{cases} \frac{{\rho_M}^{n+1} g(N-n-1,M)}{g(N,M)}, & \text{for } 0 \le n \le N-1; \\ 0, & \text{otherwise.} \end{cases} \tag{4.46}$$

Thus we have

$$\begin{aligned} E[\tilde{n}_M] &= \sum_{n=0}^{\infty} P\{\tilde{n}_M > n\} \\ &= \sum_{n=0}^{N-1} \frac{{\rho_M}^{n+1}}{g(N,M)} g(N-n-1,M) \\ &= \frac{1}{g(N,M)} \sum_{n=1}^{N} {\rho_M}^n g(N-n,M). \end{aligned} \tag{4.47}$$

Expressions for higher moments of the occupancy distribution at an arbitrary node can be derived in a similar fashion.

Given (4.47), the throughput at a given node can be specified exactly via Little's result. That is, (4.47) provides us with the average server occupancy, and we already know the average sojourn time at a server; therefore, Little's result can be used to solve for the average arrival rate to the server, which is the throughput because there is no blocking.

EXERCISE 4.7 Carefully develop the argument leading from (4.43) to (4.44).

EXERCISE 4.8 Using the recursion of (4.43) together with the initial conditions, verify the expression for $g(N, 5)$ for the special case $N = 6$ numerically for the example presented in this section.

EXERCISE 4.9 Develop an expression for throughput at node M using Little's result and (4.47).

Before closing our discussion of queueing networks, we note that when only mean occupancies (or mean sojourn times) are desired, it is possible to compute the mean values directly through an iterative technique known as *mean-value analysis*. We emphasize that mean-value analysis is an iterative technique that should not be confused with other approaches that actually yield closed-form expressions for averages. The technique is described thoroughly in Schwartz [1987], Galenbe and Pujolle [1987], Leon-Garcia [1989] and numerous other texts and papers. Although this technique has found broad application, we will not discuss it further in the volume.

We also note that a technique for solving product form networks based on *potentials*(see Çinlar [1975]) is presented in Abboud and Daigle [1997]. This method has not been fully developed for the general case, but where it has been applied, it has been found to be very fast and, in addition, requires little storage compared to other known methods.

4.2 Models Having Phase-Dependent Arrivals and Service

In Chapter 3, we discussed analyses of systems in which arrival and service rates may be state-dependent. In this section, we consider analysis of systems in which the rate at which units arrive to the server and the rate at which the units are serviced are dependent on the state of a so-called phase process. In particular, we assume that the phase (see Stern [1983]) of a system, $\{\wp(t), t \geq 0\}$, is a discrete-valued continuous-time finite Markov chain with infinitesimal generator, $\mathcal{Q}$ (Cohen [1969], Chapter 3). We define $\{\wp(t), t \geq 0\}$ to be a continuous-time Markov chain which takes on integer values between 0 and K. Thus, the dimension of the $\mathcal{Q}$-matrix is $K+1$, and

$$\frac{d}{dt}[\, P\{\wp(t)=0\} \quad P\{\wp(t)=1\} \quad \cdots \quad P\{\wp(t)=K\}\,] \tag{4.48}$$
$$= [\, P\{\wp(t)=0\} \quad P\{\wp(t)=1\} \quad \cdots \quad P\{\wp(t)=K\}\,]\, \mathcal{Q}.$$

The state diagram for the phase process is shown in Figure 4.2. As one can see from this diagram, the birth rate while the phase process is in state i, $i = 0, 1, \ldots, K$, is equal to β_i, and the rate at which the process transitions from state i to state $i-1$ is given by δ_i.

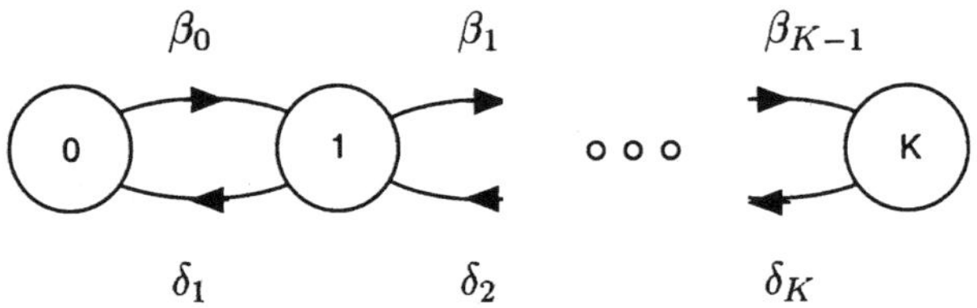

Figure 4.2. State diagram for phase process.

When $\{\wp(t), t \geq 0\}$ is in phase i, $0 \leq i \leq K$, the arrival rate of units to the server is λ_i and the service rate is μ_i. Figure 4.3 shows a partial state diagram for a queueing system having phase dependent arrival and service rates. A typical state for this system is designated by (i, j), where i specifies the current occupancy and j specifies the current phase. The process $\{(\tilde{n}(t), \wp(t)), t \geq 0\}$ is a QBD process (Neuts [1981]) on the state space $(n, i), n \geq 0, 0 \leq i \leq K$. Let

$$P_{ni} = \lim_{t \to \infty} P\{\tilde{n}(t) = n, \wp(t) = i\}.$$

In this section, we develop a number of techniques for solving for the state probabilities of QDB processes. In the first subsection, we develop a vector

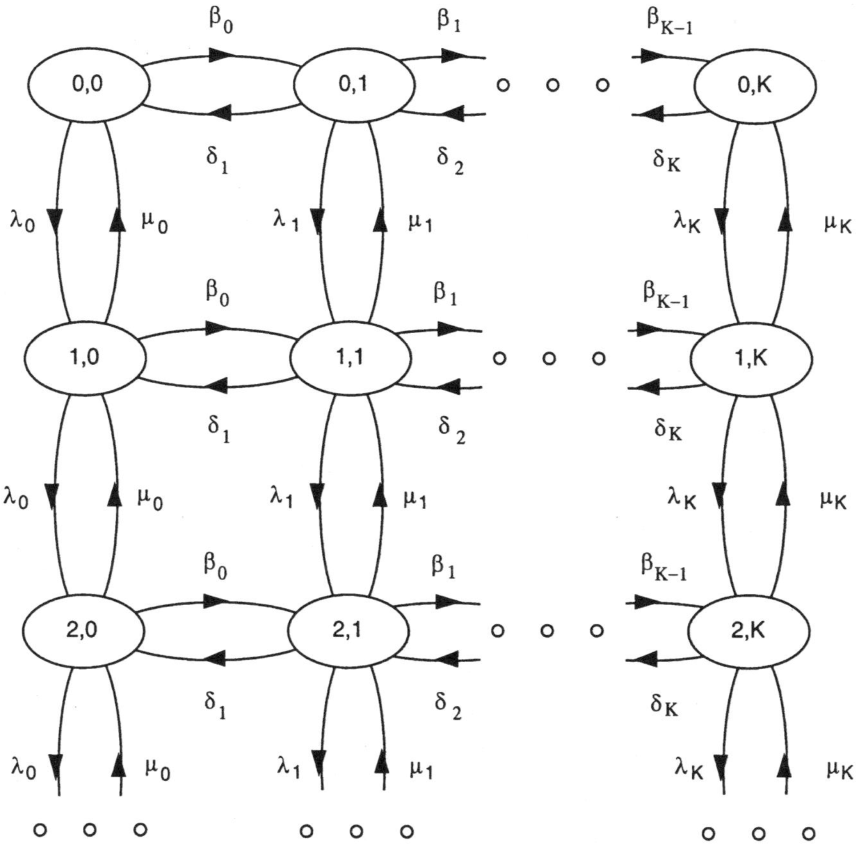

Figure 4.3. State diagram for system having phase-dependent arrival and service rates.

version of the probability generating function approach, the scalar version of which was introduced in Chapter 3. Extensive use of eigenvalue/eigenvector analysis is used to resolve the unknown vector of probabilities. In the second subsection, the matrix geometric technique is introduced and used to solve for the state probabilities of the QBD model. A technique for finding the rate matrix of the matrix geometric method, which reveals the entire solution, based on solving eigensystems is discussed in the third subsection. Finally, in the fourth subsection, a generalized state space approach is developed.

4.2.1 Probability Generating Function Approach

From the state diagram shown in Figure 4.3, it is straightforward to write the balance equations for the system. For a typical state on the interior of the diagram - that is, with $n \geq 1$ and $0 < i < K$ - we find

$$\begin{aligned}(\lambda_i + \mu_i + \beta_i + \delta_i)P_{ni} &= \lambda_i P_{n-1,i} + \beta_{i-1}P_{n,i-1} \\ &\quad + \delta_{i+1}P_{n,i+1} + \mu_i P_{n+1,i}. \qquad (4.49)\end{aligned}$$

From this balance equation, the balance equations for the states not interior to the state diagram are readily determined. We simply specialize the above equation to account for the changes due to the QBD boundary conditions . First, we consider the case $n = 0$ and $i = 0$ for which there are no transitions from state $(0, 0)$ due to "deaths," no transitions into state $(0, 0)$ due to "births," and no transitions into state $(0, 0)$ due to new arrivals. Thus, we find the balance equation for state $(0, 0)$ is

$$(\lambda_0 + \beta_0)P_{00} = \delta_1 P_{01} + \mu_0 P_{10}. \qquad (4.50)$$

Next, we consider the case $n = 0, 0 < i < K$. On this boundary, there are no transitions from state $(0, i)$ due to service completions, and no transitions into state $(0, i)$ due to new arrivals. Thus we obtain the following set of equations:

$$(\lambda_i + \beta_i + \delta_i)P_{0i} = \beta_{i-1}P_{0,i-1} + \delta_{i+1}P_{0,i+1} + \mu_i P_{1,i}. \qquad (4.51)$$

Finally, we consider the case $n = 0$ and $i = K$. On this boundary, there are no transitions from state $(0, K)$ due to service completions or "births," and no transitions into state $(0, K)$ due to either new arrivals or "deaths." Thus, the appropriate equation for this boundary is

$$(\lambda_K + \delta_K)P_{0K} = \beta_{K-1}P_{0,K-1} + \mu_K P_{1K}. \qquad (4.52)$$

Equations (4.50) through (4.52) can be rewritten in more compact form by using matrix notation. Toward this end, we define

$$P_n = [\,P_{n0} \quad P_{n1} \quad \cdots \quad P_{nK}\,].$$

Then, upon rearranging (4.50) - (4.52), we find

$$P_0(\Lambda - \mathcal{Q}) - P_1\mathcal{M} = 0, \qquad (4.53)$$

where $\Lambda = \text{diag}(\lambda_0, \lambda_1, \ldots, \lambda_K)$ is a diagonal matrix of arrival rates, and $\mathcal{M} = \text{diag}(\mu_0, \mu_1, \ldots, \mu_K)$ is a diagonal matrix of service rates.

The matrix equation (4.53) summarizes all of the required information for the states in which $n = 0$. A similar set of equations is required for the states in which $n > 0$. These equations can also be obtained from (4.49). For $i = 0$, we find

$$(\lambda_0 + \mu_0 + \beta_0)P_{n0} = \lambda_0 P_{n-1,0} + \delta_1 P_{n1} + \mu_0 P_{n+1,0}. \tag{4.54}$$

Finally, for $i = K$ with $n > 0$, by setting $P_{n,K+1} = 0$ in (4.49) and noting that the birth rate while in state (n, K) is equal to 0, we find

$$(\lambda_K + \mu_K + \delta_K)P_{nK} = \lambda_K P_{n-1,K} + \beta_{K-1}P_{n,K-1} + \mu_K P_{n+1,K}. \tag{4.55}$$

Upon rewriting (4.49), (4.54) and (4.55) in matrix form, we can readily see that

$$P_{n-1}\Lambda - P_n(\Lambda - \mathcal{Q} + \mathcal{M}) + P_{n+1}\mathcal{M} = 0, \tag{4.56}$$

where all of the terms have been previously defined.

The matrix equations (4.53) and (4.55) are analogous to the scalar balance equations derived for the M/M/1 queueing system. Formulating the probability generating function is analogous to the scalar case as well. However, the generating functions so derived will be marginal probability generating functions rather than total probability generating functions; that is, the generating functions obtained by multiplying both sides of (4.56) by z^n and summing will generate marginal distributions. These marginal distributions can then be summed to yield probability generating functions if desired.

Define

$$\mathcal{G}(z) = \sum_{n=0}^{\infty} z^n P_n \tag{4.57}$$

or, equivalently,

$$\mathcal{G}(z) = [G_0(z) \quad G_1(z) \quad \cdots \quad G_K(z)], \tag{4.58}$$

where

$$G_i(z) = \sum_{n=0}^{\infty} z^n P_{n,i},$$

for $0 \le i \le K$. Then, upon multiplying both sides of (4.56) by z^n and summing, we find

$$\begin{aligned} \mathcal{G}(z)\,\Lambda z \quad &- \quad [\mathcal{G}(z) - P_0]\,(\Lambda - \mathcal{Q} + \mathcal{M}) \\ &+ \quad [\mathcal{G}(z) - P_1 z - P_0]\,\mathcal{M}\frac{1}{z} = 0. \end{aligned} \tag{4.59}$$

After rearranging terms, we get

$$\mathcal{G}(z)\Lambda z - \mathcal{G}(z)(\Lambda - \mathcal{Q} + \mathcal{M}) + \mathcal{G}(z)\,\mathcal{M}\frac{1}{z} = P_0\mathcal{M}\left(\frac{1}{z} - 1\right) - \left[P_0(-\mathcal{Q} + \Lambda) - P_1\mathcal{M}\right]. \tag{4.60}$$

Upon comparison of the last term on the right hand side of (4.60) to (4.53), we see that this bracketed term is equal to zero. Thus, (4.60) reduces to

$$\mathcal{G}(z)\,\Lambda z - \mathcal{G}(z)\,(\Lambda - \mathcal{Q} + \mathcal{M}) + \mathcal{G}(z)\,\mathcal{M}\frac{1}{z} = P_0\,\mathcal{M}\left(\frac{1}{z} - 1\right) \tag{4.61}$$

or, equivalently,

$$\mathcal{G}(z)\,\left(\Lambda z^2 - (\Lambda - \mathcal{Q} + \mathcal{M})z + \mathcal{M}\right) = (1 - z)P_0\,\mathcal{M}. \tag{4.62}$$

To simplify the notation, define

$$\mathcal{A}(z) = \Lambda z^2 - (\Lambda - \mathcal{Q} + \mathcal{M})z + \mathcal{M}. \tag{4.63}$$

Then, we can rewrite (4.62) as

$$\mathcal{G}(z)\mathcal{A}(z) = (1 - z)P_0\,\mathcal{M}. \tag{4.64}$$

Upon solving (4.64), we find

$$\mathcal{G}(z) = \frac{1 - z}{\det\mathcal{A}(z)}P_0\mathcal{M}\ \mathrm{adj}\,\mathcal{A}(z). \tag{4.65}$$

Analogous to the M/M/1 case, in which the probability generating function approach resulted in having to resolve an unknown constant, (4.65) contains an unknown vector of coefficients. This unknown vector can be determined using exactly the same principle as that used in the scalar case. We simply observe that the vector function $\mathcal{G}(z)$ is a vector of marginal probability generating functions and is therefore bounded at least for $|z| \leq 1$ (Hunter [1983]). This means that if there are zeros in the denominator of the right-hand side of (4.65) at z_i such that $|z_i| \leq 1$, then there is also a zero in the numerator of the right-hand side of (4.65) at z_i.

Because the rows of $\mathcal{A}(1)$ sum to a zero vector, $\det\mathcal{A}(1) = 0$ and, therefore, $\det\mathcal{A}(z)$ has a $(1 - z)$ factor, which cancels the factor in the numerator. This fact, when coupled with the fact that the probabilities must sum to unity, leads to one equation in the $K + 1$ unknowns of P_0. In addition, a later exercise is to show that there are exactly K zeros of $\det\mathcal{A}(z)$ in the interval $(0, 1)$ provided that $\det\mathcal{M} \neq 0$. These K zeros lead to an additional K linear equations in the

$K+1$ unknowns of P_0. It turns out that the $K+1$ linear equations are linearly independent if det $\mathcal{M} \neq 0$ so that this linear system can be used to solve for the unknown vector P_0.

EXERCISE 4.10 Show that det $\mathcal{A}(z)$ is a polynomial, the order of which is not greater than $2(K+1)$.

More formally, let $\mathcal{F}_{\tilde{n}}(z) = \sum_{i=0}^{K} G_i(z)$ or in vector notation, $\mathcal{F}_{\tilde{n}}(z) = \mathcal{G}(z)\mathbf{e}$ where $\mathbf{e}$ is a column vector of 1s. Then, from (4.65), we find

$$\mathcal{F}_{\tilde{n}}(z) = \frac{1-z}{\det \mathcal{A}(z)} P_0 \mathcal{M} \text{ adj } \mathcal{A}(z)\mathbf{e}. \tag{4.66}$$

Based on the above discussion, we find corresponding to $z = 1$,

$$1 = \frac{1}{[\det \mathcal{A}(z)/(1-z)]_{z=1}} P_0 \mathcal{M} \text{ adj } \mathcal{A}(z)|_{z=1}\mathbf{e}. \tag{4.67}$$

To facilitate further discussion, we define $\mathcal{Z}^{(0,\infty)}$ to be the set of all z such that det $\mathcal{A}(z) = 0$. We partition the set $\mathcal{Z}^{(0,\infty)}$ into three sets: $\mathcal{Z}^{(0,1)}$ and $\mathcal{Z}^{(1,\infty)}$ containing those elements of $\mathcal{Z}^{(0,\infty)}$ less than and greater than unity, respectively, and a third set that contains only the unity element. Thus, $\mathcal{Z}^{(0,\infty)} = \mathcal{Z}^{(0,1)} \cup \{1\} \cup \mathcal{Z}^{(1,\infty)}$. The elements of $\mathcal{Z}^{(0,1)}$ are then labeled $z_0, z_1, \ldots, z_{K-1}$; the unit element is referred to as z_K; and the elements of $\mathcal{Z}^{(1,\infty)}$ are labeled $z_{K+1}, \ldots, z_n$ where n is the total number of elements in $\mathcal{Z}^{(0,\infty)}$. We assume the elements of $\mathcal{Z}^{(0,\infty)}$ are distinct, and for convenience, we order the indexing such that $z_i < z_j$ if $i < j$.

For each $z_i \in \mathcal{Z}^{(0,1)}$, we find from (4.66) and the above argument that

$$0 = P_0 \mathcal{M} \text{ adj } \mathcal{A}(z_i)\mathbf{e} \quad \text{for} \quad z_i \in \mathcal{Z}^{(0,1)}. \tag{4.68}$$

Equations (4.67) and (4.68) then form a system of $K+1$ linear equations through which P_0 may be determined. The result can then be substituted into (4.65) to obtain the marginal PGFs or into (4.66) to obtain the total PGF.

Note that we would prefer not to write explicit expressions for adj $\mathcal{A}(z)$ and det $\mathcal{A}(z)$ in order to formulate the linear system of equations. The entire problem can be formulated in terms of the eigenvalues and eigenvectors of a matrix, which can be specified directly from inspection of $\mathcal{A}(z)$; and the expressions for $\mathcal{F}_{\tilde{n}}(z)$ and $\mathcal{G}(z)$ and their corresponding probabilities can be specified in a convenient manner without the need for direct manipulation of $\mathcal{A}(z)$. However, before turning to a discussion of more advanced techniques, we present a simple numerical example.

EXAMPLE 4.2 A computer accesses a transmission line via a statistical multiplexer or packet switch. The computer acts as a source of traffic; in this case, arrivals of packets from the computer are analogous to arrivals of customers to a queue. The computer alternates between idle and busy periods, which have exponential durations with parameters β and δ, respectively. During busy periods, the computer generates packets at a Poisson rate λ. The service times on the transmission line form a sequence of independent, identically distributed exponential random variables with parameter μ. Compute the state probabilities and the occupancy distribution for the parameter values $\lambda = \beta = \delta = \mu = 1$.

Solution: In referring back to the model, we find $\beta_0 = \beta$, $\delta_1 = \delta$, $\lambda_0 = 0$, $\lambda_1 = \lambda$, and $\mu_0 = \mu_1 = \mu$. Thus we find

$$\mathcal{Q} = \begin{bmatrix} -\beta & \beta \\ \delta & -\delta \end{bmatrix} = \begin{bmatrix} -1 & 1 \\ 1 & -1 \end{bmatrix},$$

$$\mathcal{M} = \begin{bmatrix} \mu & 0 \\ 0 & \mu \end{bmatrix} = \begin{bmatrix} 1 & 0 \\ 0 & 1 \end{bmatrix},$$

and

$$\Lambda = \begin{bmatrix} 0 & 0 \\ 0 & \lambda \end{bmatrix} = \begin{bmatrix} 0 & 0 \\ 0 & 1 \end{bmatrix}.$$

Upon substitution of these definitions into (4.64), we find

$$\mathcal{A}(z) = \begin{bmatrix} 1 - 2z & z \\ z & z^2 - 3z + 1 \end{bmatrix},$$

$$\operatorname{adj} \mathcal{A}(z) = \begin{bmatrix} z^2 - 3z + 1 & -z \\ -z & 1 - 2z \end{bmatrix},$$

and

$$\begin{aligned} \det \mathcal{A}(z) &= 1 - 5z + 6z^2 - 2z^3 \\ &= 2(1-z)(1 + \frac{\sqrt{2}}{2} - z)(1 - \frac{\sqrt{2}}{2} - z). \end{aligned}$$

Thus we find that

$$\operatorname{adj} \mathcal{A}(z)\Big|_{z=1-\frac{\sqrt{2}}{2}} = \begin{bmatrix} 0.2071068 & -0.2928932 \\ -0.2928932 & 0.4142136 \end{bmatrix},$$

$$\operatorname{adj} \mathcal{A}(z)\Big|_{z=1} = \begin{bmatrix} -1 & -1 \\ -1 & -1 \end{bmatrix},$$

and

$$\left.\frac{\det \mathcal{A}(z)}{(1-z)}\right|_{z=1} = -1.$$

Substituting these numbers into (4.67) and (4.68), we find that

$$[0 \quad 1] = P_0 \begin{bmatrix} -0.0857864 & 2 \\ +0.1213204 & 2 \end{bmatrix}.$$

Thus we find that

$$P_0 = [\,0.2928932 \quad 0.2071068\,].$$

Upon substitution of this result into (4.65), we find after some algebra that

$$\begin{bmatrix} G_0(z) \\ G_1(z) \end{bmatrix}^T = \frac{1}{1 - 0.5857864z} \begin{bmatrix} 0.2928932 - 0.0857864z \\ 0.2071068 \end{bmatrix}^T.$$

After some additional algebra, this result reduces to

$$\begin{aligned} [\,G_0(z)G_1(z)\,] &= [\,0.2928932 \quad 0.2071068\,] \\ &\quad + [\,0.1464465 \quad 0.2071068\,] \sum_{n=1}^{\infty} (0.5857864)^n z^n. \end{aligned}$$

Thus, for $n \geq 1$, we find

$$[\,P_{n0} \quad P_{n1}\,] = [\,0.1464465 \quad 0.2071068\,]\,(0.5857864)^n.$$

The probability generating function for the occupancy distribution can now be computed from (4.66) or by simply summing $G_0(z)$ and $G_1(z)$. We find

$$\mathcal{F}_{\tilde{n}}(z) = 0.5 + 0.3535534 \sum_{n=1}^{\infty} (0.5857864)^n z^n.$$

From this probability generating function, we find $P_0 = 0.5$ and

$$P_n\mathrm{e} = 0.3535534 \times (0.5857864)^n \qquad \text{for } n \geq 1.$$

EXERCISE 4.11 Repeat the above numerical example for the parameter values $\beta = \delta = 2$, $\lambda = \mu = 1$. That is, the proportion of time spent in each phase is the same and the transition rate between the phases is faster than in the original example. Compare the results to those of the example by plotting curves of the respective complementary distributions. Compute the overall traffic intensity and compare.

EXERCISE 4.12 Suppose we want to use the packet switch and transmission line of the above numerical example to serve a group of users who collectively generate packets at a Poisson rate γ, independent of the computer's activity, in addition to the computer. This is a simple example of integrated traffic. Assuming that the user packets also require exponential service with rate μ, show the impact the user traffic has on the occupancy distribution of the packet switch by plotting curves for the cases of $\gamma = 0$ and $\gamma = 0.1$.

We now turn our attention to the determination of P_0 and the specification of $\mathcal{G}(z)$ and $\mathcal{F}_{\tilde{n}}(z)$ by more advanced techniques. We shall show that all of the computations required to specify these quantities can be accomplished without actually performing algebraic manipulations. With regard to (4.64), we find that[1]

$$\mathcal{G}(1)\mathcal{A}(1) = 0. \tag{4.69}$$

But from (4.63) we find that $\mathcal{A}(1) = \mathcal{Q}$. Thus we have

$$\mathcal{G}(1)\mathcal{Q} = 0. \tag{4.70}$$

That is, $\mathcal{G}(1)$ is proportional to the left eigenvector of $\mathcal{Q}$ corresponding to the eigenvalue 0 of $\mathcal{Q}$. This means that $\mathcal{G}(1)$ is the vector of ergodic probabilities of the phase process, a fact which can be readily verified by evaluation of $\mathcal{G}(1)$ using (4.57). Now, from Theorems 3.5 and 3.6, we know that the left eigenvector of $\mathcal{Q}$ corresponding to the eigenvalue zero is proportional to the rows of adj $\mathcal{Q}$, and consequently, the rows of adj $\mathcal{Q}$ are proportional to each other. But, because $\mathcal{Q}$ is the infinitesimal generator for the phase process, it can be shown that not only are the rows of adj $\mathcal{Q}$ proportional to each other, but they are also *equal* to each other. Proof of this fact is left as an exercise. Thus, if we let ϕ_K denote any row of adj $\mathcal{Q}$, we find

$$\mathcal{G}(1) = \frac{1}{\phi_K \mathbf{e}}\phi_K. \tag{4.71}$$

EXERCISE 4.13 Let $\mathcal{Q}$ denote the infinitesimal generator for a finite, discrete-valued, continuous-time Markov chain. Show that the rows of adj $\mathcal{Q}$ are equal.

It is left as an exercise to show that the sum of the columns of $\mathcal{A}(z)$ is equal to the column vector $(1-z)(\mathcal{M} - z\Lambda)\mathbf{e}$. Now, summing the columns of a matrix is an elementary transformation, and the determinant of a matrix is unaffected by elementary transformations (Noble and Daniel [1977]). Therefore,

$$\lim_{z\to 1}\left[\frac{\det \mathcal{A}(z)}{1-z}\right]$$

is given by the inner product of the last column of the cofactor matrix of $\mathcal{A}(1)$ and $(\mathcal{M} - \Lambda)\mathbf{e}$. But the last column of the cofactor matrix of $\mathcal{A}(1)$ is exactly the transpose of the last row of adj $\mathcal{Q}$, which has been defined to be ϕ_K. Therefore,

$$\lim_{z\to 1}\left[\frac{\det \mathcal{A}(z)}{1-z}\right] = \phi_K(\mathcal{M} - \Lambda)\mathbf{e}. \tag{4.72}$$

[1]We liberally use notation such as $\mathcal{G}(1)$ and $\mathcal{A}(1)$ to denote the limits of these functions as z approaches 1 without the formality of stating these are limits.

Thus, upon substituting (4.71) and (4.72) into (4.66), we find

$$1 = \frac{1}{\mathcal{G}(1)(\mathcal{M} - \Lambda)\mathbf{e}} P_0 \mathcal{M}\mathbf{e}$$

or, equivalently,

$$\mathcal{G}(1)(\mathcal{M} - \Lambda)\mathbf{e} = P_0 \mathcal{M}\mathbf{e}. \tag{4.73}$$

Thus we see that the equation corresponding to $z = 1$ can be specified without resorting to algebraic manipulation; we simply find the left eigenvector of $\mathcal{Q}$ corresponding to zero and normalize this eigenvector so that its components sum to unity to obtain $\mathcal{G}(1)$ and then use (4.73) to complete the specification.

EXERCISE 4.14 Obtain (4.73) by starting out with (4.64), differentiating both sides with respect to z, postmultiplying both sides by $\mathbf{e}$, and then taking limits as $z \to 1$.

EXERCISE 4.15 Show that the sum of the columns of $\mathcal{A}(z)$ is equal to the column vector $(1 - z)(\mathcal{M} - z\Lambda)\mathbf{e}$ so that det $\mathcal{A}(z)$ has a (1-z) factor.

Equation (4.73) has an intuitively satisfying interpretation. To see this, we rearrange (4.73) as follows:

$$\mathcal{G}(1)\Lambda\mathbf{e} = [\mathcal{G}(1) - P_0]\mathcal{M}\mathbf{e}. \tag{4.74}$$

The left side of (4.74) expresses the average rate at which units enter the service system while the right-hand side expresses the average rate at which units leave the system. Thus (4.74) is a flow balance equation. A special case of (4.74) is, for example, the relationship $\lambda = (1 - P_0)\mu$ for the M/M/1 system, which can be solved to obtain $P_0 = 1 - \lambda/\mu$.

We now turn our attention to the formulation of (4.68) by nonalgebraic techniques. Before proceeding to the details, we introduce some terminology (see Lancaster [1966]).

DEFINITION 4.1 **λ-matrix**. A λ-matrix is a matrix the elements of which are polynomials in λ.

DEFINITION 4.2 **Null value**. Let $\mathcal{A}(\lambda)$ be a λ-matrix. Then a value λ_i such that det $\mathcal{A}(\lambda_i) = 0$ is called a *null value* of $\mathcal{A}(\lambda)$. For example, $(\lambda I - A)$ is a λ-matrix, and the eigenvalues of A are null values of the λ-matrix $(\lambda I - A)$.

DEFINITION 4.3 **Null vector**. Let $\mathcal{A}(\lambda)$ be a λ-matrix, and let $X(\lambda_i)$ be a nontrivial column vector such that $\mathcal{A}(\lambda_i)X(\lambda_i) = 0$. Then $X(\lambda_i)$ is called a *null vector* of the λ-matrix $\mathcal{A}(\lambda)$ corresponding to the null value λ_i. For example, $(\lambda I - A)$ is a λ-matrix, and the eigenvectors of the matrix A are null vectors of the λ-matrix $(\lambda I - A)$.

DEFINITION 4.4 **Left null vector.** Let $\mathcal{A}(\lambda)$ be a λ-matrix, and let $X(\lambda_i)$ be a nontrivial row vector such that $X(\lambda_i)\mathcal{A}(\lambda_i) = 0$. Then $X(\lambda_i)$ is called a *left null vector* of the λ-matrix $\mathcal{A}(\lambda)$ corresponding to the null value λ_i.

From the above definitions, we see that $\mathcal{A}(\lambda)$ is a λ-matrix of degree 2 because each of the elements on the major diagonal is a polynomial of degree 2. In addition, we see that the zeros of $\det\ \mathcal{A}(\lambda)$ are equivalent to the null values of $\mathcal{A}(\lambda)$. It is left as an exercise to show that the null vector of $\mathcal{A}(\lambda)$ corresponding to the null value λ_i, $i = 0, 1, \ldots, 2K + 1$, is proportional to the columns of $\text{adj}\ \mathcal{A}(\lambda_i)$, $i = 0, 1, \ldots, 2K + 1$, respectively. Consequently, the column vectors of $\text{adj}\ \mathcal{A}(\lambda_i)$ are proportional to each other, and because the left-hand side of (4.68) is zero, we may replace $\text{adj}\ \mathcal{A}(\lambda_i)\text{e}$ in (4.68) by $X(\lambda_i)$ where $X(\lambda_i)$ is the null vector of $\mathcal{A}(\lambda)$ corresponding to λ_i. This null vector is, in turn, simply the eigenvector of $\mathcal{A}(\lambda_i)$ corresponding to the zero eigenvector of $\mathcal{A}(\lambda_i)$. Thus, if the null values of $\mathcal{A}(\lambda)$ are known exactly, then computation of the corresponding null vectors is trivial.

EXERCISE 4.16 Show that the zeros of the determinant of the λ-matrix $\mathcal{A}(z)$ are all real and nonnegative. [*Hint*: First, do a similarity transformation, transforming $\mathcal{A}(z)$ into a symmetric matrix, $\hat{\mathcal{A}}(z)$. Then, form the inner product $< X_z, \hat{\mathcal{A}}(z)X_z >$, where X_z is the null vector of $\hat{\mathcal{A}}(z)$ corresponding to z. Finally, examine the zeros of the resulting quadratic equation.]

EXERCISE 4.17 The traffic intensity for the system is defined as the probability that the server is busy at an arbitrary point in time.

1. Express the traffic intensity in terms of the system parameters and P_0.
2. Determine the average amount of time a customer spends in service using the results of part 1 and Little's result.
3. Check the result obtained in part 2 for the special case $\mathcal{M} = \mu I$.

EXERCISE 4.18 Show that $\text{adj}\ \mathcal{A}(z_i)\text{e}$ is proportional to the null vector of $\mathcal{A}(z)$ corresponding to z_i.

The null values and vectors of $\mathcal{A}(z)$ can be obtained from standard eigenvalue eigenvector routines. Towards this end, consider the system $\mathcal{A}(\sigma)X_\sigma = 0$ where σ is any null value of $\mathcal{A}(z)$, and X_σ is the corresponding null vector. Define $Y_\sigma = \sigma X_\sigma$. Then, from the definition of $\mathcal{A}(z)$ given by (4.63), we find

$$\sigma\Lambda Y_\sigma - \sigma(\Lambda - \mathcal{Q} + \mathcal{M})X_\sigma + \mathcal{M}X_\sigma = 0,$$

and, by definition,

$$Y_\sigma - \sigma X_\sigma = 0.$$

Combining these two systems, we find that

$$\left\{\begin{bmatrix} \mathcal{M} & 0 \\ 0 & I \end{bmatrix} - \sigma \begin{bmatrix} \Lambda - \mathcal{Q} + \mathcal{M} & -\Lambda \\ I & 0 \end{bmatrix}\right\} \begin{bmatrix} X_\sigma \\ Y_\sigma \end{bmatrix} = 0,$$

or, equivalently,

$$\left\{\begin{bmatrix} I & 0 \\ 0 & I \end{bmatrix} - \sigma \begin{bmatrix} \mathcal{M}^{-1}(\Lambda - \mathcal{Q} + \mathcal{M}) & -\mathcal{M}^{-1}\Lambda \\ I & 0 \end{bmatrix}\right\} \begin{bmatrix} X_\sigma \\ Y_\sigma \end{bmatrix} = 0. \quad (4.75)$$

From (4.75), we see that the null values of $\mathcal{A}(z)$ are equivalent to the inverses of the eigenvalues of the matrix

$$\mathcal{A}_E = \begin{bmatrix} \mathcal{M}^{-1}(\Lambda - \mathcal{Q} + \mathcal{M}) & -\mathcal{M}^{-1}\Lambda \\ I & 0 \end{bmatrix}, \quad (4.76)$$

where we have used the subscript E to denote *expanded*, and where the null vectors of $\mathcal{A}(z)$ are proportional to the upper and lower $(K+1)$-subvectors of the eigenvectors of this same $2(K+1)$-dimensional square matrix. Let $\phi_i = X_{z_i}$, $i = 0, 1, \cdots, K-1$ where $z_i \in \mathcal{Z}^{(0,1)}$. Then, we find that we can form the linear system of equations (4.66) and (4.67) as follows:

$$\begin{aligned} 0 &= P_0 \mathcal{M} \phi_0 \\ 0 &= P_0 \mathcal{M} \phi_1 \\ &\vdots \\ 0 &= P_0 \mathcal{M} \phi_{K-1} \\ \mathcal{G}(1)(\mathcal{M} - \Lambda)\mathbf{e} &= P_0 \mathcal{M} \mathbf{e}. \end{aligned} \quad (4.77)$$

The system (4.77) may then be solved for P_0.

Having solved for P_0, we have a formal solution for $\mathcal{G}(z)$. However, the form of $\mathcal{G}(z)$ in its present state is not suitable for manipulation, and algebraic manipulation of $\mathcal{A}(z)$ would be required to complete the specification. A reasonable approach at this point would be to expand $\mathcal{G}(z)$ using partial fraction expansions. At first glance, this would appear to be a formidable task, but a little further investigation will show that this is not the case.

As a starting point, we repeat (4.65) putting the $(1-z)$ factor in the denominator

$$\mathcal{G}(z) = \frac{1}{[\det \mathcal{A}(z)/(1-z)]} P_0 \mathcal{M} \operatorname{adj} \mathcal{A}(z). \quad (4.78)$$

Now, because Λ is not required to have full rank, the polynomial $\det \mathcal{A}(z)$ may have degree less than $2(K+1)$. If so, there will be a corresponding number of eigenvalues of $\mathcal{A}_E$ which have zero values. We therefore find that

$$\det \mathcal{A}(z) \;=\; \sum_{i=0}^{n} a_i z^i$$

$$= a_n \prod_{z_i \in \mathcal{Z}^{(0,\infty)}} (z - z_i) \tag{4.79}$$

where n is the number of eigenvalues of $\mathcal{A}_E$ having nonzero values and we recall that $\mathcal{Z}_p$ is the set of null values of $\mathcal{A}(z)$ corresponding to those n eigenvalues of $\mathcal{A}_E$. Thus we have

$$\det \mathcal{A}(z)|_{z=0} = (-1)^n a_n \prod_{z_i \in \mathcal{Z}^{(0,\infty)}} z_i. \tag{4.80}$$

But

$$\begin{aligned} \det \mathcal{A}(z)|_{z=0} &= \det \mathcal{M} \\ &= \prod_{i=0}^{K} \mu_i \end{aligned}$$

so that

$$a_n = (-1)^n \prod_{z_i \in \mathcal{Z}^{(0,\infty)}} z_i^{-1} \prod_{i=0}^{K} \mu_i. \tag{4.81}$$

Upon substituting (4.81) into (4.79), we find

$$\frac{\det \mathcal{A}(z)}{(1-z)} = \prod_{i=0}^{K} \mu_i \prod_{z_i \in \mathcal{Z}^{(0,1)} \cup \mathcal{Z}^{(1,\infty)}} (1 - z_i^{-1} z). \tag{4.82}$$

Substitution of (4.82) into (4.78) then yields

$$\mathcal{G}(z) = \frac{1}{\prod_{i=0}^{K} \mu_i \prod_{z_i \in \mathcal{Z}^{(0,1)} \cup \mathcal{Z}^{(1,\infty)}} (1 - z_i^{-1} z)} P_0 \mathcal{M} \text{ adj } \mathcal{A}(z). \tag{4.83}$$

We note that the zeros of $\det \mathcal{A}(z)$ in the interval $(0, 1)$ are canceled by the choice of P_0 so that the remaining zeros are the ones in the interval $(1, \infty)$. Recall that $\mathcal{Z}^{(1,\infty)}$ denotes the set of null values of $\mathcal{A}(z)$ in $(1, \infty)$. Thus expressing (4.83) using partial fraction expansions results in

$$\mathcal{G}(z) = C_0 + \sum_{z_{K+i} \in \mathcal{Z}^{(1,\infty)}} \frac{1}{(1 - z_{K+i}^{-1} z)} A_i, \tag{4.84}$$

where C_0 is a row vector of constants reflecting the fact that the numerator polynomials may have degree larger than that of the denominator polynomial, and A_i is a row vector representing the residue of $\mathcal{G}(z)$ corresponding to $z_{K+i} \in \mathcal{Z}^{(1,\infty)}$.

Upon multiplying both sides of (4.83) and (4.84) by $(1 - z_{K+i}^{-1}z)$ and taking limits as $z \rightarrow z_{K+i}^{-1}$, we find

$$A_i = \frac{1}{\prod_{i=0}^{K}\mu_i \prod_{z_i \in \mathcal{Z}^{(0,1)} \cup \mathcal{Z}^{(1,\infty)} \setminus \{z_{K+i}\}} (1 - z_j^{-1} z_{K+i})} P_0 \mathcal{M} \text{ adj } \mathcal{A}(z_{K+i}). \tag{4.85}$$

At this point, it is worthwhile to contemplate the difficulties of computation of A_i. At first glance, this computation may appear difficult because it involves evaluating adj $\mathcal{A}(z_i)$. However, an LU decomposition approach (Press, Flannery, Teukolsky and Vetterling [1988]) which takes into account the fact that $\mathcal{A}(z_i)$ is both tridiagonal and singular, leads to a very simple algorithm for obtaining adj $\mathcal{A}(z_i)$ as the outer product of two vectors which are easy to obtain.

Now, from (4.84), we find that

$$\begin{aligned}
\mathcal{G}(z) &= C_0 + \sum_{z_{K+i} \in \mathcal{Z}^{(1,\infty)}} A_i \sum_{j=0}^{\infty} (z_{K+i}^{-1} z)^j \\
&= C_0 + \sum_{z_i \in \mathcal{Z}^{(1,\infty)}} A_i + \sum_{z_{K+i} \in \mathcal{Z}^{(1,\infty)}} A_i \sum_{j=1}^{\infty} (z_{K+i}^{-1} z)^j \\
&= P_0 + \sum_{z_{K+i} \in \mathcal{Z}^{(1,\infty)}} A_i \sum_{j=1}^{\infty} (z_{K+i}^{-1} z)^j.
\end{aligned} \tag{4.86}$$

Thus, we need not obtain C_0 explicitly in order to compute the state probabilities. We readily find that

$$P_n = \sum_{z_{K+i} \in \mathcal{Z}^{(1,\infty)}} A_i z_{K+i}^{-n} \quad \text{for} \quad n \geq 1. \tag{4.87}$$

The marginal probabilities are obtained by summing the joint probabilities. We find that

$$P_n \mathbf{e} = \sum_{z_{K+i} \in \mathcal{Z}^{(1,\infty)}} A_i \mathbf{e} z_{K+i}^{-n} \quad \text{for} \quad j \geq 1, \tag{4.88}$$

where $P_n \mathbf{e}$ is the equilibrium probability that the occupancy is n.

EXAMPLE 4.3 In this example, we rework the previous example in which the parameter values were $\lambda = \beta = \delta = \mu = 1$, using the more advanced techniques.

Solution: From the definition of $\mathcal{A}_E$, which is given by (4.76), we find

$$\mathcal{A}_E = \begin{bmatrix} 2 & -1 & 0 & 0 \\ -1 & 2 & 0 & -1 \\ 1 & 0 & 0 & 0 \\ 0 & 1 & 0 & 0 \end{bmatrix}.$$

Then, from a standard eigenvalue/eigenvector routine, we find the eigenvalues of $\mathcal{A}_E$ to be $\{3.414214, 1.0, 0.585786, 0\}$. Upon taking inverses of those eigenvalues, we find $\mathcal{Z}^{(0,\infty)} = \{0.292893, 1.0, 1.707107\}$. Following our indexing scheme we have, $z_0 = 0.292893$, $z_1 = 1$, and $z_2 = 1.707107$.

The null vectors of $\mathcal{A}(z)$ corresponding to z_0 and z_1 are found by partitioning the matrix of eigenvectors of $\mathcal{A}_E$; the results are as follows:

$$\phi_0 = \begin{bmatrix} -0.414214 \\ 0.585786 \end{bmatrix}$$

and

$$\phi_1 = \begin{bmatrix} -0.5 \\ -0.5 \end{bmatrix},$$

respectively. From (4.71), we find

$$\mathcal{G}(1) = \begin{bmatrix} 0.5 \\ 0.5 \end{bmatrix}.$$

Thus, from (4.77), we find

$$[0 \quad 0.5] = P_0 \begin{bmatrix} -0.414214 & 0.5 \\ 0.585786 & 0.5 \end{bmatrix},$$

from which we find

$$P_0 = [0.292893 \quad 0.207107].$$

Evaluation of $\mathcal{A}(z_2)$ yields

$$\mathcal{A}(1.707107) = \begin{bmatrix} -2.414214 & 1.707107 \\ 1.707107 & -1.207107 \end{bmatrix},$$

from which we find

$$\text{adj } \mathcal{A}(1.707107) = \begin{bmatrix} -1.207107 & -1.707107 \\ -1.707107 & -2.414214 \end{bmatrix}.$$

From (4.85), we find

$$\begin{aligned} A_1 &= \frac{1}{(1 - z_1^{-1} z_2)} P_0 \mathcal{M} \text{adj } \mathcal{A}(z_2) \\ &= \frac{1}{(1 - 5.828429)} \begin{bmatrix} 0.292893 \\ 0.207107 \end{bmatrix}^T \begin{bmatrix} -1.207107 & -1.707107 \\ -1.707107 & -2.414214 \end{bmatrix} \\ &= [0.146447 \quad 0.207107], \end{aligned}$$

where the T denotes the matrix transpose operator. Thus, from (4.86), we find

$$\begin{aligned} \mathcal{G}(z) &= P_0 + A_1 \sum_{n=1}^{\infty} z_2^{-n} z^n \\ &= \begin{bmatrix} 0.292893 \\ 0.207107 \end{bmatrix}^T + \begin{bmatrix} 0.146447 \\ 0.207107 \end{bmatrix}^T \sum_{n=1}^{\infty} 0.585786^n z^n. \end{aligned}$$

Upon postmultiplication of the both sides of this expression by $\mathbf{e}$, we readily find

$$\mathcal{F}(z) = 0.5 + 0.353554 \sum_{n=1}^{\infty} 0.585786^n z^n$$

so that $P_0\mathbf{e} = 0.5$ and $P_n\mathbf{e} = 0.353554 \times 0.585786^n$ for $n \geq 1$ as before.

EXAMPLE 4.4 This example provides the solution to Exercise 4.12. As in Exercise 4.12, we increase the arrival rate in each phase by 0.1 to reflect the addition of Poisson user traffic at rate 0.1 so that the resulting parameter values are $\lambda_0 = 0.1$, $\lambda_1 = 1.1$, $\beta_0 = \delta_1 = \mu_1 = \mu_2 = 1$.

Solution: Following the procedure outlined above leads to the following results:

$$\begin{aligned} \mathcal{G}(z) &= P_0 + A_1 \sum_{n=1}^{\infty} z_2^{-n} z^n + A_2 \sum_{n=1}^{\infty} z_3^{-n} z^n \\ &= [0.234605 \quad 0.165395] + [0.13252 \quad 0.17048] \sum_{n=1}^{\infty} 0.662640^n z^n \\ &\qquad + [0.102084 \quad -0.005086] \sum_{n=1}^{\infty} 0.047568^n z^n. \end{aligned}$$

Upon postmultiplying both sides of the above equation by $\mathbf{e}$, we find

$$\mathcal{F}(z) = 0.4 + 0.303002 \sum_{n=1}^{\infty} 0.662640^n z^n + 0.096998 \sum_{n=1}^{\infty} 0.047568^n z^n.$$

so that $P_0\mathbf{e} = 0.4$ and

$$P_n\mathbf{e} = 0.303002 \times 0.662640^n + 0.096998 \times 0.047568^n \quad \text{for} \quad n \geq 1.$$

From the expression for the occupancy probabilities, it is easy to compute the complementary occupancy distribution. We find

$$P\{\tilde{n} > n\} = 0.595155 \times 0.662640^n + 0.004845 \times 0.047568^n.$$

Note that there are two null values of $\mathcal{A}(z)$ greater than unity. For large values of n, only the smaller of these has any significant effect upon the occupancy probabilities, marginal or otherwise. For example,

$$P\{\tilde{n} > 10\} \approx 9.714 \times 10^{-3}.$$

The contribution due to the larger null value is only 2.874×10^{-16}, or roughly three parts in 10^{16}. Expressing the results in the form of a geometric sum makes it easy to see the effects of each of the null values on the occupancy probabilities for all n.

The above results are now specified as a weighted sum of geometric distributions. The solution vector is then "matrix geometric." In Section 4.2.3 we describe a more direct approach, due to Neuts, to computing the ergodic probabilities when these probabilities are expressible in matrix geometric form. Although the approach is more direct from a descriptive point of view, the computation time for numerical solutions is not necessarily comparable to the current method, as will be seen later.

EXERCISE 4.19 Show that if the system described by (4.57) and (4.60) is ergodic, then there are exactly K zeros of $\det \mathcal{A}(z)$ in the interval $(0,1)$. [*Hint*: First show that this is the case if $\delta_i = 0 \ \forall\ i$. Then show that it is not possible for $\det \mathcal{A}(z)/(1-z)$ to be zero for any choice of δ_is unless $P_0 = 0$, which implies no equilibrium solution exists. The result is that the number of zeros in $(0,1)$ does not change when the δ_i change.]

EXERCISE 4.20 Beginning with (4.86) through (4.88), develop expressions for the joint and marginal complementary ergodic occupancy distributions.

EXERCISE 4.21 Develop an expression for $\text{adj}\ \mathcal{A}(z_i)$ in terms of the outer products of two vectors using LU decomposition. [*Hint*: The term in the lower right-hand corner, and consequently the last row, of the upper triangular matrix will be zero. What then is true of its adjoint?]

4.2.2 Matrix Geometric Method

Suppose there exists a matrix $\mathcal{R}$ such that

$$P_n = P_{n-1}\mathcal{R} \qquad \forall \qquad n \geq 1. \tag{4.89}$$

Then, we find by successive substitutions into (4.79) that

$$P_n = P_o\mathcal{R}^n \qquad \forall \qquad n \geq 0. \tag{4.90}$$

A solution of the form (4.90) is called a *matrix geometric* solution. The key to solving a matrix geometric system is to specify the matrix $\mathcal{R}$, which is called the rate matrix and which we shall discuss below.

Following our probability generating approach, we find from (4.89) that

$$\sum_{n=1}^{\infty} z^n P_n = \sum_{n=1}^{\infty} z^n P_{n-1}\mathcal{R},$$

so that

$$\mathcal{G}(z) - P_0 = z\mathcal{G}(z)\mathcal{R},$$

or

$$\mathcal{G}(z)\left[I - z\mathcal{R}\right] = P_0, \tag{4.91}$$

and

$$\mathcal{G}(z) = P_0\left[I - z\mathcal{R}\right]^{-1}. \tag{4.92}$$

Thus

$$\lim_{z\to 1}\mathcal{G}(z) = P_0\left[I - \mathcal{R}\right]^{-1} \tag{4.93}$$

and

$$P_0 = \mathcal{G}(1)\left[I - \mathcal{R}\right]. \tag{4.94}$$

Also, we have from (4.56) that

$$P_{n-1}\Lambda - P_n(\Lambda - \mathcal{Q} + \mathcal{M}) + P_{n+1}\mathcal{M} = 0, \quad n \geq 1.$$

Hence, upon substituting (4.89) into the above equation, we find

$$P_{n-1}\Lambda - P_{n-1}\mathcal{R}(\Lambda - \mathcal{Q} + \mathcal{M}) + P_{n-1}\mathcal{R}^2\mathcal{M} = 0, \quad n \geq 1,$$

so that

$$P_{n-1}\left[\Lambda - \mathcal{R}(\Lambda - \mathcal{Q} + \mathcal{M}) + \mathcal{R}^2\mathcal{M}\right] = 0, \quad n \geq 1. \tag{4.95}$$

Clearly, a sufficient condition for (4.95) to hold is that $\mathcal{R}$ satisfy

$$\Lambda - \mathcal{R}(\Lambda - \mathcal{Q} + \mathcal{M}) + \mathcal{R}^2\mathcal{M} = 0. \tag{4.96}$$

Thus, if one could solve (4.96) for $\mathcal{R}$, one could then use (4.94) to solve for P_0, having previously computed $\mathcal{G}(1)$ by normalizing the left eigenvector of $\mathcal{Q}$ corresponding to its zero eigenvalue, as described in Section 4.2.1. An additional check on P_0 and $\mathcal{R}$ could be obtained from the boundary condition for the system of equations as specified in (4.53). The boundary condition states that

$$P_0(\Lambda - \mathcal{Q}) - P_1\mathcal{M} = 0,$$

so that

$$P_0(\Lambda - \mathcal{Q} - \mathcal{R}M) = 0. \tag{4.97}$$

Obtaining a solution for $\mathcal{R}$ of (4.96) is not necessarily an easy task. One possibility is to specify a contraction map (Hewitt and Stromberg [1969]) on $\mathcal{R}$ based on (4.96), and then use successive approximations to obtain $\mathcal{R}$. One way to specify a contraction map is to multiply both sides of (4.65) by some positive number τ^{-1} and then add $\mathcal{R}$ to both sides of the result. This procedure yields

$$\mathcal{R} = \tau^{-1}\mathcal{R}^2\mathcal{M} + \mathcal{R}\left[I - \tau^{-1}(\mathcal{M} - \mathcal{Q} + \Lambda)\right] + \tau^{-1}\Lambda. \tag{4.98}$$

We then set

$$\mathcal{R}_i = \tau^{-1}\mathcal{R}_{i-1}^2\mathcal{M} + \mathcal{R}_{i-1}\left[I - \tau^{-1}(\mathcal{M} - \mathcal{Q} + \Lambda)\right] + \tau^{-1}\Lambda \qquad (4.99)$$

and iterate on i, starting with some suitable value for $\mathcal{R}_0$, until $\mathcal{R}_i$ converges. One possibility is to set $\mathcal{R}_0 = \text{diag}(1/(K+1), 1/(K+1), \ldots, 1/(K+1))$.

An alternative contraction map is one specified by Neuts [1981a]. It is obtained by simply solving (4.96) for $\mathcal{R}$ and then introducing subscripts. The result is

$$\mathcal{R}_j = \Lambda(\Lambda - \mathcal{Q} + \mathcal{M})^{-1} + \mathcal{R}_{j-1}^2\mathcal{M}(\Lambda + \mathcal{Q} + \mathcal{M})^{-1} \quad \text{for} \quad j \geq 1. \quad (4.100)$$

The idea is to start with $\mathcal{R}_0 = 0$ and then compute successive approximations to $\mathcal{R}$ using (4.100). Neuts [1981a] has shown that $\{\mathcal{R}_0, \mathcal{R}_1, \mathcal{R}_2, \ldots\}$ is a monotonically increasing sequence which converges to the minimal nonnegative solution to (4.96), and that this solution is the solution which uniquely provides the rate matrix $\mathcal{R}$ which satisfies (4.89).

This approach is called the *matrix geometric* approach, and it is elegantly described in Neuts [1981a]. In its most general form, Neuts describes the (QBD) process as a Markov chain on $\{(n,i), n \geq 0, 0 \leq i \leq K\}$ having an infinitesimal generator of the form

$$\tilde{\mathcal{Q}} = \begin{bmatrix} B_0 & A_0 & 0 & \cdots & \cdots \\ B_1 & A_1 & A_0 & \cdots & \cdots \\ 0 & A_2 & A_1 & A_0 & \cdots \\ 0 & 0 & A_2 & A_1 & \cdots \\ 0 & 0 & 0 & A_2 & \cdots \\ 0 & 0 & 0 & 0 & \cdots \end{bmatrix}$$

where $(B_0 + A_0)\mathbf{e} = 0$, $(B_1 + A_1 + A_0)\mathbf{e} = 0$, $(A_0 + A_1 + A_2)\mathbf{e} = 0$, and the matrix $A = A_0 + A_1 + A_2$ is a finite generator.

In Neuts' terminology, we have

$$\begin{aligned} B_0 &= -(\Lambda - \mathcal{Q}) \\ B_1 &= \mathcal{M} \\ A_0 &= \Lambda \\ A_1 &= -(\Lambda - \mathcal{Q} + \mathcal{M}) \\ A_2 &= \mathcal{M}, \end{aligned}$$

so that our system matches Neuts' definition of a QBD process.[2]

Neuts [1981] presents the following theorem relevant to the analysis of such systems:

[2] The term *quasi-birth death process* seems to have been first applied to this type of system by Evans [1967].

THEOREM 4.1 *The process $\tilde{Q}$ is positive recurrent*[3] *if and only if the minimal nonnegative solution $\mathcal{R}$ to the matrix quadratic equation*

$$\mathcal{R}^2 A_2 + \mathcal{R} A_1 + A_0 = 0$$

has all of its eigenvalues inside the unit disk and the finite system of equations

$$P_0(B_0 + \mathcal{R} B_1) = 0$$

$$P_0(I - \mathcal{R})^{-1}\mathbf{e} = 1$$

has a unique positive solution P_0.

If the matrix A $(= A_2 + A_1 + A_0)$ is irreducible, then $sp(\mathcal{R}) < 1$ if and only if $\pi A_2 \mathbf{e} > \pi A_0 \mathbf{e}$, where π is the stationary probability vector of A.[4]

The stationary probability vector $x = [P_0, P_1, \cdots]$ of $\tilde{Q}$ is given by $P_i = P_0 \mathcal{R}^i$ for $i \geq 0$.

The (equivalent) equalities

$$(\mathcal{R} A_2 - A_0)\mathbf{e} = (\mathcal{R} B_1 - B_0)\mathbf{e} = 0$$

hold. □

The equation $P_0(B_0 + \mathcal{R} B_1) = 0$ is equivalent to (4.97), and the equation $P_0(I - \mathcal{R})^{-1}\mathbf{e} = 1$ can be obtained from (4.94) by postmultiplying both sides by $\mathbf{e}$. The condition $\pi A_2 \mathbf{e} > \pi A_0 \mathbf{e}$ is equivalent to $\mathcal{G}(1)\mathcal{M}\mathbf{e} > \mathcal{G}(1)\Lambda\mathbf{e}$, which states that the maximum average rate at which service may be rendered must exceed the average arrival rate.

Matrix geometric techniques are very powerful, and their use is not limited to the analysis of QBD processes. The literature contains many applications of matrix geometric techniques to the solution of problems. An application of this method to a non-Markovian queueing system is presented in Daigle and Langford [1985,1986], and an application of this method to analysis of Ethernet-based local area networks is presented in Coyle and Liu [1985].

In addition, the results that may be obtained via matrix-geometric techniques are not limited to occupancy distributions. Ramaswami and Lucantoni [1985] discuss the application of matrix geometric techniques to obtaining stationary waiting time distributions in QBD and other systems. Additional results along these lines are given in Daigle and Lucantoni [1990]. More recently, Latouche and Ramaswami [1999] have provided a comprehensive coverage of the solution of QBD models using the matrix geometric approach. We note

[3]The phrase "the process $\tilde{Q}$" is interpreted as "the process whose infinitesimal generator is $\tilde{Q}$."

[4]The expression $sp(\mathcal{R})$ denotes the *spectral radius* of the matrix $\mathcal{R}$, which is defined as the magnitude of the largest eigenvalue of $\mathcal{R}$. A plot of $sp(\mathcal{R})$ as a function of overall traffic intensity is sometimes called the *caudal characteristic curve* for the system.

that it is also possible to obtain such distributions using the PGF approach described in Section 4.2.1, but the development is more cumbersome.

EXAMPLE 4.5 Provide the solution in matrix geometric form to the problem solved in Example 4.4.

Solution: Upon substituting the parameters from Example 4.4 into (4.99), we find

$$\mathcal{R}_i = \tau^{-1}\mathcal{R}_{i-1}^2 + \mathcal{R}_{i-1}\left[I - \tau^{-1}\begin{pmatrix} 2.1 & -1.0 \\ -1 & 3.1 \end{pmatrix}\right] + \tau^{-1}\begin{bmatrix} 0.1 & 0.0 \\ 0.0 & 1.1 \end{bmatrix},$$

which will be solved iteratively for $\mathcal{R}$.

We arbitrarily choose $\tau = 3$, and the iteration process yields

$$\mathcal{R} = \begin{bmatrix} 0.070500 & 0.029500 \\ 0.460291 & 0.639709 \end{bmatrix}.$$

We next determine the equilibrium phase probabilities by any of the methods previously described, all of which are very simple to apply. We find

$$\mathcal{G}(1) = [0.5 \quad 0.5].$$

Substituting this result into (4.94) yields

$$\begin{aligned} P_0 &= [0.5 \quad 0.5]\left[I - \begin{pmatrix} 0.070500 & 0.029500 \\ 0.460291 & 0.639709 \end{pmatrix}\right] \\ &= [0.234605 \quad 0.165395]. \end{aligned}$$

The solution expressed in matrix geometric form as in (4.90) is then

$$\begin{aligned} P_n &= P_0\mathcal{R}^n \\ &= [0.234605 \quad 0.165395]\begin{bmatrix} 0.070500 & 0.029500 \\ 0.460291 & 0.639709 \end{bmatrix}. \end{aligned}$$

We see from this example that the behavior of the tail probabilities is somewhat less obvious than they are in the form of the previous example, where the geometric quantities are expressed in scalar form.

EXERCISE 4.22 Solve for the equilibrium state probabilities for Example 4.5 using the matrix geometric approach. Specify the results in terms of the matrix $\mathcal{R}$. Determine numerically the range of values of τ for which (4.98) converges. Also, verify numerically that the results are the same as those obtained above.

EXERCISE 4.23 Solve Exercise 3.22 using the matrix geometric approach. Evaluate the relative difficulty of using the matrix geometric approach to that of using the probability generating function approach.

4.2.3 Rate Matrix Computation via Eigenanalysis

We now turn our attention to a discussion of the relationship between the probability generating function and the matrix geometric approaches discussed in the previous subsections. To begin our discussion, let $\nu \neq 0$ be an eigenvalue of $\mathcal{R}$, and let V_ν be a left eigenvector of $\mathcal{R}$ corresponding to ν. Then, by the defining relationship between eigenvalues and eigenvectors, we have

$$V_\nu \mathcal{R} = V_\nu \nu. \tag{4.101}$$

Also, upon premultiplying (4.96) by V_ν, we find that

$$V_\nu \left[\Lambda - \mathcal{R}(\Lambda - \mathcal{Q} + \mathcal{M}) + \mathcal{R}^2 \mathcal{M}\right] = 0.$$

Substitution of (4.101) into the previous equation yields

$$V_\nu \left[\Lambda - \nu(\Lambda - \mathcal{Q} + \mathcal{M}) + \nu^2 \mathcal{M}\right] = 0. \tag{4.102}$$

Then, upon substituting $\sigma = 1/\nu$ into (4.102) and multiplying both sides by σ^2, we find

$$V_\nu \left[\Lambda \sigma^2 - (\Lambda - \mathcal{Q} + \mathcal{M})\sigma + \mathcal{M}\right] = 0. \tag{4.103}$$

But this latter expression is exactly

$$V_\nu \mathcal{A}(\sigma) = 0. \tag{4.104}$$

This means that if V_ν is a left eigenvector of $\mathcal{R}$ corresponding to ν, then V_ν is a left null vector of $\mathcal{A}(z)$ corresponding to its null value $1\,/\nu$. It is obvious from (4.90) that, for a stable system, all of the eigenvalues of $\mathcal{R}$ have magnitudes less than unity. Therefore, the null values of $\mathcal{A}(z)$ which are of interest are those that have magnitudes greater than unity, that is, those in the set $\mathcal{Z}^{(1,\infty)}$. Analogous to the case of Section 4.2.2, these eigenvalues and left eigenvectors can be found via a standard eigenanalysis of the matrix

$$\mathcal{A}'_E = \begin{bmatrix} \mathcal{M}^{-1}(\Lambda - \mathcal{Q}^T + \mathcal{M}) & -\mathcal{M}^{-1}\Lambda \\ I & 0 \end{bmatrix}, \tag{4.105}$$

where the only difference between (4.76) and (4.105) is that the matrix $\mathcal{Q}$ is transposed.

Now, the matrix $\mathcal{A}'_E$ may have zero eigenvalues. In fact, it is easy to show (Daigle and Lucantoni [1990]) that the number of zero eigenvalues of this matrix is exactly the same as the number of terms that are zero on the major diagonal of Λ. First we use an elementary transformation to transform $\mathcal{A}(z)$ into a symmetric λ-matrix. Then, we can write the quadratic form of the transformed symmetric λ-matrix, which has the same null values as the original matrix, can be written as

$$V_\nu \hat{\mathcal{A}}(\sigma) V_\nu^T = \ell\sigma^2 - (\ell - q + m)\sigma + m = 0,$$

where ℓ, m, and q are nonnegative, positive, and nonpositive, respectively. The discriminant of the solution of this quadratic equation is readily found to be $(\ell - m)^2 - 2q(\ell + m) + q^2$, which is always nonnegative. Therefore, all null values of $\mathcal{A}(z)$ are real. It is also easy to see that the null values of $\mathcal{A}(z)$ are nonnegative. But, as the value of ℓ becomes smaller and smaller, the value of the null value corresponding to V_ν becomes larger and larger, and its inverse becomes smaller and smaller. In the limit, the null value becomes infinity and its inverse becomes zero. In effect, the matrix $\mathcal{A}$ has one less null value, but the inverse of this "null value at infinity" shows up as a zero eigenvalue of $\mathcal{A}_E$.

From (4.100), it is easy to see that if $\lambda_i = 0$ for some i, then the corresponding row of the $\mathcal{R}$ matrix will be zero; this can be seen by doing successive substitutions in (4.100) starting with $\mathcal{R}_0 = 0$. Thus, if we denote the $(K+1) \times 1$ column vector whose ith element is 1 with all other elements being 0 by $\mathbf{e}_i$, then it is easy to see that $\mathbf{e}_i$ is a left eigenvector of $\mathcal{R}$.

Now, suppose there are n values of i for which $\lambda_i = 0$. Then there will be n rows of $\mathcal{R}$ which will be identically 0. Define $\mathcal{T}$ to be the elementary transformation such that $\mathcal{T}\Lambda\mathcal{T}$ is a diagonal matrix in which the 0 values of λ_i appear as the first n diagonal elements. Then, the first n rows of the matrix $\mathcal{TRT}$ will be identically 0, and the row vectors $\mathbf{e}_i$, $0 \le i < n$ will be left eigenvectors of this matrix. Next, we denote the matrix formed by the collection of the remaining left eigenvectors of $\mathcal{TRT}$ by $\hat{\mathcal{V}}$, and partition this matrix into the matrix $\left[\hat{\mathcal{V}}_1 \hat{\mathcal{V}}_2\right]$, where $\hat{\mathcal{V}}_1$ contains the first n columns of $\hat{\mathcal{V}}$. Then, we can verify that

$$\begin{bmatrix} I & 0 \\ \hat{\mathcal{V}}_1 & \hat{\mathcal{V}}_2 \end{bmatrix} \mathcal{TRT} = \begin{bmatrix} 0 & 0 \\ 0 & \hat{\mathcal{N}} \end{bmatrix} \begin{bmatrix} I & 0 \\ \hat{\mathcal{V}}_1 & \hat{\mathcal{V}}_2 \end{bmatrix}, \tag{4.106}$$

where $\hat{\mathcal{N}}$ is the diagonal matrix of the nonzero eigenvalues of $\mathcal{R}$. The form of (4.106) indicates that if the matrix of left eigenvectors spans the $K+1$ dimensional eigenspace, then the matrix $\hat{\mathcal{V}}_2$ is nonsingular. Thus we find

$$\mathcal{R} = \mathcal{T} \begin{bmatrix} 0 & 0 \\ \hat{\mathcal{V}}_2^{-1}\hat{\mathcal{N}}\hat{\mathcal{V}}_1 & \hat{\mathcal{V}}_2^{-1}\hat{\mathcal{N}}\hat{\mathcal{V}}_2 \end{bmatrix} \mathcal{T}. \tag{4.107}$$

The implication of the above is that a zero-valued eigenvalue of $\mathcal{R}$ having multiplicity greater than one simplifies, rather than complicates, computation of $\mathcal{R}$.

The computation of the matrix of left eigenvectors of $\mathcal{R}$ is quite straightforward using standard eigenanalysis packages. First, we formulate the matrix $\mathcal{A}_E'$ and obtain its eigenvalues and corresponding eigenvectors. We then select the set of eigenvalues that are less than unity together with their corresponding eigenvectors. The last $K + 1$ elements of the eigenvector of $\mathcal{A}'_E$ corresponding to each eigenvalue of $\mathcal{A}'_E$ which is less than unity are then transformed by $\mathcal{M}^{-1}$ to yield the elements of the left eigenvector of $\mathcal{R}$ corresponding to the same eigenvalue. If the diagonal matrix whose diagonal elements are the eigenvalues of $\mathcal{R}$ is denoted by $\mathcal{N}$, and the matrix of corresponding left eigenvectors is denoted by $\mathcal{V}$, then we compute $\mathcal{R}$ from

$$\mathcal{R} = \mathcal{V}^{-1}\mathcal{N}\mathcal{V}. \tag{4.108}$$

Once we know $\mathcal{R}$, we may solve for P_0 using (4.94) and then compute P_n for $n \geq 1$ via (4.90). Note that the computational effort required to compute P_0 via (4.108) and (4.94) is roughly equal to that required to compute P_0 via (4.77). Note also that computation of a particular power of $\mathcal{R}$ is readily accomplished by making use of the identity

$$\mathcal{R}^n = \mathcal{V}^{-1}\mathcal{N}^n\mathcal{V}.$$

We now turn to the computation of the survivor function and the moments of the occupancy distribution. With regard to the survivor functions, define the joint (occupancy, phase) survivor function as

$$\Sigma_n = \sum_{m=n+1}^{\infty} P_m. \tag{4.109}$$

Then due to (4.90), we find

$$\begin{aligned}
\Sigma_n &= \sum_{m=n+1}^{\infty} P_0\mathcal{R}^m \\
&= P_0[I - \mathcal{R}]^{-1}\mathcal{R}^{n+1} \\
&= \mathcal{G}(1)\mathcal{R}^{n+1}.
\end{aligned} \tag{4.110}$$

Thus we can readily see that the values of the vector Σ_n for successive values of n can be obtained via a postmultiplication by $\mathcal{R}$. The marginal survivor function for the queue occupancy can then be obtained by summing the elements of Σ_n or, equivalently, by postmultiplication by $\mathbf{e}$. Also, the terms of the conditional survivor functions for the queue occupancy can be obtained by dividing the ith element of Σ_n by the ith element of $\mathcal{G}(1)$.

By using the final form of (4.110) and the well known result that the expected value of a nonnegative random variable is given by the integral of its survivor function, we find

$$\begin{aligned} E[\tilde{n}] &= \sum_{n=0}^{\infty} \Sigma_n \mathbf{e} \\ &= \mathcal{G}(1)[I - \mathcal{R}]^{-1}\mathcal{R}\mathbf{e}. \end{aligned}$$

But, because our technique for computing $\mathcal{R}$ yields the eigenvalues and eigenvectors, the expectation can be computed using these quantities. In particular, by using (4.108) and modest algebraic manipulation, we find

$$\begin{aligned} E[\tilde{n}] &= \mathcal{G}(1)\mathcal{V}^{-1} \\ &\quad \operatorname{diag}\left[\, \nu_0/(1-\nu_0) \quad \nu_1/(1-\nu_1) \quad \cdots \quad \nu_K/(1-\nu_K) \,\right] \mathcal{V}\mathbf{e}. \end{aligned} \tag{4.111}$$

Computational forms for higher factorial moments can be easily derived along the same lines. The resulting formulae are as follows:

$$\begin{aligned} \mathcal{G}^{(n)}(1)\mathbf{e} &= \left. \frac{d^n}{dz^n}\mathcal{G}(z) \right|_{z=1} \mathbf{e} \\ &= n!\mathcal{G}(1)\mathcal{V}^{-1} \operatorname{diag}[\, (\nu_0/[1-\nu_0])^n \\ &\quad \cdots (\nu_K/[1-\nu_K])^n \,]\mathcal{V}\mathbf{e}. \end{aligned} \tag{4.112}$$

Formulae for the above quantities based on the partial fraction expansion representation of the occupancy distribution can be readily developed, the most complicated operation being the summing of a geometric series.

EXERCISE 4.24 Prove the result given by (4.112) for the n-th factorial moment of $\tilde{n}$.

An alternate matrix geometric approach for solving QBD models, based on the notion of *complete level crossing information*, is described by Beuerman and Coyle [1989]. Beuerman and Coyle first expand the state space so that the resulting model has *complete level crossing information*, and then they describe a technique for obtaining an alternate rate matrix, W, that is completely specified analytically. In addition, Zhang and Coyle [1989] describe a procedure, based on transform analysis, to determine the time-dependent state probabilities for QBD processes. These results will not be commented upon further here, but the reader interested in solutions to QBD models is encouraged to consult the references.

4.2.4 Rate Matrix Computation via Generalized State-Space Methods

An interesting alternative for determining the rate matrix is the so-called generalized state-space approach discussed in Akar, Oğuz, and Sohraby [1998]. For the case of a QBD process, we shall see that a natural state vector for the system of balance equations is found by concatenating two level probability vectors. Specifically, we will define in a natural way the state vector $y_i = [\,P_i \quad P_{i+1}\,]$, where P_i and P_{i+1} are two successive level probability vectors. Again, in a very straightforward manner we will see that successive state vectors are related through two matrices; specifically, $y_i E = y_{i-1} A$, where E and A are easily specified square matrices. We will then solve for the state vectors via a process that involves reduction of the pair (A, E) to generalized Schur form, which is accomplished via the widely available QZ algorithm, which is discussed in Golub and Van Loan [1996].

Recall from our earlier discussion that the QBD process is a Markov chain on $\{(n,i), n \geq 0, 0 \leq i \leq K\}$ having an infinitesimal generator of the form

$$\tilde{Q} = \begin{bmatrix} B_0 & A_0 & 0 & \cdots & \cdots \\ B_1 & A_1 & A_0 & \cdots & \cdots \\ 0 & A_2 & A_1 & A_0 & \cdots \\ 0 & 0 & A_2 & A_1 & \cdots \\ 0 & 0 & 0 & A_2 & \cdots \\ 0 & 0 & 0 & 0 & \cdots \end{bmatrix},$$

where $(B_0 + A_0)\mathbf{e} = 0$, $(B_1 + A_1 + A_0)\mathbf{e} = 0$, $(A_0 + A_1 + A_2)\mathbf{e} = 0$, and the matrix $A = A_0 + A_1 + A_2$ is a finite generator.

From the generator, $\tilde{Q}$, we immediately have

$$\begin{aligned} 0 &= P_0 B_0 + P_1 B_1, \text{ and} \\ 0 &= P_{i-1} A_0 + P_i A_1 + P_{i+1} A_2 \quad \text{for} \quad i \geq 1. \end{aligned} \tag{4.113}$$

From the previous equation, we then have

$$P_{i+1} A_2 = [\,P_{i-1} \quad P_i\,] \begin{bmatrix} -A_0 \\ -A_1 \end{bmatrix} \quad \text{for} \quad i \geq 1.$$

Then, by making the simple statement $P_i = P_i$ and combining this simple statement with the previous equation, it follows that

$$[\,P_i \quad P_{i+1}\,] \begin{bmatrix} I & 0 \\ 0 & A_2 \end{bmatrix} = [\,P_{i-1} \quad P_i\,] \begin{bmatrix} 0 & -A_0 \\ I & -A_1 \end{bmatrix} \quad \text{for} \quad i \geq 1. \tag{4.114}$$

If we now define

$$\begin{aligned} y_i &= [\,P_i \quad P_{i+1}\,] \\ E &= \begin{bmatrix} I & 0 \\ 0 & A_2 \end{bmatrix}, \text{ and} \\ A &= \begin{bmatrix} 0 & -A_0 \\ I & -A_1 \end{bmatrix}, \end{aligned} \tag{4.115}$$

we find

$$y_i E = y_{i-1} A. \tag{4.116}$$

In order to solve (4.116) for the state probabilities, we shall first decouple the matrix equations (4.116) into two sets of equations; those representing stable modes and those representing unstable modes. We shall accomplish this decoupling by reducing the pair (A, E) to generalized Schur form. Before doing so, we introduce a minimum of terminology and theory from linear algebra.

DEFINITION 4.5 . **Generalized eigenvectors, eigenvalues, and spectrum.** A scalar value λ such that $AS = \lambda ES$ for a non-trivial vector S is called a *generalized eigenvalue of A with respect to E*. The vector S is said to be the *generalized eigenvector of A with respect to E* corresponding to the generalized eigenvalue λ. The set of all generalized eigenvalues is called the *spectrum*, and it is denoted by $\lambda(A, E)$.

The following theorem is a composite of various theorems that appear in Demmel [1997], Golub and Van Loan [1996], and Akar, Oğuz, and Sohraby [1998] to which the reader is referred for more details.

THEOREM 4.2 *Suppose A and E are both real matrices and have spectrum $\lambda(A, E)$. Suppose further that $\lambda(A, E)$ is partitioned into two sets, say $\lambda_u(A, E)$ and $\lambda_s(A, E)$ such that $\lambda_u(A, E) \cap \lambda_s(A, E) = \emptyset$. Then, there exist matrices, Q and Z, such that*

$$Q^T EZ = \begin{bmatrix} E_{uu} & E_{us} \\ 0 & E_{ss} \end{bmatrix} \quad \text{and} \quad Q^T AZ = \begin{bmatrix} A_{uu} & A_{us} \\ 0 & A_{ss} \end{bmatrix},$$

where all matrices are real, Q and Z are orthogonal, meaning $Q^{-1} = Q^T$ and $Z^{-1} = Z^T$, E_{uu} and E_{ss} are upper triangular, and A_{uu} and A_{ss} are block upper triangular, meaning that their diagonal elements are either 1×1 or 2×2 blocks, depending upon whether the eigenvalues are real or occur in complex conjugate pairs. The row dimensions of E_{uu} and A_{uu} and E_{ss} and A_{ss} are $n_u = \text{card } (\lambda_u(A, E))$ and $n_s = \text{card } (\lambda_s(A, E))$, respectively. □

The decomposition indicated in Theorem 4.2 can be accomplished using the routine *dgges()* from the LAPACK Users' Guide [1999] or its C-language version CLAPACK, the latter of which is broadly distributed in C-language development packages. The routine *dgges()* allows the user to write and specify a function that determines whether or not a particular generalized eigenvalue is in the set $\lambda_u(A, E)$. In our particular case, we define $\lambda_u(A, E)$ to be the set of all *unstable* generalized eigenvalues, by which we mean $||\lambda|| \geq 1$ for all $\lambda \in \lambda_u(A, E)$. By default, we then have $||\lambda|| < 1$ for all $\lambda \in \lambda_s(A, E)$.

The set $\lambda_u(A, E)$ contains the set $\mathcal{Z}^{(0,1)}$ defined earlier in the context of very specific assignments for the matrices B_0, B_1, A_0, A_1, and A_2. Let n_u and n_s denote the number of elements of $\lambda_u(A, E)$ and $\lambda_s(A, E)$, respectively. Then, as in the special case we considered earlier, $n_u = n_s = K + 1$ for stable

QBDs. Based on these definitions of $\lambda_u(A, E)$ and $\lambda_s(A, E)$, we use *dgges()* to find the corresponding matrices Q and Z as defined in Theorem 4.2.

For each $j \geq 0$ define $y_j = u_j Q^T$. Then, substitution into (4.116) and postmultiplication by Z yields

$$u_j Q^T E Z = u_{j-1} Q^T A Z. \tag{4.117}$$

Let $u_{j,u}$ and $u_{j,s}$ denote the first n_u and the last n_s elements of u_j so that $u_j = [\, u_{j,u} \quad u_{j,s} \,]$. Then, by choice of Q and Z, we have the equations

$$u_{j,u} E_{uu} = u_{j-1,u} A_{uu}, \tag{4.118}$$

and

$$u_{j,s} E_{ss} = u_{j-1,u} A_{us} + u_{j-1,s} A_{ss}. \tag{4.119}$$

From (4.118), the equations for the unstable generalized eigenvalues, we find

$$u_{j,u} = u_{0,u} [A_{uu} E_{uu}^{-1}]^j.$$

But, since $\|\lambda\| \geq 1$ for $\lambda \in \lambda_u(A, E)$, $\sum_{j=1}^{\infty} u_{j,u}$ grows without bound if $u_{0,u} > 0$. Therefore, we must choose $u_{0,u} = 0$.

On the other hand, $\|\lambda\| < 1$ for $\lambda \in \lambda_s(A, E)$. Therefore, $u_{0,s}$ may have a non-zero value. Thus

$$u_{j,s} = u_{j-1,s} [A_{ss} E_{ss}^{-1}],$$

where $u_{j,s} \neq 0$. This leads to $u_j = [\, 0 \quad u_{j,s} \,]$ for $j \geq 0$.

To facilitate further exploration of the stable part of the solution, define the following partitions of Q and Q^T:

$$Q = [\, Q_u \quad Q_s \,] = \begin{bmatrix} Q_{uu} & Q_{us} \\ Q_{su} & Q_{ss} \end{bmatrix} \quad \text{and} \quad Q^T = \begin{bmatrix} L_u \\ L_s \end{bmatrix} = \begin{bmatrix} L_{uu} & L_{us} \\ L_{su} & L_{ss} \end{bmatrix},$$

where Q_{su}, Q_{ss}, L_{su}, and L_{ss} are all $(K+1) \times (K+1)$. Then,

$$y_j = u_j Q^T = [\, 0 \quad u_{j,s} \,] \begin{bmatrix} L_u \\ L_s \end{bmatrix} = u_{j,s} L_s = u_{j-1,s} [A_{ss} E_{ss}^{-1}] L_s. \tag{4.120}$$

Now, $y_j = u_j Q^T$ implies $u_j = y_j Q$. Therefore, we have $[\, u_{ju} \quad u_{js} \,] = y_j \, [\, Q_u \quad Q_s \,]$ so that $u_{js} = y_j Q_s$. Upon using this result in (4.120), we then find

$$y_j = y_{j-1} Q_s [A_{ss} E_{ss}^{-1}] L_s. \tag{4.121}$$

But, because $y_i = [\, P_i \quad P_{i+1} \,]$, (4.121) is equivalent to

$$[\, P_i \quad P_{i+1} \,] = [\, P_{i-1} \quad P_i \,] \begin{bmatrix} Q_{su} \\ Q_{ss} \end{bmatrix} [A_{ss} E_{ss}^{-1}] \, [\, L_{su} \quad L_{ss} \,]. \tag{4.122}$$

Thus, from the first matrix equation of (4.122), we have

$$P_i = [\,P_{i-1} \quad P_i\,]\begin{bmatrix} Q_{su} \\ Q_{ss} \end{bmatrix}[A_{ss}E_{ss}^{-1}]\,L_{su}.$$

Upon solving the previous equation for P_i, we have

$$P_i = P_{i-1}Q_{su}A_{ss}E_{ss}^{-1}L_{su}\left[I - Q_{ss}A_{ss}E_{ss}^{-1}L_{su}\right]^{-1}. \tag{4.123}$$

Now define

$$R = Q_{su}A_{ss}E_{ss}^{-1}L_{su}\left[I - Q_{ss}A_{ss}E_{ss}^{-1}L_{su}\right]^{-1}. \tag{4.124}$$

Then, (4.123) reduces to

$$P_i = P_{i-1}R, \quad \text{for} \quad i \geq 1$$

so that

$$P_i = P_0 R^i, \quad \text{for} \quad i \geq 0, \tag{4.125}$$

which is in standard matrix geometric form.

As before, upon summing the probabilities, we find

$$\sum_{i=0}^{\infty} P_i = P_0 \sum_{i=0}^{\infty} R^i = P_0\,[I - R]^{-1} \quad \text{or} \quad P_0 = \left[\sum_{i=0}^{\infty} P_i\right][I - R]\,.$$

But, in the solution to Exercise 4.25 it will be seen that

$$\sum_{i=0}^{\infty} P_i = \pi,$$

where π is the stationary vector of the stochastic matrix $A_0 + A_1 + A_2$. Thus, we have

$$P_0 = \pi\,[I - R]\,. \tag{4.126}$$

We now have a complete specification for the solution of all of the state probabilities. In summary, the procedure is as follows:

1 From the problem statement, determine A_0, A_1, and A_2. There is no need to specify B_0 and B_1 as the information is already contained in A_0, A_1, and A_2.

2 Determine the matrices A and E as defined in (4.115).

3 Perform the decomposition indicated in Theorem 4.2 using the LAPACK routine *dgges()* to find Q^T, E_{ss}, and A_{ss}.

4 Determine R from (4.124).

5 Determine the stationary vector, π, of $A_0 + A_1 + A_2$.

6 Compute P_0 using (4.126).

7 Compute the desired level probabilities using (4.125).

EXERCISE 4.25 Beginning with (4.113), show that

$$\sum_{i=0}^{\infty} P_i = \pi.$$

[Hint: First, sum the elements of the right hand side of (4.113) from $i = 1$ to $i = \infty$. This will yield $0 = [\sum_{i=0}^{\infty} P_i][A_0 + A_1 + A_2] + \theta(P_0, P_1)$. Next, use the fact that, because $\tilde{Q}$ is stochastic, the sum of the elements of each of the rows of $\tilde{Q}$ must be a zero matrix to show that $\theta(P_0, P_1) = 0$. Then complete the proof in the obvious way.]

EXAMPLE 4.6 Consider the system of Example 4.4, where the parameter values are $\lambda_0 = 0.1$, $\lambda_1 = 1.1$, $\beta_0 = \delta_1 = \mu_1 = \mu_2 = 1$ and the QBD matrices are as follows:

$$A_0 = \begin{bmatrix} 0.1 & 0.0 \\ 0 & 1.1 \end{bmatrix}, \; A_1 = \begin{bmatrix} -2.1 & 1.0 \\ 1 & -3.1 \end{bmatrix}, \text{ and } A_2 = \begin{bmatrix} 1.0 & 0.0 \\ 0.0 & 1.0 \end{bmatrix}.$$

We wish to find the rate matrix, R, and the 0-level probability vector, P_0 via generalized state-space methods.

Solution: From (4.115), we have

$$E = \begin{bmatrix} 1.0 & 0.0 & 0.0 & 0.0 \\ 0.0 & 1.0 & 0.0 & 0.0 \\ 0.0 & 0.0 & 1.0 & 0.0 \\ 0.0 & 0.0 & 0.0 & 1.0 \end{bmatrix} \quad \text{and} \quad A = \begin{bmatrix} 0.0 & 0.0 & -0.1 & 0.0 \\ 0.0 & 0.0 & 0.0 & -1.1 \\ 1.0 & 0.0 & 2.1 & -1.0 \\ 0.0 & 1.0 & -1.0 & 3.1 \end{bmatrix}.$$

Schur decomposition via *dgges()* then yields:

$$Q^T = \begin{bmatrix} 0.015987 & -0.249442 & -0.557910 & 0.791365 \\ -0.062789 & -0.565609 & 0.744934 & 0.348162 \\ 0.856779 & -0.429783 & -0.137423 & -0.249661 \\ -0.511597 & -0.658142 & -0.339005 & -0.436112 \end{bmatrix},$$

$$A_{uu} = \begin{bmatrix} 3.489792 & -0.732157 \\ 0.000000 & 1.000000 \end{bmatrix}, \; E_{uu} = \begin{bmatrix} 1.000000 & 0.000000 \\ 0.000000 & 1.000000 \end{bmatrix},$$

$$A_{ss} = \begin{bmatrix} 0.047568 & 0.348278 \\ 0.000000 & 0.662640 \end{bmatrix}, \text{ and } E_{ss} = \begin{bmatrix} 1.000000 & 0.000000 \\ 0.000000 & 1.000000 \end{bmatrix}.$$

From Q^T of the Schur decomposition, we can readily extract the following:

$$Q_{su} = \begin{bmatrix} 0.856779 & -0.511597 \\ -0.429783 & -0.658142 \end{bmatrix}, \quad Q_{ss} = \begin{bmatrix} -0.137423 & -0.339005 \\ -0.249661 & -0.436112 \end{bmatrix},$$

and

$$L_{su} = \begin{bmatrix} 0.856779 & -0.429783 \\ -0.511597 & -0.658142 \end{bmatrix}.$$

Note that there is no need to invert Q^{-1} to find Q because Q is orthogonal by construction. We may now use (4.124) to compute R. We find

$$R = \begin{bmatrix} 0.070500 & 0.029500 \\ 0.460291 & 0.639709 \end{bmatrix}.$$

We preiously had $\pi = [0.500000 \quad 0.500000]$. Substitution of R and π into (4.126) yields

$$P_0 = [0.234605 \quad 0.165395].$$

We may now compute any desired level probabilities by using (4.125).

We also note that by dividing the diagonal elements of the A matrices by the diagonal elements of the E matrices of the Schur decomposition, we can readily find the generalized eigenvalues to be $\{3.489792, 1.0, 0.047568, 0.662640\}$. Note that the eigenvalues whose moduli are less than 1 are partitioned from those whose moduli is at least 1.

The example is now complete.

Reduction to generalized Schur form is accomplished via the so-called QZ algorithm, which is described in detail in Golub and Van Loan [1996]. They report that the QZ algorithm requires 46 n^3 floating point operations, including computation of Q^{-1}, and that the the speed of the algorithm is not affected by any rank deficiency of E.

We shall see later that the type of formulation and the solution methodology presented here is useful in a much broader context.

4.3 Service-Time Distributions of the Phase Type and Other Variations

At this point we have just begun to scratch the surface concerning systems having matrix geometric solutions. In this section, we make minor modifications to the model of the preceding section and show how more general service-time distributions may be considered via matrix geometric methods. In particular, our modifications serve to introduce distributions of the *phase type*, a broad class of distributions which are covered in detail in Neuts [1981a].

The essence of our modification is that we allow both the level and the phase of the process to change simultaneously. In particular, we express the infinitesimal generator of the phase process as

$$\mathcal{Q} = \mathcal{S} + \mathcal{S}^0 b, \tag{4.127}$$

where $\mathcal{S}$ is a nonsingular $(K+1) \times (K+1)$ matrix representing phase changes within the same level, $\mathcal{S}^0$ is a nonnegative $(K+1)$-column vector equal to $-\mathcal{S}e$, b is a nonnegative $(K+1)$-row vector such that $be = 1$, and $\mathcal{S}^0 b$ is a $(K+1) \times (K+1)$ matrix representing phase changes that result in a level decrease, that is, a service completion.

Since $\mathcal{S}^0 b$ represents phase changes that are simultaneously level decreases, the net of effect of the corresponding transitions of this nature is service completion. Thus we let

$$\mathcal{M} = \mathcal{S}^0 b. \tag{4.128}$$

In addition, we restrict the arrival process so that the arrival rate is independent of the phase so that

$$\Lambda = \lambda I. \tag{4.129}$$

Then, substituting (4.127) through (4.129) into (4.53) and (4.56) we get

$$P_0(\lambda I - \mathcal{S} - \mathcal{S}^0 b) - P_1 \mathcal{S}^0 b = 0, \tag{4.130}$$

$$P_{n-1}\lambda I - P_n(\lambda I - \mathcal{S}) + P_{n+1}\mathcal{S}^0 b = 0, \quad \text{for} \quad n \geq 1. \tag{4.131}$$

We now modify the behavior of the system so that the state of the system following a service completion from level 1 is always $(0, 0)$, and the state of the system following an arrival from level 0 is $(1, i)$ with probability b_i, $i = 0, 1, \ldots, K$. Then, we find that

$$P_0 = [\, p_{00} \quad 0 \quad \ldots \quad 0 \,],$$

equations analogous to (4.130) and (4.131) for this special case are

$$P_{00}\lambda - P_1 \mathcal{S}^0 = 0, \tag{4.132}$$

$$P_{00}\lambda b - P_1(\lambda I - \mathcal{S}) + P_2 \mathcal{S}^0 b = 0, \tag{4.133}$$

and

$$P_{n-1}\lambda I - P_n(\lambda I - \mathcal{S}) + P_{n+1}\mathcal{S}^0 b = 0, \quad \text{for} \quad n \geq 2. \tag{4.134}$$

Neuts [1981a], pp. 83-86 shows, using very elementary arguments, that the system of equations (4.131) through (4.134) has the unique solution

$$P_{00} = 1 - \rho,$$

$$P_n = (1-\rho) b \mathcal{R}^n, \quad \text{for} \quad n \geq 1, \tag{4.135}$$

where

$$\mathcal{R} = \lambda \left[\lambda I - \lambda \mathbf{e} b - \mathcal{S} \right]^{-1}, \tag{4.136}$$

and

$$\rho = -\lambda b \mathcal{S}^{-1} \mathbf{e}. \tag{4.137}$$

Note that the matrix $\mathcal{R}$ is given explicitly, so that the state distribution can be determined in closed form involving only a matrix inversion. This demonstrates that it is worthwhile to attempt to find an analytic solution to a problem even though it may at first appear difficult.

A little thought reveals that the system we have just analyzed is a queueing system having Poisson arrivals, identically, but not exponentially distributed, service times, and infinite waiting capacity. In short, the system is a special case of the M/G/1 queueing system. The special case is the one in which the service-time distribution is of the *phase type*.

For service-time distributions of the phase type, each time service is begun the phase of the process is initiated in phase i with probability b_i, $i = 0, 1, \ldots, K$, with $b\mathbf{e} = 1$. Changes in the phase process are then governed by the infinitesimal generator

$$\tilde{Q} = \begin{bmatrix} \mathcal{S} & \mathcal{S}^0 \\ 0 & 0 \end{bmatrix}, \tag{4.138}$$

where $\mathcal{S}$ and $\mathcal{S}^0$ are defined as above, 0 is a $(K+1)$-row vector of zeros, and 1 is a scalar. A transition to phase $K+1$ represents absorption of the process, or, equivalently, the end of the current service time.

In forming our system of equations, we represent absorption by a level change, or service completion, so that phase $(K+1)$ is not needed. The column vector $\mathcal{S}^0$ represents the rate at which the process changes levels, and the vector b is the vector of probabilities determining the phase of the process immediately following a service completion. Equivalently, b is the vector of probabilities determining the phase of the process at the beginning of the service time, and consequently, it is needed to specify the phase of the process following an arrival from the empty state, only one of which is needed since the joint probability $P_{0i} = 0$ for $i \neq 0$.

EXERCISE 4.26 With $\tilde{Q}$ defined as in (4.138) and b any vector of probability masses such that $b\mathbf{e} = 1$, show that the matrix $\mathcal{S} + \mathcal{S}^0 b$ is always an infinitesimal generator.

If the infinitesimal generator matrix $\mathcal{Q} = \mathcal{S} + \mathcal{S}^0 b$ is irreducible, then Neuts [1981a], pp. 48-51 shows that the resulting service-time distribution will be

$$F_{\tilde{x}}(x) = 1 - b e^{\mathcal{S}x} \mathbf{e}, \tag{4.139}$$

and the moments of the distribution are readily shown to be given by

$$E[\tilde{x}^n] = (-1)^n n! \, b \, \mathcal{S}^{-n} \, \mathbf{e}. \tag{4.140}$$

A broad class of distributions can be described by judicious choice of the terms of $\mathcal{Q}$. For example, these terms can be easily chosen so that the resulting distribution is exponential, or Erlang-k. However, the representation of a given distribution is not unique. For example, two representations for the exponential distribution with parameter μ follow.

First we choose to let $\mathcal{Q}$ be a scalar. Then, $\mathcal{S} = -\mu$, $\mathcal{S}^0 = \mu$, and $b = 1$. Second , we choose

$$\mathcal{S} = \begin{bmatrix} -(\mu+\alpha) & \alpha \\ \beta & -(\mu+\beta) \end{bmatrix}, \quad \mathcal{S}^0 = \begin{bmatrix} \mu \\ \mu \end{bmatrix}, \quad b = [\, p_0 \quad p_1 \,], \qquad (4.141)$$

where α and β are any nonnegative constants, and p_0 and p_1 are any two nonnegative numbers such that $p_0 + p_1 = 1$. The interested reader is urged to consult Neuts [1981a] and also the more recent book, Latouche and Ramaswami [1999], for superb treatment of distributions of the phase type and queueing systems having interarrival arrival and service time distributions of the phase type. We also point out that there are readily available software packages on the World Wide Web that facilitate definition of phase-type distributions from empirical data.

EXERCISE 4.27 Consider a single-server queueing system having Poisson arrivals. Suppose upon entering service, each customer initially receives a type 1 service increment. Each time a customer receives a type 1 service increment, the customer leaves the system with probability $(1-p)$ or else receives a type 2 service increment followed by an additional type 1 service increment. Suppose type 1 and type 2 service increment times are each drawn independently from exponential distributions with parameters μ_1 and μ_2, respectively. Define the phase of the system to be 1 if a customer in service is receiving a type 2 service increment. Otherwise, the system is in phase 0. Define the state of the system to be the 0 when the system is empty and by the pair (i, j) where $i > 0$ is the system occupancy and $j = 0, 1$ is the phase of the service process. Define $P_i = [\, P_{i0} \quad P_{i1} \,]$ for $i > 0$ and P_0, a scalar. Draw the state diagram, and determine the matrix $\mathcal{Q}$, the infinitesimal generator for the continuous-time Markov chain defining the occupancy process for this system.

EXERCISE 4.28 Suppose

$$\mathcal{S} = \begin{bmatrix} -\mu & \mu \\ 0 & -\mu \end{bmatrix} \quad \mathcal{S}^0 = \begin{bmatrix} 0 \\ \mu \end{bmatrix}, \quad \text{and} \quad P_t(0) = [\, 1 \quad 0 \,].$$

Find $F_{\tilde{x}}(t) = P\{\tilde{x} \le t\}$ and $f_{\tilde{x}}(t)$, and identify the form of $f_{\tilde{x}}(t)$. [*Hint*: First solve for $P_0(t)$, then for $P_1(t)$, and then for $P_2(t) = P_a(t)$. There is never a need to do matrix exponentiation.]

EXERCISE 4.29 Starting with (4.139) as given, prove the validity of (4.140).

Modifications to the basic model of phase-dependent arrival and service rates can be made to model both independent, identically distributed arrival and service times of the phase type; such models are classified as $PH/PH/1$. Neuts [1981a] covers the analysis of $PH/PH/1$ models, which have matrix geometric solutions, as well as numerous other models. We defer further coverage of models having the matrix geometric solution to Chapter 7 and turn to the analysis of other models in the next chapter.

4.4 Supplementary Problems

4-1 Let $\mathcal{Q}$ be an $(m+1)$-square matrix representing the infinitesimal generator for a continuous-time Markov chain with state space $\{0, 1, \ldots, m\}$. Let

$$\tilde{\mathcal{Q}} = \begin{bmatrix} \mathcal{T} & \mathcal{T}^0 \\ 0 & 0 \end{bmatrix},$$

where $\mathcal{T}$ is an m-square matrix, $\mathcal{T}^0$ is an $m \times 1$ column vector, and the remaining terms are chosen to conform, be a matrix obtained by replacing any row of $\mathcal{Q}$ by a row of zeros and then exchanging rows so that the final row is a vector of zeros. Let $P(t) = [\, P_t(t) \quad P_a(t)\,]$ denote the state probability vector for the Markov chain for which $\tilde{\mathcal{Q}}$ is the infinitesimal generator, with $P_t(t)$ a row vector of dimension m and $P_a(t)$ a scalar.

(a) Argue that if $\mathcal{Q}$ is the infinitesimal generator for an irreducible Markov chain, then the states $0, 1, \ldots, m-1$ of the modified chain are all transient, and state m is an absorbing state.

(b) Prove that if $\mathcal{Q}$ is the infinitesimal generator for an irreducible Markov chain, then the matrix $\mathcal{T}$ must be nonsingular. [*Hint*: Solve for $P_t(t)$, then prove by contradiction. Make use of the fact that if $\mathcal{T}$ is singular, then $\mathcal{T}$ has a zero eigenvalue.]

(c) Show that $P_a(t) = 1 - P_t(0) \exp\{\mathcal{T}t\}\mathbf{e}$. [*Hint*: Use the fact that $P_t(t)\mathbf{e}$ is the probability that the state of the modified Markov chain is in the set $\{0, \ldots, m-1\}$ at time t.]

(d) Let $\tilde{x}$ be the time required for the modified Markov chain to reach state m given an initial probability vector $P(0) = [\, P_t(0) \quad 0\,]$, that is, with $P_t(0)\mathbf{e} = 1$. Argue that $P\{\tilde{x} \le t\} = P_a(t)$; that is,

$$P\{\tilde{x} \le t\} = 1 - P_t(0) \exp\{\mathcal{T}t\}\mathbf{e}.$$

(e) Argue that if $\mathcal{Q}$ is the infinitesimal generator for an irreducible Markov chain, then the matrix $\tilde{\mathcal{T}} = \mathcal{T} + \mathcal{T}^0 P_t(0)$ is the infinitesimal generator for an irreducible Markov chain with state space $\{0, 1, \ldots, m-1\}$.

4-2 Consider an m-server queueing system having Poisson arrivals. Suppose upon entering service, each customer initially receives a type 1 service increment. Each time a customer receives a type 1 service increment, the customer leaves the system with probability $(1 - p)$ or else receives a type 2 service increment followed by an additional type 1 service increment. Suppose type 1 and type 2 service increment times are each drawn independently from exponential distributions with parameters μ_1 and μ_2, respectively. With the service process defined as in Problem 4-1, suppose there are m servers. Define the phase of the system to be j if there are j customers receiving a type 2 service increment, $j = 0, 1, \cdots, m$. Define the state of the system to be 0 when the system is empty and by the pair (i, j) where $i \geq 0$ is the system occupancy and $j = 0, \cdots, i$ is the phase of the service process. Define $P_i = [\, P_{i0} \quad P_{i1} \quad \cdots \quad P_{i,\min\{i,m\}} \,]$ for $i > 0$ and P_0, a scalar.

(a) Draw the state transition diagram for the special case of $m = 3$.

(b) Write the matrix balance equations for the special case of $m = 3$.

(c) Write the matrix balance equations for the case of general values of m.

(d) Determine the matrix $\mathcal{Q}$, the infinitesimal generator for the continuous-time Markov chain defining the occupancy process for this system.

(e) Comment on the structure of the matrix $\mathcal{Q}$ relative to that for the phase-dependent arrival and service rate queueing system and to the M/PH/1 system. What modifications in the solution procedure would have to be made to solve this problem? [*Hint*: See Neuts [1981a], pp. 24-26.]

Chapter 5

THE BASIC M/G/1 QUEUEING SYSTEM

In the previous chapters we made extensive use of the memoryless properties of the exponential distribution to study the dynamics of the M/M/1 and other queueing systems, as well as the service-time distributions and interarrival time distributions, which were exponentially distributed. Due to the memoryless property of the exponential distribution, the evolution of such systems from any point in time forward is independent of past history. Thus, the memoryless property allowed us to specify the state of the system at an arbitrary point in time and to write equations describing the system dynamics conveniently.

If the service system has service times that are drawn from a general distribution, then the memoryless property is lost, and it is then necessary to choose observation times carefully in order that the state of the system at the observation times can be easily specified. That is, if we choose the observation times carefully, we may be able to specify the state of the system conveniently, and further, we may succeed in having the evolution of the process from that point forward be independent of past history. Suppose, for example, that we choose our observation times as those instants in time when a customer has just completed service. At those points in time, both the arrival process, which is memoryless, and the service process, which is not necessarily memoryless, start over again. Thus, in order to determine the future evolution of the system, it is necessary to know only the number of customers left in the system immediately following customer departures.

Define $\{\tilde{q}_n, n \geq 1\}$ to be the number of customers left in the system by the nth departing customer. Then, according to our previous observations, the process $\{\tilde{q}_n, n \geq 1\}$ is Markovian. We call $\{\tilde{q}_n, n \geq 1\}$ an embedded Markov chain, which we introduced in Chapter 2. We say that we have "embedded a Markov chain at the points of customer departure." Following our notation of the previous chapters, we denote the number of customers in the system, including the one in service, if any, by $\tilde{n}(t)$. The process $\{\tilde{n}(t), t \geq 0\}$ does not have the Markovian property. That is, unlike the M/M/1 case, the future evolution of the process depends upon the length of time the system has been in the present state. However, the process does have the Markov property at instants of time just after customer departures. Thus, the process $\{\tilde{n}(t), t \geq 0\}$ is called a *semi-Markov process*.

The analytical tools used to study queueing systems of this type are fundamentally the same as those used to study Markovian models, but their application is quite different. The notion of balance equations in non-Markovian systems, for example, can still be applied, but the application is much more difficult (see Daigle and Whitehead [1985]). Thus, different approaches are used to examine non-Markovian systems. In this chapter, we present basic tools that are useful in the analysis of non-Markovian systems. Our presentation is accomplished through the development of many of the classical results for the M/G/1 system and some of its variants, in addition to presentation of some nontraditional approaches and results.

In Section 5.1 we begin our study of the M/G/1 queueing system with a classical development of the Pollaczek-Khintchine transform equation, or probability generating function, for the occupancy distribution. In the same section, we develop the Laplace-Stieltjes transforms for the ergodic waiting time, sojourn time, and busy period distributions.

In Section 5.2, we address inversion of the occupancy distribution's probability generating function, which was developed in the first section. Three methods are presented. The first method is based upon Fourier analysis (Daigle [1989]), and the second approach, due to Keilson and Servi [1989], is recursive. The second approach appears to be useful when only a few terms of the distribution are required; while the first appears to be more appropriate when the entire distribution is desired. The third approach is based on generalized state space methods, which were used in Chapter 4 to determine the equilibrium probabilities for QBD processes. A number of practical issues regarding a variety of approximations are addressed using the generalized state space approach. All of the approaches are applicable to systems other than the ordinary M/G/1.

We next turn our attention to the direct computation of average waiting and sojourn times for the M/G/1 queueing system. Our development follows that for the M/M/1 system to the point at which the consequences of not having the Markovian property surfaces. At this point, a little renewal theory is introduced so that the analysis can be completed. Additional insight into the properties of the M/G/1 system are also introduced at this point. Following completion of the waiting- and sojourn-time development, we introduce alternating renewal theory and use a basic result of alternating renewal theory to compute the average length of the M/G/1 busy period directly. The results of this section play a key role in the analysis of queueing systems with priority, which we address in Chapter 6.

We conclude the chapter with a set of supplementary exercises.

5.1 M/G/1 Queueing System Transform Equations

In this section we examine the behavior of the ordinary M/G/1 queueing system. We develop the probability generating function for the occupancy distribution, and then we use this result to develop the Laplace-Stieltjes transforms for the ergodic waiting-time, sojourn-time, and busy-period distributions.

Recall that we have defined $\{\tilde{q}_n, n \geq 1\}$ to be the number of customers left in the system by the nth departing customer. Now, it is easy to see that the number of customers left by the $(n+1)$st departing customer is equal to the number of customers who arrive during the $(n+1)$st service plus either zero or one fewer than the number left by the nth departing customer, whichever is greater. Thus, we define $\tilde{v}_n$ to be the number of arrivals that occur during the nth customer's service. Then, we find that

$$\tilde{q}_{n+1} = (\tilde{q}_n - 1)^+ + \tilde{v}_{n+1} \tag{5.1}$$

where

$$(a)^+ = \max\{a, 0\}. \tag{5.2}$$

We will solve (5.1) by making use of probability generating functions and Laplace-Stieltjes transforms. In particular, we will develop an expression for the probability generating functions for the sequence of random variables $\tilde{q}_n$, and then we will obtain the ergodic probability generating function by taking limits.

Observation of (5.1) reveals that $\tilde{q}_{n+1}$ is the sum of two independent random variables, $(\tilde{q}_n - 1)^+$ and $\tilde{v}_{n+1}$. Each of these random variables has a generic form, which will occur in later analysis. In addition, we will constantly be encountering sums of independent random variables. Therefore, before proceeding to the analysis of (5.1), we present the following three theorems, which will be useful both in the current analysis and later analyses.

THEOREM 5.1 *Let $\tilde{x}$ and $\tilde{y}$ be two independent, nonnegative, integer-valued random variables, then*

$$\mathcal{F}_{\tilde{x}+\tilde{y}}(z) = \mathcal{F}_{\tilde{x}}(z)\mathcal{F}_{\tilde{y}}(z),$$

where $\mathcal{F}_{\tilde{x}}(z) \stackrel{\Delta}{=} E[z^{\tilde{x}}]$.

Proof The proof follows directly from the fact that $z^{\tilde{x}}$ and $z^{\tilde{y}}$ are independent random variables and consequently $E[z^{\tilde{x}}z^{\tilde{y}}] = E[z^{\tilde{x}}]E[z^{\tilde{y}}]$. □

THEOREM 5.2 *Let $\tilde{x} \sim F_{\tilde{x}}(x)$ denote a random period of time, and let $F^*_{\tilde{x}}(s)$ denote the Laplace-Stieltjes transform of $F_{\tilde{x}}(x)$; i.e., $F^*_{\tilde{x}}(x) = \int_0^\infty e^{-st} dF_{\tilde{x}}(t)$. Further, let $\{\tilde{n}(t), t \geq 0\}$ be a Poisson process with rate λ, and let $\tilde{y}$ denote the number of events from $\{\tilde{n}(t), t \geq 0\}$ that occur during the period of time $\tilde{x}$. Then, $\mathcal{F}_{\tilde{y}}(z) = F^*_{\tilde{x}}(\lambda[1-z])$. That is, the probability generating function for the number of events from a Poisson process that occur during a random period of time is given by the Laplace-Stieltjes transform for the distribution of the length of the period of time with the transform variable, s, evaluated at the point $\lambda[1-z]$.*

Proof

$$\mathcal{F}_{\tilde{y}}(z) = \int_0^\infty E[z^{\tilde{y}} | \tilde{x} = x] dF_{\tilde{x}}(x).$$

But,

$$\begin{aligned} E[z^{\tilde{y}} | \tilde{x} = x] &= \sum_{y=0}^{\infty} E[z^{\tilde{y}} | \tilde{x} = x, \tilde{y} = y] P\{\tilde{y} = y | \tilde{x} = x\} \\ &= \sum_{y=0}^{\infty} z^y \frac{(\lambda x)^y}{y!} e^{-\lambda x} \\ &= e^{-\lambda(1-z)x}. \end{aligned}$$

Thus we have

$$\begin{aligned} \mathcal{F}_{\tilde{y}}(z) &= \int_0^\infty e^{-\lambda(1-z)x} dF_{\tilde{x}}(x) \\ &= \int_0^\infty e^{-sx} dF_{\tilde{x}}(x) \Big|_{s=\lambda(1-z)}. \end{aligned}$$

That is,

$$\mathcal{F}_{\tilde{y}}(z) = F^*_{\tilde{x}}(\lambda[1-z]).$$

□

EXERCISE 5.1 With $\tilde{x}$, $\{\tilde{n}(t), t \geq 0\}$, and $\tilde{y}$ defined as in Theorem 5.2, show that $E[\tilde{y}(\tilde{y}-1)\cdots(\tilde{y}-n+1)] = \lambda^n E[\tilde{x}^n]$.

THEOREM 5.3 *Let $\tilde{x}$ be a nonnegative integer valued random variable with probability generating function $\mathcal{F}_{\tilde{x}}(z)$. Then*

$$\mathcal{F}_{(\tilde{x}-1)^+}(z) = \left(1 - \frac{1}{z}\right) P\{\tilde{x} = 0\} + \frac{1}{z} \mathcal{F}_{\tilde{x}}(z).$$

□

EXERCISE 5.2 Prove Theorem 5.3.

We now turn to the analysis of (5.1). As before, we let $\tilde{x}_n$ denote the service time of the nth customer and we assume $\{\tilde{x}_n, n \geq 1\}$ to be a sequence of independent, identically distributed random variables with mean

$$\frac{1}{\mu} = \int_0^\infty x \, dF_{\tilde{x}}(x)$$

where $\tilde{x}$ represents a generic $\tilde{x}_i$. Clearly, $\tilde{v}_{n+1}$ is independent of $(\tilde{q}_n - 1)^+$ because the number of arrivals during the $(n+1)$st service time does not depend on the number of customers left in the system by the nth departing customer. Therefore, by Theorem 5.1,

$$\mathcal{F}_{\tilde{q}_{n+1}}(z) = \mathcal{F}_{(\tilde{q}_n - 1)^+}(z)\mathcal{F}_{\tilde{v}_{n+1}}(z).$$

But, according to Theorem 5.2, we find that

$$\mathcal{F}_{\tilde{v}_{n+1}}(z) = F^*_{\tilde{x}_{n+1}}(\lambda[1-z]),$$

and by Theorem 5.3,

$$\mathcal{F}_{(\tilde{q}_n - 1)^+} = \left(1 - \frac{1}{z}\right) P\{\tilde{q}_n = 0\} + \frac{1}{z}\mathcal{F}_{\tilde{q}_n}(z).$$

Thus we have

$$\mathcal{F}_{\tilde{q}_{n+1}}(z) = \left[\left(1 - \frac{1}{z}\right) P\{\tilde{q}_n = 0\} + \frac{1}{z}\mathcal{F}_{\tilde{q}_n}(z)\right] F^*_{\tilde{x}_{n+1}}(\lambda[1-z]). \quad (5.3)$$

In the limit as $n \to \infty, \tilde{q}_n \to q$, $\tilde{x}_n \to \tilde{x}$, so

$$\mathcal{F}_{\tilde{q}}(z) = \left[\left(1 - \frac{1}{z}\right) P\{\tilde{q} = 0\} + \frac{1}{z}\mathcal{F}_{\tilde{q}}(z)\right] F^*_{\tilde{x}}(\lambda[1-z]). \quad (5.4)$$

Upon solving (5.4), we find that

$$\mathcal{F}_{\tilde{q}}(z) = \frac{(1-z)P\{\tilde{q} = 0\}F^*_{\tilde{x}}(\lambda[1-z])}{F^*_{\tilde{x}}(\lambda[1-z]) - z}. \quad (5.5)$$

It remains to specify $P\{\tilde{q} = 0\}$, the probability that a departing customer leaves no customers in the system. It is straightforward to determine this unknown probability by using the facts that $F^*_X(0) = 1$, $\mathcal{F}_X(1) = 1$, and other properties of the Laplace-Stieltjes transform and probability generating function that we have previously discussed. By taking limits on both sides of (5.5), applying L'Hôpital's rule, and then using the properties of Laplace transforms and probability generating functions, we find that

$$P\{\tilde{q} = 0\} = 1 - \frac{\lambda}{\mu} = 1 - \rho, \tag{5.6}$$

where $\rho = \lambda/\mu$ is the server utilization as defined in Chapter 3. Thus we find that

$$\mathcal{F}_{\tilde{q}}(z) = \frac{(1-z)(1-\rho)F^*_{\tilde{x}}(\lambda[1-z])}{F^*_{\tilde{x}}(\lambda[1-z]) - z}. \tag{5.7}$$

EXERCISE 5.3 Starting with (5.5), use the properties of Laplace transforms and probability generating functions to establish (5.6).

EXERCISE 5.4 Establish (5.6) directly by using Little's result.

Now, in the M/G/1 queueing system, the probability that a departing customer leaves n customers in the system is the same as the probability that an arriving customer finds n customers in the system when the system is in stochastic equilibrium. A little thought will show that this must be true in order that an equilibrium distribution exist. In addition, we have pointed out earlier that the Poisson arrival's view of the system is exactly the same as that of a random observer. Thus we find that

$$P\{\tilde{n} = n\} = P\{\tilde{q} = n\} = P\{\tilde{q}' = n\},$$

where $\tilde{q}'$ is the number of customers found in the system by an arbitrary arrival when the system is in stochastic equilibrium, and $\tilde{n}$ is the number of customers found in the system by an arbitrary random observer. Thus we find

$$\mathcal{F}_{\tilde{n}}(z) = \frac{(1-z)(1-\rho)F^*_{\tilde{x}}(\lambda[1-z])}{F^*_{\tilde{x}}(\lambda[1-z]) - z}. \tag{5.8}$$

DEFINITION 5.1 **Squared coefficient of variation.** For any nonnegative random variable, $\tilde{x}$ with $E[\tilde{x}] > 0$, the *squared coefficient of variation* for $\tilde{x}$ is defined to be the quantity

$$C_{\tilde{x}}^2 = \frac{Var(\tilde{x})}{E^2[\tilde{x}]}. \tag{5.9}$$

EXERCISE 5.5 **Batch Arrivals.** Suppose arrivals to the system occur in batches of size $\tilde{b}$, and the batches occur according to a Poisson process at rate λ. Develop an expression equivalent to (5.5) for this case. Be sure to define each of the variables carefully.

EXERCISE 5.6 Using the properties of the probability generating function, show that

$$\begin{aligned} E[\tilde{n}] &= \rho + \frac{\lambda\rho}{1-\rho} \frac{E[\tilde{x}^2]}{2E[\tilde{x}]} \\ &= \rho \left(1 + \frac{\rho}{1-\rho} \frac{C_{\tilde{x}}{}^2 + 1}{2}\right). \end{aligned} \tag{5.10}$$

[*Hint*: The algebra will be greatly simplified if (5.8) is first rewritten as

$$\mathcal{F}_{\tilde{n}}(z) = \alpha(z)/\beta(z),$$

where

$$\alpha(z) = (1-\rho)F_{\tilde{x}}^*(\lambda[1-z]),$$

and

$$\beta(z) = 1 - \frac{1 - F_{\tilde{x}}^*(\lambda[1-z])}{1-z}.$$

Then, in order to find

$$\lim_{z\to 1} \frac{d}{dz}\mathcal{F}_{\tilde{n}}(z),$$

first find the limits as $z \to 1$ of $\alpha(z)$, $\beta(z)$, $d\alpha(z)/dz$, and $d\beta(z)/dz$, and then substitute these limits into the formula for the derivative of a ratio. Alternatively, multiply both sides of (5.8) to clear fractions and then differentiate and take limits.]

EXERCISE 5.7 Let $\delta_n = 1$ if $\tilde{q}_n = 0$ and let $\delta_n = 0$ if $\tilde{q}_n > 0$ so that $\tilde{q}_{n+1} = \tilde{q}_n - 1 + \delta_n + \tilde{v}_{n+1}$. Starting with this equation, find $E[\delta_\infty]$ and $E[\tilde{n}]$. Interpret $E[\delta_\infty]$. [*Hint*: To find $E[\tilde{n}]$, start off by squaring both sides of the equation for $\tilde{q}_{n+1}$.]

5.1.1 Sojourn Time for the M/G/1 System

Recall $s_n \to \tilde{s}$ is the total amount of time spent in system by an arbitrary customer. Thus

$$F_{\tilde{s}}^*(s) = \int_0^\infty e^{-st} dF_{\tilde{s}}(t)$$

is the Laplace-Stieltjes transform of the distribution of the total amount of time that an arbitrary customer spends in the system. From Theorem 5.2, we therefore find that the probability generating function for the number of customers that arrive during the time a customer spends in the system is given by $F_{\tilde{s}}^*(\lambda[1-z])$. But, for a FCFS system, the number of customers that arrive while a customer is in the system is exactly the same as the number of customers left behind by that customer. So

$$\mathcal{F}_{\tilde{q}}(z) = F^*_{\tilde{s}}(\lambda[1 - z]). \tag{5.11}$$

From this we conclude that

$$F^*_{\tilde{s}}(s) = \mathcal{F}_{\tilde{q}}(z)\Big|_{z=1-s/\lambda}. \tag{5.12}$$

Thus, from (5.8), we find that

$$F^*_{\tilde{s}}(s) = \frac{(1-z)(1-\rho)F^*_{\tilde{x}}(\lambda[1-z])}{F^*_{\tilde{x}}(\lambda[1-z]) - z}\Big|_{z=1-s/\lambda}. \tag{5.13}$$

After a little algebra, (5.13) reduces to the Pollaczek-Khintchine transform equation for the sojourn time, which is

$$F^*_{\tilde{s}}(s) = \frac{(1-\rho)sF^*_{\tilde{x}}(s)}{s - \lambda[1 - F^*_{\tilde{x}}(s)]}. \tag{5.14}$$

An alternate presentation of (5.14) which we will find useful later is

$$F^*_{\tilde{s}}(s) = \frac{(1-\rho)F^*_{\tilde{x}}(s)}{1 - \rho\{[1 - F^*_{\tilde{x}}(s)]/sE[\tilde{x}]\}}. \tag{5.15}$$

In principle, $F^*_{\tilde{s}}(s)$ can be inverted to obtain $d/dtP\{\tilde{s} \leq t\}$, which is the density of the sojourn time. This can be done fairly easily if $F^*_{\tilde{x}}(s)$ is rational (that is, it is a ratio of polynomials) by using partial fraction expansions. The above expression can also be differentiated to obtain moments; for example,

$$E[\tilde{s}] = -\frac{d}{ds}F^*_{\tilde{s}}(s)|_{s=0}.$$

EXERCISE 5.8 Using (5.14) and the properties of the Laplace transform, show that

$$\begin{aligned} E[\tilde{s}] &= \frac{\rho}{1-\rho}\frac{E[\tilde{x}^2]}{2E[\tilde{x}]} + E[\tilde{x}] \\ &= \left(\frac{\rho}{1-\rho}\frac{C_{\tilde{x}}^2 + 1}{2} + 1\right) E[\tilde{x}]. \end{aligned} \tag{5.16}$$

Combine this result with that of Exercise 5.6 to verify the validity of Little's result applied to the M/G/1 queueing system. [*Hint*: Use (5.15) rather than (5.14) as a starting point, and use the hint for Exercise 5.6.]

5.1.2 Waiting Time for the M/G/1 System

Recall that $w_n \to \tilde{w}$ refers to the amount of time a customer spends in the queue waiting for service to begin. This means that $\tilde{s} = \tilde{w} + \tilde{x}$. Because the service time for a customer does not depend upon the amount of time the customer waits for service to begin, $\tilde{w}$ and $\tilde{x}$ are independent and consequently, $F^*_{\tilde{s}}(s) = F^*_{\tilde{w}}(s)F^*_{\tilde{x}}(s)$. It therefore follows from (5.14) and (5.15) that

$$\begin{aligned} F^*_{\tilde{w}}(s) &= \frac{(1-\rho)s}{s - \lambda[1 - F^*_{\tilde{x}}(s)]} \\ &= \frac{(1-\rho)}{1 - \rho\{[1 - F^*_{\tilde{x}}(s)]/(sE[\tilde{x}])\}}. \end{aligned} \tag{5.17}$$

EXERCISE 5.9 Using (5.17) and the properties of the Laplace transform, show that

$$E[\tilde{w}] = \frac{\rho}{1-\rho}\frac{C_{\tilde{x}}^2 + 1}{2}E[\tilde{x}]. \tag{5.18}$$

Combine this result with the result of Exercise 5.6 to verify the validity of Little's result when applied to the waiting line for the M/G/1 queueing system.

Platzman, Ammons, and Bartholdi [1988] describe an approximate method for inverting transforms such as those for the waiting and sojourn time. Experience has shown that this method works quite well, especially for the tail of the distribution.

5.1.3 Busy Period for the M/G/1 Queueing System

In this section, we will determine the Laplace-Stieltjes transform for the distribution of the length of the busy period for the M/G/1 queueing system. As before, we let $\tilde{y}$ denote the length of an M/G/1 busy period, and let $F^*_{\tilde{y}}(s)$ denote the Laplace-Stieltjes transform of the distribution of $\tilde{y}$; that is, $F^*_{\tilde{y}}(s) = E[e^{-s\tilde{y}}]$. Further, denote the length of service time for the first customer in the busy period by $\tilde{x}$, and let $\tilde{v}$ denote the number of arrivals during the service time of this customer. Then

$$F^*_{\tilde{y}}(s) = \int_0^\infty E[e^{-s\tilde{y}} \mid \tilde{x} = x]dF_{\tilde{x}}(x). \tag{5.19}$$

Also,

$$E[e^{-s\tilde{y}} \mid \tilde{x} = x] = \sum_{v=0}^{\infty} E[e^{-s\tilde{y}} \mid \tilde{x} = x, \tilde{v} = v]P\{\tilde{v} = v \mid \tilde{x} = x\}.$$

Now, if $\tilde{v}$ customers arrive during x then at the end of x there will be $\tilde{v}$ customers in the system, none of whom have begun service. Since order of service does not affect the length of the busy period, the remainder of the busy period has length $\tilde{y}_0 + \tilde{y}_1 + \tilde{y}_2 + \cdots + \tilde{y}_v$ where $\tilde{y}_j$ denotes the length of the sub-busy period due to the j-th customer who arrived during x and $\tilde{y}_0 = 0$ with probability 1. Thus

$$\begin{aligned} E[e^{-s\tilde{y}} \mid \tilde{x} = x] &= \sum_{v=0}^{\infty} E[e^{-s(x+\sum_{i=0}^{v} \tilde{y}_i)}] \frac{(\lambda x)^v}{v!} e^{-\lambda x} \\ &= \sum_{v=0}^{\infty} e^{-sx} E\left[\prod_{i=0}^{v} e^{-s\tilde{y}_i}\right] \frac{(\lambda x)^v}{v!} e^{-\lambda x}. \end{aligned}$$

But, for $i \geq 1$, the $\tilde{y}_i$'s are independent, identically distributed random variables with common distribution $F_{\tilde{y}}$, so

$$E\left[\prod_{i=0}^{v} e^{-s\tilde{y}_i}\right] = [F_{\tilde{y}}^*(s)]^v,$$

where we have used the fact that $E[e^{-s\tilde{y}_0}] = E[e^{-s0}] = 1$. Thus

$$\begin{aligned} E[e^{-s\tilde{y}} \mid \tilde{x} = x] &= \sum_{v=0}^{\infty} e^{-sx} \frac{[F_{\tilde{y}}^*(s)]^v (\lambda x)^v}{v!} e^{-\lambda x} \\ &= e^{-[s+\lambda-\lambda F_{\tilde{y}}^*(s)]x}. \end{aligned} \tag{5.20}$$

Substitution of (5.20) into (5.19) then yields

$$\begin{aligned} F_{\tilde{y}}^*(s) &= \int_0^{\infty} e^{-[s+\lambda-\lambda F_{\tilde{y}}^*(s)]x} dF_{\tilde{x}}(x) \\ &= F_{\tilde{x}}^*[s + \lambda - \lambda F_{\tilde{y}}^*(s)]. \end{aligned} \tag{5.21}$$

Equation (5.21) is a functional relationship defining $F_{\tilde{y}}^*(s)$. This functional relationship can be used to determine moments for the length of the busy period.

EXERCISE 5.10 Using properties of the Laplace transform, show that

$$E[\tilde{y}] = \frac{E[\tilde{x}]}{1 - \rho}. \tag{5.22}$$

We point out in passing that busy-period analysis is an extremely important tool in the analysis of priority queueing systems. Often delays can be specified entirely in terms of busy periods, as will be shown later on. One variant of the M/G/1 system which is useful in analysis of complicated systems is the M/G/1 with exceptional first service. In this system, the service time for the first customer in each busy period is drawn from the service-time distribution $F_{\tilde{x}_e}(x)$, and the remaining service times in each busy period are drawn from the general distribution $F_{\tilde{x}}(x)$. The length of the busy period for this system is denoted by $\tilde{y}_e$. It is left as an exercise to show that

$$E[\tilde{y}_e] = \frac{E[\tilde{x}_e]}{1-\rho}, \tag{5.23}$$

where we retain the definition $\rho = \lambda E[\tilde{x}]$. Thus the expected length of the busy period is proportional to the expected length of the first service time. In addition, it is easy to see that if $F_{\tilde{x}_e}(x) = F_{\tilde{x}}(x)$, then (5.23) reduces to (5.22).

EXERCISE 5.11 For the ordinary M/G/1 queueing system determine $E[\tilde{y}]$ without first solving for $F_{\tilde{y}}(s)$. [*Hint*: Condition on the length of the first customer's service and the number of customers that arrive during that period of time.]

EXERCISE 5.12 **M/G/1 with Exceptional First Service.** A queueing system has Poisson arrivals with rate λ. The service time for the first customer in each busy period is drawn from the service-time distribution $F_{\tilde{x}_e}(x)$, and the remaining service times in each busy period are drawn from the general distribution $F_{\tilde{x}}(x)$. Let $\tilde{y}_e$ denote the length of the busy period for this system. Show that

$$E[\tilde{y}_e] = \frac{E[\tilde{x}_e]}{1-\rho},$$

where $\rho = \lambda E[\tilde{x}]$.

EXERCISE 5.13 For the M/G/1 queueing system with exceptional first service as defined in the previous exercise, show that $F^*_{\tilde{y}_e}(s) = F^*_{\tilde{x}_e}(s + \lambda - \lambda F^*_{\tilde{y}}(s))$.

EXERCISE 5.14 Comparison of the formulas for the expected waiting time for the M/G/1 system and the expected length of a busy period for the M/G/1 system with the formula for exceptional first service reveals that they both have the same form; that is, the expected waiting time in an ordinary M/G/1 system is the same as the length of the busy period of an M/G/1 system in which the expected length of the first service is given by

$$E[\tilde{x}_e] = \rho\frac{E[\tilde{x}^2]}{2E[\tilde{x}]}.$$

Explain why these formulas have this relationship. What random variable must $\tilde{x}_e$ represent in this form? [*Hint*: Consider the operation of the M/G/1 queueing system under a nonpreemptive, LCFS, service discipline and apply Little's result, taking into account that an arriving customer may find the system empty.]

5.2 Ergodic Occupancy Distribution for M/G/1

We have previously pointed out that the ergodic occupancy distribution can be calculated by using (5.8) and the properties the probability generating functions. Because this process requires differentiation of fractions and numerous applications of L'Hôpital's rule, the process of generating the occupancy distribution in this manner is tedious at best. In this section, we present three alternative methods of computing the occupancy distribution. The first method is based on discrete Fourier transform analysis, and the second approach, due to Keilson and Servi [1989], is based upon a recursion. The third approach is based on generalized state-space methods, which are commonly applied in the control theory area (Akar, Oğuz, and K. Sohraby [1998]).

5.2.1 Discrete Fourier Transform Approach to Ergodic Occupancy Computation

In this subsection, we describe an approach to inversion of the probability generating function for the occupancy distribution which is based upon Fourier analysis (Daigle [1989]). We first show that a tracing of the PGF around the unit circle describes the *characteristic function* (Feller [1971]) for the occupancy distribution in the form of a complex Fourier series in which the coefficients are the probability masses. Approximate values for these Fourier coefficients are then expressed via finite Riemann sums, where the resulting approximations are a finite set of $K+1$ discrete Fourier transform (DFT) coefficients. It turns out that, if K is properly chosen, then the $K+1$ DFT coefficients can be used to obtain approximations for all of the probability masses. A finite number of coefficients is sufficient because the tail of the occupancy distribution decreases geometrically; this fact can be used both to convert the DFT coefficients into probability estimates and to generate the tail probabilities. We also present an algorithm for choosing an appropriate value of K.

Briefly stated, our objective in this section is as follows. Starting with the definition of the PGF,

$$\mathcal{F}_{\tilde{n}}(z) = E[z^{\tilde{n}}] = \sum_{n=0}^{\infty} p_n z^n, \tag{5.24}$$

we will develop a simple computational technique based upon DFTs to compute

$$p_n = P\{\tilde{n} = n\} \qquad \text{for} \qquad n \geq 0$$

from $\mathcal{F}_{\tilde{n}}(z)$. The methodology developed here is useful for inversion of PGFs for a large class of distributions, namely, those having geometrically decreasing tails.

To begin our development, we define

$$\Psi_{\tilde{n}}(\alpha) = \mathcal{F}_{\tilde{n}}\left(e^{-j2\pi\alpha}\right) = \sum_{n=0}^{\infty} p_n e^{-j2\pi n\alpha}, \tag{5.25}$$

where $j = \sqrt{-1}$ and α is a real variable. That is, $\Psi_{\tilde{n}}(\alpha)$ is the *Fourier-Stieltjes integral* (Feller [1971]), or *characteristic function*, of the cumulative distribution function $F_{\tilde{n}}$. We note that $\Psi_{\tilde{n}}(\alpha)$ is always periodic in α with period 1, and that $\Psi_{\tilde{n}}(\alpha)$ is expressed in (5.25) as a complex Fourier series (Churchill and Brown [1987]), that is, a Fourier series in which the basis set (Hewitt and Stromberg [1969]) is $\{\phi_n(\alpha), n = 0, \pm 1, \pm 2, \ldots\}$, where $\phi_n(\alpha) = e^{-j2\pi n\alpha}$. Indeed, the Fourier coefficients are simply the probability masses.

As usual, the Fourier coefficients are given by the integral, averaged over one period, of the product of the function in question and the complex conjugate of the basis function for the coefficient in question; that is,

$$p_n = \int_0^1 \Psi_{\tilde{n}}(\alpha)\phi_n^*(\alpha)d\alpha, \tag{5.26}$$

where $\phi_n^*(\alpha)$ denotes the complex conjugate of $\phi_n(\alpha)$. To perform the integration indicated in (5.26) numerically, we partition the interval [0,1] into $K+1$ equal subintervals. We denote the approximate value of the integral thus obtained by $c_{n,K}$, and we find that

$$\begin{aligned} c_{n,K} &= \sum_{k=0}^{K} \Psi_{\tilde{n}}(\alpha_k)\phi_n^*(\alpha_k)\Delta\alpha \qquad (5.27) \\ &= \frac{1}{K+1}\sum_{k=0}^{K} \mathcal{F}_{\tilde{n}}\left(e^{-j[(2\pi k)/(K+1)]}\right) e^{j[(2\pi nk)/(K+1)]}, \end{aligned}$$

where $\alpha_k = k/(K+1)$ and $\Delta\alpha = 1/(K+1)$. The right-hand side of (5.27) is the *inverse discrete Fourier transform* (IDFT) (Nussbaumer [1982]) of the finite sequence $\{\mathcal{F}_{\tilde{n}}(e^{(j2\pi k)/(K+1)}), 0 \le k \le K\}$. It is easily verified by substituting $n + m(K+1)$ for n that the resulting $c_{n,K}$ sequence is periodic with period $K+1$. However, we will think of the $c_{n,K}$ as being defined for $0 \le n \le K$ only.

The exact relationship between the probability masses, p_n and $c_{n,K}$ for $0 \le n \le K$, can be obtained by substituting the definition of $\Psi_{\tilde{n}}(\alpha)$ into (5.27) and performing the indicated summation. We find that

$$c_{n,K} = \frac{1}{K+1}\sum_{k=0}^{K}\sum_{\ell=0}^{\infty} p_\ell e^{-j[(2\pi k\ell)/(K+1)]}\, e^{j[(2\pi nk)/(K+1)]}$$

$$
\begin{aligned}
&= \frac{1}{K+1}\sum_{k=0}^{K}\sum_{\ell=0}^{\infty} p_\ell e^{-j\{[2\pi k(\ell-n)]/(K+1)\}} \qquad (5.28)\\
&= \frac{1}{K+1}\sum_{\ell=0}^{\infty} p_\ell \sum_{k=0}^{K} e^{-j\{[2\pi k(\ell-n)]/(K+1)\}}.
\end{aligned}
$$

The final summation on the right-hand side of (5.28) is the sum of a finite number of terms from a geometric series. It is easily shown that the sum is $K+1$ if $\ell = (K+1)m + n$ and zero otherwise. Therefore,

$$
c_{n,K} = p_n + \sum_{m=1}^{\infty} p_{m(K+1)+n} \qquad \text{for} \qquad 0 \le n \le K. \tag{5.29}
$$

Clearly, $\sum_{n=0}^{K} c_{n,K} = \sum_{n=0}^{\infty} p_n$. Thus the $c_{n,K}$ sum to unity for each K. As an example, if we choose $K = 255$, then

$$
\begin{aligned}
c_{0,255} &= p_0 + p_{256} + p_{512} + p_{768} + \cdots \\
c_{1,255} &= p_1 + p_{257} + p_{513} + p_{769} + \cdots \\
&\vdots \\
c_{255,255} &= p_{255} + p_{511} + p_{767} + p_{1023} + \cdots .
\end{aligned}
$$

EXERCISE 5.15 Show that the final summation on the right-hand side of (5.28) is $K+1$ if $\ell = (K+1)m + n$ and zero otherwise.

The lack of equality between $c_{n,K}$ and p_n in this particular form is called *aliasing*, and it is a direct result of approximating the integral (5.25) by a finite sum. It will be pointed out later in this subsection that tail probabilities in queueing systems decrease geometrically. Thus it is clear by observation of (5.29) that the error due to aliasing can be reduced to any desired degree by increasing K. But this practice results in round-off error, and, in addition, still does not yield tail probabilities.

We now develop an approach that takes advantage of the fact that the tail probabilities decay geometrically. This approach simultaneously addresses aliasing error and round-off error, and, in addition, provides values for p_n for $n \ge K+1$.

From (5.8), we find

$$
\mathcal{F}_{\tilde{n}}(z) = \frac{(1-\rho)(z-1)}{z - F^*_{\tilde{x}}(\lambda[1-z])} F^*_{\tilde{x}}(\lambda[1-z]). \tag{5.30}
$$

It can be shown that the denominator of the right-hand side of (5.30) has only one zero for $z > 1$; call this zero z_0. Then, the geometric rate at which the tail probabilities decrease is given by the inverse of z_0. This fact follows directly from the Laurent series expansion (Churchill [1960]) of $\mathcal{F}_{\tilde{n}}(z)$ about its singularity at z_0. Indeed, the *principle part* of $\mathcal{F}_{\tilde{n}}(z)$ is given by the quantity $b_1/(z-z_0)$ where

$$\begin{aligned} b_1 &= \lim_{z\to z_0}(z-z_0)\mathcal{F}_{\tilde{n}}(z) \\ &= \left[\frac{(1-\rho)(z-1)F_{\tilde{x}}^*(\lambda[1-z])}{1-d/dzF_{\tilde{x}}^*(\lambda[1-z])}\right]_{z=z_0}. \end{aligned} \tag{5.31}$$

Thus the principle part of $\mathcal{F}_{\tilde{n}}(z)$ is given in power series form by the expression

$$-b_1\sum_{n=0}^{\infty} {z_0}^{-(n+1)} z^n.$$

It turns out that for large n, the values of the coefficients of z^n in the Laurent series expansion of $\mathcal{F}_{\tilde{n}}(z)$ are dominated by the principle part. Thus the principle part can be used to obtain very close estimates of the tail probabilities as in Woodside and Ho [1987]. Two obvious disadvantages of this method of computing tail probabilities are that the singularity z_0 must usually be found numerically, and the derivative of $F_{\tilde{x}}^*(\lambda[1-z])$ must be evaluated at z_0. Sometimes neither of these operations is straightforward, but so long as it is possible to evaluate the denominator of (5.30), it is easy to bound the difference between z_0^{-1} and the actual decay rate, r_0, as determined by the algorithm defined below.

Remark. Discussions of Laurent series expansions are sometimes quite difficult to follow. The Laurent series in this particular case may be related to the Taylor series in the following way. First, define a function $\xi(z)$ such that $(z-z_0)\mathcal{F}_{\tilde{n}}(z) = \xi(z)$. Then $\xi(z)$ has no singularities and has a Taylor series. That is,

$$\xi(z) = \sum_{i=0}^{\infty} \frac{\xi^{(i)}(z_0)}{i!}(z-z_0)^i,$$

where $\xi^{(i)}(z_0)$ denotes the ith derivative of $\xi(z)$ evaluated at z_0. Now divide the Taylor series of $\xi(z)$, term by term, by the quantity $z-z_0$; the result is the Laurent series for $\mathcal{F}_{\tilde{n}}(z)$. More on Laurent series can be found in Churchill [1960].

| EXERCISE 5.16 Argue the validity of the expression for b_1 in (5.31).

EXERCISE 5.17 Show that the denominator of the right-hand side of (5.30) for the probability generating of the occupancy distribution has only one zero for $z > 1$. [*Hint*: From Theorem 5.2, we know that $F_{\tilde{x}}^*(\lambda[1-z])$ is the probability generating function for the number of arrivals during a service time. Therefore, $F_{\tilde{x}}^*(\lambda[1-z])$ can be expressed as a power series in which the coefficients are probabilities and therefore nonnegative. The function and all of its derivatives are therefore nonnegative for nonnegative z, and so on. Now compare the functions $f_1(z) = z$ and $f_2(z) = F_{\tilde{x}}^*(\lambda[1-z])$, noting that the expression $\mathcal{F}_{\tilde{n}}(z)$ can have poles neither inside nor on the unit circle (Why?).]

Let n_g denote the occupancy level at and above which the tail probabilities are geometrically decreasing to within computational accuracy, and let r_0 denote the geometric rate of decay; that is, $r_0 = z_0^{-1}$. Then, with $K \geq n_g$, for each n, $0 \leq n \leq K$, the sequence $\{p_{m(K+1)+n}, m \geq 1\}$, is geometrically decreasing at rate ${r_0}^{(K+1)}$. Thus, in general,

$$p_{i+n} \approx p_i {r_0}^n \quad \text{for} \quad i \geq n_g, \tag{5.32}$$

and, in particular, with $i = K + 1$,

$$p_{K+1+n} \approx p_K {r_0}^{n+1}. \tag{5.33}$$

By using (5.32) in (5.29), one can find after a moderate amount of algebraic manipulation that

$$r_0 \approx \frac{c_{0,K} - p_0}{c_{K,K}}, \tag{5.34}$$

and

$$p_n \approx c_{n,K} - (c_{0,K} - p_0){r_0}^n \qquad \text{for} \qquad 1 \leq n \leq K. \tag{5.35}$$

Finally, from (5.24), we have that

$$p_0 = \mathcal{F}_{\tilde{n}}(0). \tag{5.36}$$

Thus, to the extent that K has been chosen sufficiently large so that the tail probabilities are actually decaying geometrically, we see that (5.33) through (5.36) can be used to obtain approximations for p_n, $n \geq 0$ in which aliasing has been removed, and only round-off error affects the results. Having selected a value of K, we first compute the $c_{n,K}$, $0 \leq n \leq K$ using either (5.27) or, more likely, a fast Fourier transform (FFT) algorithm (Nussbaumer [1982]), then compute p_0 using (5.36). Next compute r_0 using (5.34). Finally, compute the probability masses using (5.35) for $1 \leq n \leq K$ and (5.33) for $n > K$.

EXERCISE 5.18 Starting with (5.29) and (5.32), establish the validity of (5.33) through (5.36).

We now turn our attention to specifying a method of choosing K. There are two conflicting objectives. On the one hand, the larger the value of K, the more likely it is that the probability masses are geometrically decreasing at least starting with p_K, thus satisfying the assumption leading to (5.35). On the other hand, the larger the selected value of K, the larger the round-off error will be in the computations leading to the specifications of the $c_{n,K}$. Since the FFT yields round-off errors of equal magnitude for all coefficients, the most serious affect of this round-off error will be upon the accuracy of $c_{K,K}$. This will obviously lead to inaccuracies in the computation of r_0, which is the key to the generation of tail probabilities. Thus the appropriate choice of K is the minimum for which the tail probabilities are geometrically decreasing.

The value for K should be chosen algorithmically so that one may passively use the computational technique. Towards this end, we note that under the assumption that the quantity n_g is large enough so that the tail probabilities are decreasing, it follows that

$$c_{n,K} = \frac{p_n}{1 - {r_0}^{K+1}} \quad \text{for} \quad n_g \leq n \leq K. \tag{5.37}$$

That is, in our choice of K, we insist that K be sufficiently large to assure that the tail probabilities are geometrically decreasing beginning at n_g, which is in turn less than K. If so, then we also have that $p_{n+1} = r_0 p_n$ for all $n \geq n_g$. Thus, from (5.35), it is readily seen that $c_{n,K}/c_{n-1,K} = r_0$ for $n_g < n \leq K$. We therefore have two ways of computing r_0: first, using (5.35) and second, taking ratios of successive coefficients. An algorithm for choosing K based on these observations is as follows:

1. For a candidate K, let $n_g = K - \lfloor K/4 \rfloor$.
2. Compute $c_{n,K}, 0 \leq n \leq K$ using (5.27) and r_0 using (5.34).
3. Compute $r_{n,K} = c_{n,K}/c_{n-1,K}$ for $n_g < n \leq K$.
4. Compute $a_K = \max_{n_g < n \leq K} |(r_0 - r_{n,K})/r_0|$.
5. Let $K' = 2(K+1) - 1$.
6. Compute $a_{K'}$ as in (a), (b), (c), and (d).
7. If $a_{K'} < a_K$, then replace K' by K and repeat (e) and (f), else use the computations based on the current value of K in the final results.

The values a_K are a measure of the maximum deviation of the calculated ratios for the last one-fourth of the coefficients from the computed value of the ratio r_0 based on (5.35). Thus, a_K may be viewed as a measure of the accuracy of the assumption that K is large enough to assure geometrically decreasing tail probabilities.

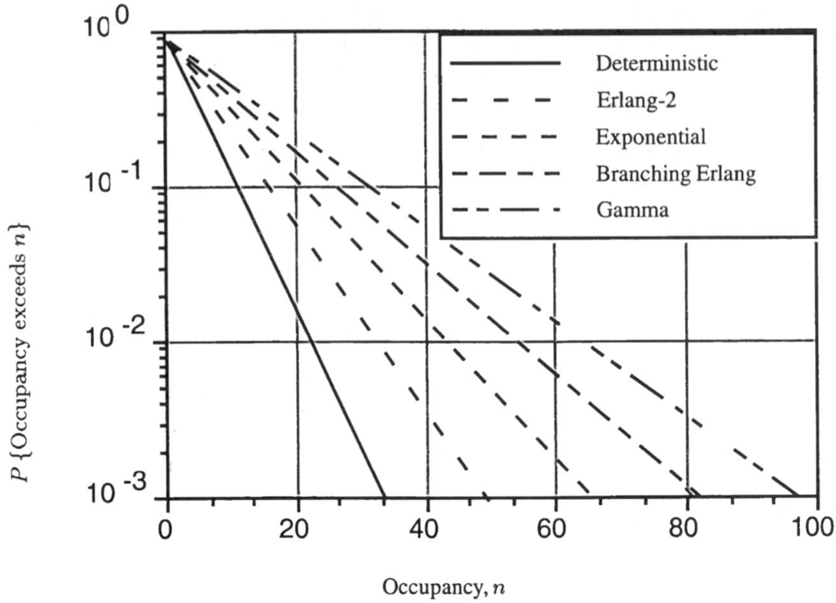

Figure 5.1. Survivor functions with deterministic, Erlang-2, exponential, branching Erlang and gamma service-time distributions at $\rho = 0.9$.

We note that the above development was discussed in terms of the M/G/1 queueing system, but that the techniques are applicable to the inversion of any PGF for which the tail probabilities are geometrically decreasing. Queueing systems having these properties include the G/G/c system as reported in Tijms [1986] and many priority queueing systems which we will discuss later. We also note that once the ergodic occupancy distribution is known for the case of infinite capacity, it is straightforward to obtain the ergodic occupancy distribution for the case of finite waiting room by methods outlined in Cooper [1972, 1981] and Keilson and Servi [1989].

The above technique is more fully described and evaluated with respect to computation of M/G/1 occupancy distributions in Daigle [1989], which shows that satisfactory results can be obtained with very few coefficients, especially if traffic intensity is high. An example of the type of results that might be obtained using the techniques is given below.

DEFINITION 5.2 **Erlang-k distribution.** The distribution of $\tilde{x}$ is said to be *Erlang-k* with mean $1/\mu$ if $\tilde{x}$ is the sum of k independent exponentially distributed random variables each of which has mean $1/(k\mu)$.

EXAMPLE 5.1 Compare the survivor functions for the occupancy distribution of the M/G/1 queueing system for the following service-time distributions:

- M/D/1, the ordinary M/G/1 queueing system having deterministic, unit, service time. The squared coefficient of variation of the service-time distribution is 0, and the Laplace-Stieltjes transform of the service-time distribution is $F^*_{\tilde{x}}(s) = e^{-s}$.

- M/E$_2$/1, the ordinary M/G/1 queueing system having Erlang-2, unit mean, service times. The squared coefficient of variation of the service-time distribution is 0.5, and the Laplace-Stieltjes transform (LST) of the service-time distribution is $F^*_{\tilde{x}}(s) = [2/(s+2)]^2$.

- M/M/1, the ordinary M/G/1 queueing system having exponential, unit mean, service time. The squared coefficient of variation of the service-time distribution is 1.0, and the LST of the service-time distribution is $F^*_{\tilde{x}}(s) = 1/(s+1)$.

- M/B$_{1.5}$/1, the ordinary M/G/1 queueing system having a two-phase, unit mean, branching Erlang, service-time distribution. The particular branching Erlang distribution used here is the one given in Chandy and Sauer [1981] for specifying distributions whose squared coefficient of variation is greater that 1. The LST for that distribution is

$$F^*_{\tilde{x}}(s) = \frac{\mu_1}{s+\mu_1} \frac{vs+\mu_2}{s+\mu_2}$$

  where

$$\mu_1 = 1 + \sqrt{1 - \frac{2}{1+C_{\tilde{x}}{}^2}},$$

$$\mu_2 = 1 - \sqrt{1 - \frac{2}{1+C_{\tilde{x}}{}^2}},$$

$$v = C_{\tilde{x}}{}^2 \mu_2,$$

  and $C_{\tilde{x}}{}^2$ is the squared coefficient of variation of the distribution of $\tilde{x}$. In our example, the squared coefficient of variation was chosen to be 1.5.

- M/G$_2$/1, the ordinary M/G/1 queueing system having service times drawn from a gamma distribution. The LST for the gamma distribution is $F^*_{\tilde{x}}(s) = (1+\beta s)^{-\alpha}$ (Hogg and Craig [1978]). In this case, $E[\tilde{x}] = \alpha\beta$ and $Var(\tilde{x}) = \alpha\beta^2$, so that the parameter values for which the service-time distribution has unit mean and a squared coefficient of variation of 2, are $\beta = 2$ and $\alpha = 0.5$.

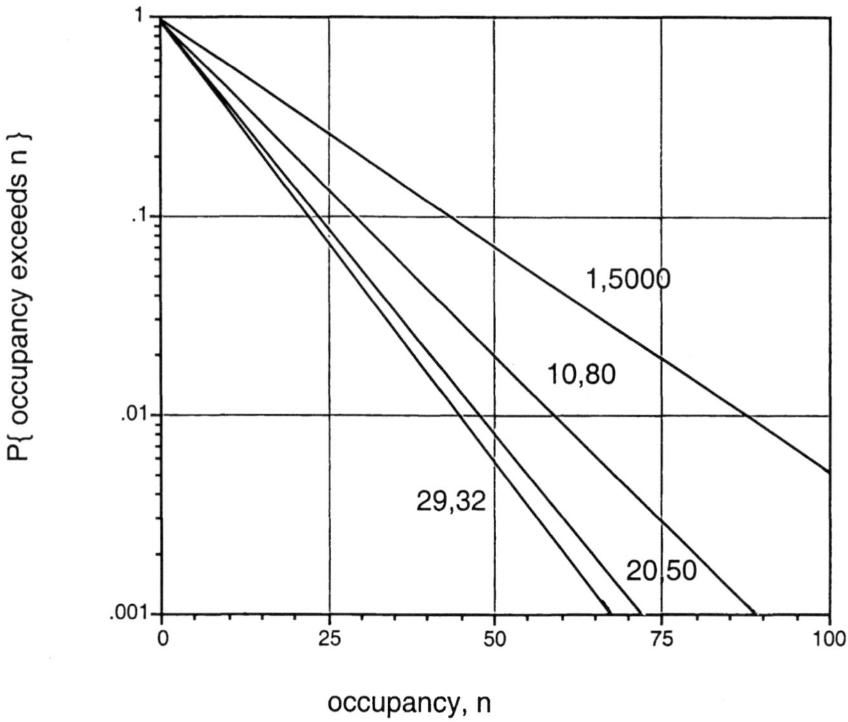

Figure 5.2. Survivor functions for system occupancy with message lengths drawn from truncated geometric distributions at $\rho = 0.95$.

Solution: The results are shown in Figure 5.1, which shows the occupancy probabilities are greatly affected by the form of the service-time distribution. For example, the probability that the queue length exceeds 60 is about 10 times as large for the case in which service times are drawn from the gamma distribution than it is for the case in which service times are drawn from the exponential distribution.

EXAMPLE 5.2 **Message Lengths having Truncated Geometric Distribution.** In analyzing communication systems, it is common to represent the distribution of message lengths as geometric. For example, the number of characters in a typed line may be represented as having a geometric distribution with mean 30. It is clear that the distribution cannot actually be geometric because the number of characters in a typed line on an 80-character screen cannot exceed 80 characters. Consider a communication system in which messages are transmitted over a communication line having a capacity of 2400 bits/sec, or equivalently, 300 characters per second. Suppose the message lengths are drawn from a geometric distribution having a mean of 30 characters, but truncated at a and b characters on the lower and upper ends of the distribution, respectively. That is, message lengths are drawn from a distribution characterized as follows:

$$P\{\tilde{m} = m\} = k\theta(1-\theta)^{m-1} \quad \text{for} \quad a \leq m \leq b,$$

where $\tilde{c}$ is the number of characters in a message and k is a normalizing constant. We wish to determine the survivor function for several different values of a and b at a traffic utilization of 95%, assuming a transmission capacity of 30 characters/sec.

Solution: Curves showing the desired results and additional results are presented in Figure 5.2. The pair of numbers shown beside each curve gives the points at which the geometric distributions is truncated. Note that the survivor function is an increasing function of the spread between the lower and upper truncation points. For example, if $a = 29$ and $b = 32$, then the survivor function is very nearly that of the M/D/1 system. On the other hand, if $a = 1$ and $b = 5000$, then the service-time distribution is very nearly geometric, and we expect the survivor function to approach that of the M/M/1 system.

EXERCISE 5.19 Approximate the distribution of the service time for the previous example by an exponential distribution with an appropriate mean. Plot the survivor function for the corresponding M/M/1 system at 95% utilization. Compare the result to those shown in Figure 5.2.

An important consideration in the design of communication systems is the blocking probability. As stated in Chapter 3, the blocking probability is defined as the proportion of the customers seeking admission to the queueing system who are denied admission. We assume a finite waiting room of capacity K. As in the case of the general birth-death model covered in Chapter 3, ergodicity exists for the finite K model even when it does not exist for the case of infinite waiting room. For the special case in which ergodicity exists for the unbounded waiting room case, Keilson and Servi [1989] present a simple relationship between the ergodic occupancy probabilities and the blocking probabilities for the finite case. The following is their result, which we state without proof:

$$P_B(K) = \frac{1-\rho}{1-\rho S_{K-1}} S_{K-1}, \tag{5.38}$$

where ρ has the usual definition and $S_{K-1} = P\{\tilde{n} > K-1\}$. Table 5.1 illustrates the drastic difference between $P\{\tilde{n} > K-1\} = P\{\tilde{n} \geq K\}$ and the blocking probability at $\rho = 0.9$. Again, the reader is cautioned against using survivor function and blocking probabilities interchangeably.

Table 5.1. Blocking probabilities versus occupancy probabilities for various service time distributions.

System	$K = 10$		$K = 30$	
	S_{K-1}	$P_B(K)$	S_{K-1}	$P_B(K)$
M/D/1	0.17795	0.021188	0.002825	0.000283
M/E_2/1	0.29806	0.040733	0.018246	0.001855
M/M/1	0.38742	0.059482	0.047101	0.004919
M/$B_{1.5}$/1	0.45117	0.075961	0.085055	0.009211
M/G_2/1	0.50514	0.092622	0.145434	0.016734

In case ergodicity does not exist for the unbounded waiting room case (that is, $\rho \geq 1$), the blocking probability is more difficult to compute. The interested reader is referred to Langford [1990] and also Niu and Cooper [1993] for a treatment of the more general case.

5.2.2 Recursive Approach to Ergodic Occupancy Computation

In this section, we present an alternate method of computing the occupancy distribution that is based on the paper by Keilson and Servi [1989], although our approach is somewhat different and the results are stated in a slightly different form.

To begin our development, we rewrite (5.14) as follows:

$$F_{\tilde{s}}^*(s) = \frac{(1-\rho)s}{s - \lambda[1 - F_{\tilde{x}}^*(s)]} F_{\tilde{x}}^*(s).$$

We will show in the next section that the expression

$$\frac{1 - F_{\tilde{x}}^*(s)}{s\, E[\tilde{x}]}$$

is the Laplace-Stieltjes transform of the distribution of a random variable, which we will denote by $\tilde{x}_r$ and refer to as the residual life of $\tilde{x}$. Therefore, we define

$$F_{\tilde{x}_r}^*(s) = \frac{1 - F_{\tilde{x}}^*(s)}{s\, E[\tilde{x}]}. \tag{5.39}$$

From (5.15) and (5.39), we find that the Laplace-Stieltjes transform of the sojourn time distribution can be written as

$$F_{\tilde{s}}^*(s) = \frac{(1-\rho)F_{\tilde{x}}^*(s)}{1-\rho F_{\tilde{x}_r}^*(s)}. \tag{5.40}$$

In turn, Equation (5.40) can be rewritten as

$$F_{\tilde{s}}^*(s) = (1-\rho)F_{\tilde{x}}^*(s)\sum_{i=0}^{\infty}\left[\rho F_{\tilde{x}_r}^*(s)\right]^i. \tag{5.41}$$

After some minor algebra, this equation can be rearranged as

$$F_{\tilde{s}}^*(s) = (1-\rho)F_{\tilde{x}}^*(s) + \rho F_{\tilde{x}_r}^*(s)F_{\tilde{s}}^*(s). \tag{5.42}$$

Using (5.42) as a starting point, we will specify a convenient method of generating the occupancy distribution for the M/G/1 system.

We have previously argued that $\mathcal{F}_{\tilde{n}}(z) = F_{\tilde{s}}^*(\lambda[1-z])$; that is, the probability generating function for the occupancy distribution is given by the Laplace-Stieltjes transform for the sojourn-time distribution evaluated at the point $\lambda(1-z)$. For consistency of notation, we will specify the probability generating function for the number of arrivals that occur from a Poisson process with rate λ during a random period of time $\tilde{x}$ by

$$F_{\tilde{x}}^*(\lambda[1-z]) = \sum_{i=0}^{\infty} z^i P_{\tilde{x},i}. \tag{5.43}$$

Then, from (5.42), we find

$$\sum_{i=0}^{\infty} z^i P_{\tilde{s},i} = (1-\rho)\sum_{i=0}^{\infty} z^i P_{\tilde{x},i} + \rho\left[\sum_{i=0}^{\infty} z^i P_{\tilde{x}_r,i}\right]\left[\sum_{i=0}^{\infty} z^i P_{\tilde{s},i}\right],$$

or equivalently,

$$\sum_{i=0}^{\infty} z^i P_{\tilde{s},i} = (1-\rho)\sum_{i=0}^{\infty} z^i P_{\tilde{x},i} + \rho\sum_{i=0}^{\infty} z^i\left[\sum_{n=0}^{i} P_{\tilde{x}_r,n}P_{\tilde{s},i-n}\right]. \tag{5.44}$$

Upon matching the coefficients of z^i in (5.44), we find that

$$P_{\tilde{s},i} = (1-\rho)P_{\tilde{x},i} + \rho\sum_{n=0}^{i} P_{\tilde{x}_r,n}P_{\tilde{s},i-n}. \tag{5.45}$$

Equation (5.45) may then be solved for $P_{\tilde{s},i}$. We find

$$P_{\tilde{s},i} = \frac{(1-\rho)P_{\tilde{x},i} + \rho\sum_{n=1}^{i} P_{\tilde{x}_r,n}P_{\tilde{s},i-n}}{1-\rho P_{\tilde{x}_r,0}}. \tag{5.46}$$

Now, it is straightforward to show that

$$P_{\tilde{x}_r,i} = \frac{1}{\rho}\left[1 - \sum_{n=0}^{i} P_{\tilde{x},n}\right]. \tag{5.47}$$

Substituting (5.47) into (5.46), we find

$$P_{\tilde{s},0} = 1 - \rho, \tag{5.48}$$

and for $i \geq 1$,

$$P_{\tilde{s},i} = \frac{(1-\rho)P_{\tilde{x},i} + \sum_{n=1}^{i}\left(1 - \sum_{m=0}^{n} P_{\tilde{x},n}\right) P_{\tilde{s},i-n}}{P_{\tilde{x},0}}. \tag{5.49}$$

EXERCISE 5.20 Starting with (5.39), demonstrate the validity of (5.47).

Clearly, $P_{\tilde{s},i}$ can be determined recursively from (5.49). The number of terms of the form $P_{\tilde{x},i}$ that must be computed is limited to the number of occupancy probabilities that the analyst is interested in computing for the particular problem at hand. As we have noted in Section 5.1, $P_{\tilde{s},i}$ is equivalent to $P_i = P\{\tilde{n} = i\}$. Thus we find

$$P_i = \frac{(1-\rho)P_{\tilde{x},i} + \sum_{n=1}^{i}\left(1 - \sum_{m=0}^{n} P_{\tilde{x},n}\right) P_{i-n}}{P_{\tilde{x},0}}. \tag{5.50}$$

In addition, as we have seen earlier in this subsection, the ratio P_i/P_{i-1} converges to a constant as i increases for all $F_{\tilde{x}}(x)$ of practical interest. Thus, (5.50) offers a practical method for calculating occupancy probabilities so long as the $P_{\tilde{x},i}$ can be computed readily. In general, the computation of these quantities is straightforward if the Laplace-Stieltjes transform for the service-time distribution is rational or if the service-time distribution can be adequately approximated by a discrete distribution. On the other hand, for more general distributions, this task, in and of itself, is more difficult to accomplish than is the direct computation of the occupancy distribution using the methods presented in the previous section.

EXERCISE 5.21 Evaluate $P_{\tilde{x},i}$ for the special case in which $\tilde{x}$ has the exponential distribution with mean $1/\mu$. Starting with (5.50), show that the ergodic occupancy distribution for the M/M/1 system is given by $P_i = (1-\rho)\rho^i$, where $\rho = \lambda/\mu$.

EXERCISE 5.22 Evaluate $P_{\tilde{x},i}$ for the special case in which $P\{\tilde{x} = 1\} = 1$. Use (5.50) to calculate the occupancy distribution. Compare the complementary occupancy distribution ($P\{N > i\}$) for this system to that of the M/M/1 system with $\mu = 1$.

EXERCISE 5.23 Evaluate $P_{\tilde{x},i}$ for the special case in which $P\{\tilde{x} = \frac{1}{2}\} = P\{\tilde{x} = \frac{3}{2}\} = \frac{1}{2}$. Use (5.50) to calculate the occupancy distribution. Compare the complementary occupancy distribution ($P\{N > i\}$) for this system to that of the M/M/1 system with $\mu = 1$.

5.2.3 Ergodic Occupancy Distributions via Generalized State-Space Approach

As discussed in Section 5.1, the process $\{\tilde{q}_n, n \geq 1\}$, which denotes the number of customers left in the system by the nth departing customer, is a Markov chain embedded at points of customer departure. Our objective is to present a linear algebra-based approach for obtaining the distribution of the number of customers left in the system by a departing customer for a variation of the M/G/1 system. We specifically consider the cases in which the probability generating function of the number of arrivals that occur during service periods can be expressed either as a finite polynomial or a ratio of polynomials. Such cases include, but are not limited to, the M/G/1 system whose service time distribution has a rational Laplace-Stieltjes transform.

Although we take a somewhat different approach to the formulation of the various problems in this class, our solution takes advantage of the ideas presented in Akar, Oğuz, and Sohraby [1998], where more complex systems, such as those discussed in Chapter 7, are considered. Specifically, the problem formulation results in a simple system of finite matrix equations. This system has both stable and unstable modes, and Schur decomposition is used as a starting point for decoupling the system into stable and unstable subsystems. The solutions are then expressed in a specialized matrix geometric form, similar to that for the QDB systems discussed in Chapter 3.

We consider both the cases of simple and multiple boundary conditions. Although the development itself is somewhat tedious, we show that in the end, all of the problems in this class can be solved in a uniform way. Furthermore, all of the steps involved in formulating and solving the problem are easy to implement on a computer using a combination of routines from LAPACK and simple program development.

We first address the case in which the probability generating function of the number of arrivals that occur during service periods is a finite polynomial. We consider the broader case in which the first service is exceptional. That is, the first service of the busy period has distribution $F_{\tilde{x}_e}(x)$ and all remaining service times have distribution $F_{\tilde{x}}(x)$. The finite equations characterizing the system are developed directly from the equations for the equilibrium departure probabilities, namely, $\pi = \pi\mathcal{P}$, $\pi\mathrm{e} = 1$. The solution procedure described in Akar, Oguz, and Sohraby [1998] is then described and used to solve the problem.

Next, we consider the case in which the probability generating functions of the number of arrivals that occur during exceptional and ordinary service periods are a rational functions. In this case, the equations that must be solved are identical in form to those for the more elementary case, but the equations are specified from the probability generating function of the occupancy distribution as seen by departing customers. The finite polynomial case is then a special case wherein the denominator is identically 1. In fact, when the finite case is posed in terms of the rational case, the form of the solution is identical to that of the less general case.

Finally, the more general case in which there are multiple boundaries and the probability generating functions of the number of arrivals that occur during service periods are either finite polynomials or ratios of polynomials is considered. In particular, there are C boundary probabilities, $\pi_0, \pi_1, \ldots, \pi_{C-1}$. Again, in this case, the equations that must be solved and the form of the solution are identical in form to those for the more elementary cases.

At the end of the section, we discuss the practical issue of implementing the solution procedures on a computer.

Recall that the recursive equation for the queue upon the departure of the nth customer is as follows:

$$\tilde{q}_{n+1} = (\tilde{q}_n - 1)^+ + \tilde{v}_{n+1}. \tag{5.51}$$

Therefore,

$$P\{\tilde{q}_{n+1} = j\} = P\{(\tilde{q}_n - 1)^+ + \tilde{v}_{n+1} = j\}. \tag{5.52}$$

Upon conditioning on the value of $\tilde{q}_n$, we find

$$P\{\tilde{q}_{n+1} = j\} = \sum_{i=0}^{\infty} P\{(\tilde{q}_n - 1)^+ + \tilde{v}_{n+1} = j | \tilde{q}_n = i\} P\{\tilde{q}_n = i\}. \tag{5.53}$$

But, $P\{(\tilde{q}_n - 1)^+ + \tilde{v}_{n+1} = j | \tilde{q}_n = i\}$ is just the probability that there are $j - (i-1)^+$ arrivals during the service time of the $(n+1)$st customer; that is,

$$P\{(\tilde{q}_n - 1)^+ + \tilde{v}_{n+1} = j | \tilde{q}_n = i\} = P\{\tilde{v}_{n+1} = j - (i-1)^+ | \tilde{q}_n = i\}, \tag{5.54}$$

where the conditional argument is retained to account for the fact that the distribution of the number of arrivals may, in fact, depend upon $\tilde{q}_n$. As an example, in the case of exceptional first service, which was introduced in Exercise 5.12, the first service of every busy period has the distribution $F_{\tilde{x}_e}(x)$ rather than the distribution $F_{\tilde{x}}(x)$, which is common for all other services. Therefore $\tilde{v}_{n+1}$ is computed on the basis of $F_{\tilde{x}_e}(x)$ whenever $\tilde{q}_n = 0$.

We now consider the case where the first service of a busy period may be exceptional but all other service times are drawn from a common distribution.

For this case, we define

$$\begin{aligned} b_j &= P\{\tilde{v}_{n+1} = j | \tilde{q}_n = 0\} \\ a_j &= P\{\tilde{v}_{n+1} = j | \tilde{q}_n \neq 0\}. \end{aligned} \tag{5.55}$$

Then, upon combining (5.54) and (5.54), we find

$$P\{\tilde{q}_{n+1} = j\} = b_j P\{\tilde{q}_n = 0\} + \sum_{i=1}^{j+1} a_{j+1-i} P\{\tilde{q}_n = i\}. \tag{5.56}$$

Upon passing to the limit, we have

$$\pi_j = P\{\tilde{q} = j\} = b_j P\{\tilde{q} = 0\} + \sum_{i=1}^{j+1} a_{j+1-i} P\{\tilde{q} = i\}. \tag{5.57}$$

Upon rewriting (5.57) in matrix form, we find

$$\pi = \pi \mathcal{P}, \tag{5.58}$$

where $\pi = [\pi_0 \quad \pi_1 \quad \pi_2 \quad \cdots]$, and

$$\mathcal{P} = \begin{bmatrix} b_0 & b_1 & b_2 & b_3 & \cdots \\ a_0 & a_1 & a_2 & a_3 & \cdots \\ 0 & a_0 & a_1 & a_2 & \ddots \\ 0 & 0 & a_0 & a_1 & \ddots \\ 0 & 0 & 0 & a_0 & \ddots \\ \vdots & \vdots & \vdots & \ddots & \ddots \end{bmatrix}.$$

Any matrix that has the same form as $\mathcal{P}$ is said to be a matrix of the M/G/1 type.

We now find $\mathcal{F}_n(z)$ in terms of the probability generating functions of the sequences $\{b_i, i = 0, 1, \cdots\}$ and $\{a_i, i = 0, 1, \cdots\}$. Define

$$\mathcal{F}_{\tilde{a}}(z) = \sum_{n=0}^{\infty} a_i z^i \quad \text{and} \quad \mathcal{F}_{\tilde{b}}(z) = \sum_{n=0}^{\infty} b_i z^i.$$

Upon multiplying both sides of (5.58) by z^j and summing over j from 0 to ∞, or equivalently, upon multiplying both sides of (5.58) by diag $(1, z, z^2, \cdots)$ and then postmultiplying by $\mathbf{e}$, we find

$$\mathcal{F}_{\tilde{q}}(z)\left[z - \mathcal{F}_{\tilde{a}}(z)\right] = \pi_0 \left[z\,\mathcal{F}_{\tilde{b}}(z) - \mathcal{F}_{\tilde{a}}(z)\right]. \tag{5.59}$$

The above equation yields the Pollaczek-Khintchine transform equation for the queue length distribution in an M/G/1 system with exceptional first service.

EXERCISE 5.24 Beginning with (5.59), suppose $F_{\tilde{x}_e}(x) = F_{\tilde{x}}(x)$. Show that (5.59) reduces to the standard Pollaczek-Khintchine transform equation for the queue length distribution in an ordinary M/G/1 system.

EXERCISE 5.25 Beginning with (5.59), use the fact that $\lim_{z\to 1} \mathcal{F}_{\tilde{q}}(z) = 1$ to show that

$$\pi_0 = \frac{1 - \mathcal{F}_{\tilde{a}}'(1)}{1 - \mathcal{F}_{\tilde{a}}'(1) + \mathcal{F}_{\tilde{b}}'(1)}. \tag{5.60}$$

We now define a linear algebra-based approach for solving (5.59) to obtain $\{\pi_0, \pi_1, \ldots\}$. We first consider the case where $\mathcal{F}_{\tilde{a}}(z)$ and $\mathcal{F}_{\tilde{b}}(z)$ are finite polynomials, and then we consider the case where $\mathcal{F}_{\tilde{a}}(z)$ and $\mathcal{F}_{\tilde{b}}(z)$ are rational functions in a more generalized setting, which will be defined below. We will see that the solution in the former case can be accomplished entirely without the use of generating functions.

Suppose $\mathcal{F}_{\tilde{a}}(z)$ and $\mathcal{F}_{\tilde{b}}(z)$ are polynomials of degree m. Then, a_j, $j \in \{0, 1, \ldots, m\}$ represents the probability of j arrivals during an ordinary service time. For stability, we require the average number of arrivals during a service time to be strictly less than unity. Thus, $a_0 > 0$. In addition, we require $a_m > 0$ so that we actually have a polynomial of degree m. The coefficients of $\mathcal{F}_{\tilde{b}}(z)$ are less restricted.

From (5.58), we find

$$\begin{bmatrix} \pi_0 & \pi_1 & \ldots & \pi_m \end{bmatrix} = \pi_0 \begin{bmatrix} b_0 & b_1 & \ldots & b_m \end{bmatrix} + \begin{bmatrix} \pi_1 & \pi_2 & \ldots & \pi_{m+1} \end{bmatrix} \begin{bmatrix} a_0 & a_1 & a_2 & \cdots & a_m \\ 0 & a_0 & a_1 & \cdots & a_{m-1} \\ 0 & 0 & a_0 & \cdots & a_{m-2} \\ \vdots & \ddots & \ddots & \ddots & \vdots \\ 0 & 0 & 0 & \cdots & a_0 \end{bmatrix}. \tag{5.61}$$

Define $z_0 = \begin{bmatrix} \pi_1 & \pi_2 & \ldots & \pi_{m+1} \end{bmatrix}$. Then, upon rearranging (5.61), we find

$$z_0 D = \pi_0 N, \tag{5.62}$$

where $N = \begin{bmatrix} b_0 - 1 & b_1 & \ldots & b_m \end{bmatrix}$ and

$$D = \begin{bmatrix} -a_0 & 1 - a_1 & -a_2 & \cdots & -a_m \\ 0 & -a_0 & 1 - a_1 & \ddots & -a_{m-1} \\ 0 & 0 & -a_0 & \ddots & -a_{m-2} \\ \vdots & \ddots & \ddots & \ddots & \vdots \\ 0 & 0 & 0 & \cdots & -a_0 \end{bmatrix}.$$

Because $a_0 > 0$, D is nonsingular. Thus, we can solve (5.62) to obtain

$$\begin{bmatrix} \pi_1 & \pi_2 & \ldots & \pi_{m+1} \end{bmatrix} = \pi_0 N D^{-1}. \tag{5.63}$$

But, we have already shown in Exercise 5.25 that

$$\pi_0 = \frac{1 - \mathcal{F}_{\tilde{a}}'(1)}{1 - \mathcal{F}_{\tilde{a}}'(1) + \mathcal{F}_{\tilde{b}}'(1)}. \tag{5.64}$$

Therefore, we have a complete specification of $z_0 = [\, \pi_1 \quad \pi_2 \quad \dots \quad \pi_{m+1} \,]$.

EXERCISE 5.26 Argue rigorously that in order for the M/G/1 queue to be stable, we must have $a_0 > 0$.

EXERCISE 5.27 Verify (5.62).

We turn now to the computation of the remaining probabilities. First, we note that for any $j \geq 1$,

$$\pi_{m+j} = \sum_{i=j+1}^{m+1+j} \pi_i a_{m+j+1-i}.$$

For example,

$$\pi_{m+1} = \sum_{i=2}^{m+2} \pi_i a_{m+2-i}, \quad \text{and} \quad \pi_{m+2} = \sum_{i=3}^{m+3} \pi_i a_{m+3-i}.$$

Define $y_j = [\, \pi_{j+1} \quad \pi_{j+2} \quad \dots \quad \pi_{j+m-1} \quad \pi_{j+m} \,]$. Then, in matrix form, this equation becomes

$$y_j \begin{bmatrix} -a_m \\ -a_{m-1} \\ \vdots \\ -a_2 \\ 1-a_1 \end{bmatrix} = y_{j+1} \begin{bmatrix} 0 \\ 0 \\ \vdots \\ 0 \\ a_0 \end{bmatrix}. \tag{5.65}$$

Also, by putting the simple statements $\pi_{j+i} = \pi_{j+i}$ for $i = 1, 2, \dots, m-1$ in matrix form, it is easy to see that the $m-1$ equations become

$$y_j \begin{bmatrix} 0 & 0 & \cdots & 0 \\ 1 & 0 & \cdots & 0 \\ 0 & 1 & \ddots & 0 \\ \vdots & \ddots & \ddots & 0 \\ 0 & 0 & \cdots & 1 \end{bmatrix} = y_{j+1} \begin{bmatrix} 1 & 0 & \cdots & 0 \\ 0 & 1 & \ddots & 0 \\ \vdots & \ddots & \ddots & 0 \\ 0 & 0 & \cdots & 1 \\ 0 & 0 & \cdots & 0 \end{bmatrix}.$$

Combining the previous expression with (5.65), we find

$$y_{i+1} E = y_i A, \tag{5.66}$$

where $E = \text{diag }(1, 1, \dots, 1, a_0)$, and

$$A = \begin{bmatrix} 0 & 0 & \cdots & \cdots & -a_m \\ 1 & 0 & \ddots & \ddots & -a_{m-1} \\ 0 & 1 & \ddots & \ddots & -a_{m-2} \\ \vdots & \ddots & \ddots & \ddots & \vdots \\ 0 & 0 & \cdots & 1 & 1-a_1 \end{bmatrix}.$$

We note that $y_1 = [\pi_2 \quad \pi_3 \quad \dots \quad \pi_m \quad \pi_{m+1}]$. Therefore, we have already solved for y_1 in terms of π_0. Since E is always nonsingular, it is now possible to solve for

$$\pi_{j+2} = y_1[AE^{-1}]^j e_0, \quad \text{for} \quad j = 1, 2, \dots, \tag{5.67}$$

where e_j denotes the column vector in which the jth element is 1 and the remaining elements are 0.

EXAMPLE 5.3 Consider a discrete time service system having unit service. Suppose the number of arrivals during the service period has the binomial distribution with parameter $(N, p) = (3, 0.3)$ for all service periods. Determine the matrices A, E, D, and N.

Since the arrival process is common to all service periods, $b_i = a_i$ for all i. In addition, $a_i = 0$ for $i \geq 4$. The binomial probabilities are $(a_0, a_1, a_2, a_3) = (0.343, 0.441, 0.189, 0.027)$. Given these values, the following results are readily obtained from (5.62) and (5.66):

$$A = \begin{bmatrix} 0 & 0 & -0.027 \\ 1 & 0 & -0.189 \\ 0 & 1 & 0.559 \end{bmatrix}, \; D = \begin{bmatrix} -0.343 & 0.550 & -0.189 & -0.027 \\ 0 & -0.343 & 0.550 & -0.189 \\ 0 & 0 & -0.343 & 0.550 \\ 0 & 0 & 0 & -0.343 \end{bmatrix},$$

$$E = \text{diag }(1, 1, 0.343), \text{ and } N = [-0.657 \quad 0.441 \quad 0.189 \quad 0.027].$$

While (5.67) is correct, it is not in the simplest possible form, and indeed the form is not suitable for general computations, such as computing moments. The fundamental reason is that the solution contains unstable modes; that is, all eigenvalues of the matrix AE^{-1} are not strictly less than unity. To address this problem, we would like to decouple the matrix equations (5.66) into two sets of equations; those representing stable modes and those representing unstable modes. Before doing this, however, we introduce the more general class of problems in which there may be multiple boundaries and where the generating functions may be ratios of polynomials rather than merely finite polynomials.

Suppose our queueing system has C boundary conditions. One situation in which this would occur would be in a system that serves a random number of units in a service period. A specific case of interest is a slotted wireless transmission system, where the link quality varies from time slot to time slot, as discussed in Chapter 1. In such a case, the arrival process to the system might be stationary, but the queue transition probabilities would depend upon the number of units served. In some cases, the number of units served can be assumed to be independent from slot to slot, and the number of packets transmitted over a time slot may vary between 0 and some maximum number, say, C. In such cases, the $\mathcal{P}$ matrix would then have the form

$$\mathcal{P} = \begin{bmatrix} b_{00} & b_{01} & b_{02} & b_{03} & \cdots \\ b_{10} & b_{11} & b_{12} & b_{13} & \cdots \\ \cdots & \cdots & \cdots & \cdots & \cdots \\ b_{C-1,0} & b_{C-1,1} & b_{C-1,2} & b_{C-1,3} & \cdots \\ a_0 & a_1 & a_2 & a_3 & \cdots \\ 0 & a_0 & a_1 & a_2 & \ddots \\ 0 & 0 & a_0 & a_1 & \ddots \\ 0 & 0 & 0 & a_0 & \ddots \\ \vdots & \ddots & \ddots & \ddots & \ddots \end{bmatrix},$$

where, again, $\pi = \pi\mathcal{P}$.

We now find $\mathcal{F}_n(z)$ in terms of the probability generating functions of the sequences $\{b_{ji}, i = 0, 1, \cdots, j = 0, 1, \ldots, C-1\}$ and $\{a_i, i = 0, 1, \cdots\}$. Define

$$\mathcal{F}_{\tilde{a}}(z) = \sum_{n=0}^{\infty} a_i z^i \quad \text{and} \quad \mathcal{F}_{\tilde{b},j}(z) = \sum_{n=0}^{\infty} b_{ji} z^i.$$

Upon multiplying both sides of (5.58) by diag $(1, z, z^2, \cdots)$ and then post-multiplying by e, we find

$$\mathcal{F}_{\tilde{q}}(z)\left[z^C - \mathcal{F}_{\tilde{a}}(z)\right] = \sum_{j=0}^{C-1} \pi_j \left[\mathcal{F}_{\tilde{b},j}(z) - z^j \mathcal{F}_{\tilde{a}}(z)\right]. \tag{5.68}$$

Now, (5.68) has a vector of unknown probabilities, $[\,\pi_0 \quad \pi_1 \ \cdots \ \pi_{C-1}\,]$; we call these boundary probabilities. Once the boundary probabilities are known, the transform equation is well-defined, and, in principle, all of the probabilities can be computed.

We now wish to transform (5.68) into an equation that has no boundary probabilities on its left hand side. By beginning with

$$\mathcal{F}_{\tilde{q}}^{(1)}(z) = \frac{1}{z}\left[\mathcal{F}_{\tilde{q}}(z) - \pi_0\right],$$

and forming iteratively

$$\mathcal{F}_{\tilde{q}}^{(i)}(z) = \frac{1}{z}\left[\mathcal{F}_{\tilde{q}}^{(i-1)}(z) - \pi_{i-1}\right],$$

for $i = 2, 3, \ldots, C$, it is straightforward to show that (5.68) reduces to

$$\mathcal{F}_{\tilde{q}}^{(C)}(z)\left[z^C - \mathcal{F}_{\tilde{a}}(z)\right] = \sum_{j=0}^{C-1} \pi_j \left[\mathcal{F}_{\tilde{b},j}(z) - z^j\right]. \tag{5.69}$$

EXERCISE 5.28 For $i = 1, 2, \ldots$, show that

$$\mathcal{F}_{\tilde{q}}^{(i)}(z) = \sum_{j=0}^{\infty} \pi_{i+j} z^j.$$

EXERCISE 5.29 Define

$$\mathcal{F}_{\tilde{q}}^{(1)}(z) = \frac{1}{z}\left[\mathcal{F}_{\tilde{q}}(z) - \pi_0\right] \text{ and } \mathcal{F}_{\tilde{q}}^{(i+1)}(z) = \frac{1}{z}\left[\mathcal{F}_{\tilde{q}}^{(i)}(z) - \pi_i\right], i \geq 1.$$

Starting with (5.68), substitute a function of $\mathcal{F}_{\tilde{q}}^{(1)}(z)$ for $\mathcal{F}_{\tilde{q}}(z)$, then a function of $\mathcal{F}_{\tilde{q}}^{(2)}(z)$ for $\mathcal{F}_{\tilde{q}}^{(1)}(z)$, and continue step by step until a function of $\mathcal{F}_{\tilde{q}}^{(C)}(z)$ is substituted for $\mathcal{F}_{\tilde{q}}^{(C-1)}(z)$. Show that at each step, one element of

$$\sum_{j=0}^{C-1} \pi_j z^j \mathcal{F}_{\tilde{a}}(z)$$

is eliminated, resulting in (5.69).

Now, suppose

$$\mathcal{F}_{\tilde{a}}(z) = \frac{u(z)}{w(z)} \quad \text{and} \quad \mathcal{F}_{\tilde{b},j}(z) = \frac{v_j(z)}{w(z)} \text{ for } j = 1, 2, \cdots, C-1. \tag{5.70}$$

Then after replacing $\mathcal{F}_{\tilde{a}}(z)$ and $\mathcal{F}_{\tilde{b},j}(z)$ by their ratio forms, we have

$$\mathcal{F}_{\tilde{q}}^{(C)}(z)\left[z^C w(z) - u(z)\right] = \sum_{j=0}^{C-1} \pi_j \left[v_j(z) - z^j w(z)\right]. \tag{5.71}$$

EXAMPLE 5.4 Consider the M/M/1 system with exceptional first service. Then,

$$F_{\tilde{x}_e}^*(s) = \frac{\mu_e}{s + \mu_e} \quad \text{and} \quad F_{\tilde{x}}^*(s) = \frac{\mu}{s + \mu}.$$

Thus, the corresponding probability generating functions for the numbers of arrivals over periods of exceptional and ordinary service times are

$$\mathcal{F}_{\tilde{b}}(z) = \frac{\mu_e}{\lambda(1-z) + \mu_e} \quad \text{and} \quad \mathcal{F}_{\tilde{a}}(z) = \frac{\mu}{\lambda(1-z) + \mu}.$$

Thus, $w(z) = (\lambda + \mu - \lambda z)(\lambda + \mu_e - \lambda z)$, $u(z) = \mu(\lambda + \mu_e - \lambda z)$, and $v(z) = \mu_e(\lambda + \mu - \lambda z)$.

Now define $d(z) = z^C w(z) - u(z)$, ν_d as the degree of $d(z)$, and d_i as the coefficient of z^i in $d(z)$. Similarly, define $n_j(z) = v_j(z) - z^j w(z)$ for $j = 1, 2, \cdots, C-1$, ν_n as the maximal degree of $n_j(z)$ over all j, and $n_{j,i}$ as the coefficient of z^i in $n_j(z)$. Note that some of the coefficients of $d(z)$ and $n_j(z)$ may be zero, but $d_{\nu_d} \neq 0$ and $n_{j,\nu_n}(z) \neq 0$ for at least one value of j. Finally, define

$$\nu = \max\{\nu_d, \nu_n + 1\}. \tag{5.72}$$

By making these substitutions in (5.71), we obtain

$$\mathcal{F}_{\tilde{q}}^{(C)}(z)d(z) = \sum_{j=0}^{C-1} \pi_j n_j(z), \tag{5.73}$$

where

$$d(z) = \sum_{i=0}^{\nu} d_i z^i = \sum_{i=0}^{\nu} [w_{i-C} - u_i]\, z^i,$$

and

$$n_j(z) = v_j(z) - z^j w(z) = \sum_{i=0}^{\nu-1} [v_{j,i} - w_{i-C}]\, z^i.$$

We note also that the definition of ν assures that the polynomials $n_j(z)$ have degree at most $\nu - 1$. The coefficients $d(z)$ and $n_j(z)$ are then computed as follows:

$$\begin{aligned} d_i &= w_{i-C} - u_i, \text{ for } i = 0, 1, \ldots, \nu, \qquad (5.74)\\ n_{j,i} &= v_{b,j,i} - w_{i-j}, \text{ for } i = 0, 1, \ldots, \nu - 1, \text{ and } j = 0, 1, \ldots, C-1. \end{aligned}$$

Since, as shown in Exercise 5.28,

$$\mathcal{F}_{\tilde{q}}^{(i)}(z) = \sum_{j=0}^{\infty} \pi_{i+j} z^j,$$

we can rewrite (5.69) as

$$\left[\sum_{i=0}^{\infty} \pi_{C+i} z^i\right] \left[\sum_{i=0}^{\nu} d_i z^i\right] = \sum_{j=0}^{C-1} \pi_j \sum_{i=0}^{\nu-1} [v_{j,i} - w_{i-C}]\, z^i.$$

Upon reordering the summations of the previous equation, we find

$$\sum_{i=0}^{\infty} \left[\sum_{k=0}^{\min\{i,\nu\}} \pi_{C+i-k} d_k\right] z^i = \sum_{i=0}^{\nu-1} \left[\sum_{j=0}^{C-1} \pi_j n_{j,i}\right] z^i. \tag{5.75}$$

By matching the coefficients of z^i for $i = 0, 1, \ldots, \nu - 1$ on the left and right sides of (5.75), we obtain the following matrix equation

$$y_0 D = x_0 N, \tag{5.76}$$

where $x_0 = [\,\pi_0 \quad \pi_1 \quad \ldots \quad \pi_{C-1}\,]$, $y_0 = [\,\pi_C \quad \pi_{C+1} \quad \ldots \quad \pi_{C+\nu-1}\,]$,

$$D = \begin{bmatrix} d_0 & d_1 & d_2 & \cdots & d_{\nu-1} \\ 0 & d_0 & d_1 & \ddots & d_{\nu-2} \\ 0 & 0 & d_0 & \ddots & d_{\nu-3} \\ \vdots & \vdots & \ddots & \ddots & \cdots \\ 0 & 0 & 0 & \cdots & d_0 \end{bmatrix},$$

and

$$N = \begin{bmatrix} n_{01} & n_{02} & \cdots & n_{0,\nu-1} \\ n_{11} & n_{12} & \cdots & n_{1,\nu-1} \\ \vdots & \vdots & \cdots & \vdots \\ n_{C-1,1} & n_{C-1,2} & \cdots & n_{C-1,\nu-1} \end{bmatrix}.$$

EXAMPLE 5.5 Consider a time division multiplexing system having 3 packet slots per frame. During each frame, suppose the number of packet arrivals that occur is binomially distributed with parameters $(N, p) = (5, 0.54)$. Specify ν, D, N, E, and A.

Solution: So long as there are three or less packets in the system, the number of packets left at the of a frame will be simply the number of packets that arrive during the frame. If there are more than 3 packets present at the beginning of a time slot, then the number of packets remaining at the end of the frame will be the number that arrive during the frame time plus the number of packets in excess of three at the beginning of the frame. Thus, $w(z) = 1$, and

$$v_{j,i} = u_i = P\{\tilde{v}_{n+1} = i\} = a_i.$$

Since $C = 3$ and the degrees of $u(z)$ and $v(z)$ are both 5, $\nu_d = 5$ and $\nu_n = 5$; thus, $\nu = 6$. Thus, D, E, and A are all 6×6 matrices and N is a 3×6 matrix. The numerical results are summarized in Table 5.2 From the table, the arrays D, N, E, and A can be determined. For example,

$$N = \begin{bmatrix} -0.979 & 0.121 & 0.284 & 0.333 & 0.196 & 0.046 \\ 0.021 & -0.879 & 0.284 & 0.333 & 0.196 & 0.046 \\ 0.021 & 0.121 & -0.716 & 0.333 & 0.196 & 0.046 \end{bmatrix}.$$

For $i \geq \nu$, the coefficient of z^i on the right hand side of (5.75) is zero. Thus, we have, for $i = \nu + j$,

$$\sum_{k=0}^{\nu+j} \pi_{C+\nu+j-k} d_k = 0 \quad \text{for} \quad j \geq 0,$$

Table 5.2. Parameters for Example 5.5.

i	u_i	d_i	$v_{0,i}$	$v_{1,i}$	$v_{2,i}$
0	0.021	-0.021	-0.979	0.021	0.021
1	0.121	-0.121	0.121	-0.879	0.121
2	0.284	-0.284	0.284	0.284	-0.716
3	0.333	0.667	0.333	0.333	0.333
4	0.196	-0.196	0.196	0.196	0.196
5	0.046	-0.046	0.046	0.046	0.046

or

$$\pi_{C+\nu+j} d_0 = \sum_{k=1}^{\nu} \pi_{C+\nu+j-k} d_k \quad \text{for} \quad j \geq 0.$$

If we now define $y_j = [\, \pi_{C+j} \quad \pi_{C+j+1} \quad \cdots \quad \pi_{C+j+\nu-1} \,]$, the previous equation can be written in vector form as

$$y_{j-1} \begin{bmatrix} -d_{\nu-1} \\ -d_{\nu-2} \\ \vdots \\ -d_2 \\ -d_1 \end{bmatrix} = y_j \begin{bmatrix} 0 \\ 0 \\ \vdots \\ 0 \\ d_0 \end{bmatrix} \text{ for } j \geq 0. \tag{5.77}$$

By putting the simple statements $\pi_{j+i} = \pi_{j+i}$ for $i = 1, 2, \ldots, m - 1$ into the previous equation, we have the following:

$$y_j E = y_{j-1} A \text{ for } j \geq 1, \tag{5.78}$$

where $E = \text{diag }(1, 1, \ldots, 1, d_0)$, and

$$A = \begin{bmatrix} 0 & 0 & \cdots & \cdots & -d_\nu \\ 1 & 0 & \ddots & \ddots & -d_{\nu-1} \\ 0 & 1 & \ddots & \ddots & -d_{\nu-2} \\ \vdots & \ddots & \ddots & \ddots & \vdots \\ 0 & 0 & \cdots & 1 & -d_1 \end{bmatrix}.$$

Thus, in order to solve for all the probabilities we need only to solve the system of equations defined by (5.76) and (5.78):

$$y_0 D = x_0 N \quad \text{and} \quad y_j E = y_{j-1} A \text{ for } j \geq 1.$$

Note that (5.76) and (5.78) are identical in form to (5.62) and (5.66); the only difference is that π_0 of (5.62) is replaced by x_0 of (5.76). But, if $C = 1$, then $x_0 = \pi_0$ so that the two forms are identical. We therefore choose to solve the general case first and then do examples for the case $C = 1$.

EXERCISE 5.30 Suppose $\mathcal{F}_{\tilde{a}}(z)$ and $\mathcal{F}_{\tilde{b}}(z)$ are each polynomials of degree m as discussed in the first part of this section. Define $w(z) = 1$. Find ν, D, N, A, and E using (5.72), (5.73), and (5.78). Compare the results to those presented in (5.62) and (5.66).

We now turn to the solution procedure based on the generalized state space approach. We will use generalized Schur decomposition as we did in solving QBD systems in an earlier chapter. For continuity, we repeat the following theorem:

THEOREM 5.4 **Generalized Schur Decomposition.** *Suppose A and E are both real matrices with spectrum $\lambda(A, E)$ and $\lambda(A, E)$ is partitioned into two sets, say $\lambda_u(A, E)$ and $\lambda_s(A, E)$ such that $\lambda_u(A, E) \cap \lambda_s(A, E) = \emptyset$. Then, there exist (non-singular) orthogonal matrices, Q and Z, such that*

$$Q^T E Z = \begin{bmatrix} E_{uu} & E_{us} \\ 0 & E_{ss} \end{bmatrix} \quad \text{and} \quad Q^T A Z = \begin{bmatrix} A_{uu} & A_{us} \\ 0 & A_{ss} \end{bmatrix},$$

where all matrices are real, E_{uu} and E_{ss} are upper triangular, and A_{uu} and A_{ss} are block upper triangular, meaning that their diagonal elements are either 1×1 or 2×2 blocks, depending upon whether the eigenvalues are real or occur in complex conjugate pairs. The row dimensions of E_{uu} and A_{uu} and E_{ss} and A_{ss} are $n_u = \text{card}\ (\lambda_u(A, E))$ and $n_s = \text{card}\ (\lambda_s(A, E))$, respectively. □

The generalized Schur decomposition of Theorem 5.4 is carried out efficiently by using the so-called QZ algorithm, which is described in detail in Golub and Van Loan [1996]. In turn, the QZ algorithm is implemented in the routine *dgges()* of LAPACK (Anderson [1999]).

As in the solution procedure for QBD processes, we first define the partitions $\lambda_u(A, E)$ and $\lambda_s(A, E)$ as the unstable and stable sets of generalized eigenvalues of A with respect to E. Next, we define $y_j = u_j Q^T$. We then substitute $u_j Q^T$ and $u_{j-1} Q^T$ for y_j and y_{j-1} in (5.78) and postmultiply both sides of the result by Z to obtain

$$\begin{bmatrix} u_{j,u} & u_{j,s} \end{bmatrix} \begin{bmatrix} E_{uu} & E_{us} \\ 0 & E_{ss} \end{bmatrix} = \begin{bmatrix} u_{j-1,u} & u_{j-1,s} \end{bmatrix} \begin{bmatrix} A_{uu} & A_{us} \\ 0 & A_{ss} \end{bmatrix}, \tag{5.79}$$

where $u_{j,u}$ and $u_{j,s}$ represent the unstable and stable parts of u_j, respectively, and whose dimensions are n_u and n_s, respectively. As in the QDB case, we recognize that $u_{j,u}$ must be 0 for all j in order to have a stable solution. Thus, (5.79) implies $u_{j,s} E_{ss} = u_{j-1,s} A_{ss}$. This leads to

$$u_{j,s} = u_{j-1,s} A_{ss} E_{ss}^{-1}. \tag{5.80}$$

Because $u_{j,u} = 0$ for all j, $u_jQ^T = [0 \quad u_{j,s}]Q^T$. Therefore, it is convenient to partition Q^T so that we may write $y_j = u_{j,s}L_s$, where L_s is the matrix containing the last n_s rows of Q^T. Now substitute $y_{j-1}Q_s$ for $u_{j-1,s}$ in (5.80) and then postmultiply the result by L_s to obtain

$$y_j = y_{j-1}Q_sA_{ss}E_{ss}^{-1}L_s. \tag{5.81}$$

Equivalently,

$$\begin{aligned} y_j &= y_0\left[Q_sA_{ss}E_{ss}^{-1}L_s\right]^j, \\ y_j &= y_0Q_s\left[A_{ss}E_{ss}^{-1}\right]^j L_s, \text{ and} \\ \pi_{C+j} &= g\left[A_{ss}E_{ss}^{-1}\right]^j H \quad \text{for all} \quad j \geq 0, \end{aligned} \tag{5.82}$$

where $g = y_0Q_s$, H is defined as the first column of L_s, and the second step of the previous equation results from the fact that $L_sQ_s = I$. Thus, (5.82) specifies all level probabilities for levels greater than C in terms of y_0Q_s, while $y_0 = [\pi_C \quad \pi_{C+1} \quad \cdots \quad \pi_{C+\nu-1}]$. Thus, to complete the solution, it remains only to specify y_0.

From (5.76), we have

$$[x_0 \quad y_0]\begin{bmatrix} -N \\ D \end{bmatrix} = 0.$$

In addition, because Q is orthogonal, $y_j = u_jQ^T$ implies $u_j = y_jQ$ so that $y_jQ_u = u_{j,u}$, where Q_u is the first n_u columns of Q. Thus, we must have

$$y_jQ_u = 0 \text{ for all } j \geq 0.$$

We thus have from the previous two equations

$$[x_0 \quad y_0]\begin{bmatrix} -N & 0 \\ D & Q_u \end{bmatrix} = 0. \tag{5.83}$$

We now define

$$x_0 = k_0\hat{x}_0 \quad \text{and} \quad y_0 = k_0\hat{y}_0, \tag{5.84}$$

where $\hat{x}_0\mathrm{e} = 1$. Equation 5.83 can then be rewritten as

$$[\hat{x}_0 \quad \hat{y}_0]\begin{bmatrix} -\hat{N} & \mathrm{e} & 0 \\ \hat{D} & 0 & Q_u \end{bmatrix} = [0 \quad 1 \quad 0], \tag{5.85}$$

where $\hat{N}$ and $\hat{D}$ are the matrices N and D with their last columns deleted. Note that (5.85) yields numerical values for $\hat{x}_0$ and $\hat{y}_0$.

In terms of $\hat{y}_0$, (5.82) becomes

$$\pi_{C+j} = k_0 \hat{y}_0 Q_s \left[A_{ss} E_{ss}^{-1}\right]^j H \quad \text{for all} \quad j \geq 0.$$

Since the individual probabilities must sum to unity, we then have

$$1 = x_0 \mathbf{e} + \sum_{i=0}^{\infty} \pi_{C+j} \mathbf{e},$$

or equivalently,

$$\begin{aligned} 1 &= \hat{k}_0 \left[\hat{x}_0 + \hat{y}_0 Q_s \left\{\sum_{i=0}^{\infty} \left[A_{ss} E_{ss}^{-1}\right]^j\right\} H\right] \mathbf{e} \\ &= \hat{k}_0 \left[\hat{x}_0 + \hat{y}_0 Q_s \left\{I - \left[A_{ss} E_{ss}^{-1}\right]\right\}^{-1} H\right] \mathbf{e}. \end{aligned} \tag{5.86}$$

Thus, we can determine k_0 from (5.86).

After finding k_0, we can substitute its value into (5.84) to determine x_0 and y_0, noting that x_0 contains the vectors $\pi_0, \pi_1, \ldots, \pi_{C-1}$. We can then use y_0 in (5.82) to determine π_{C+i} for $i \in \{0, 1, \ldots\}$, which results in a complete solution for the level probabilities.

In summary, the generalized state-space solution to multiple-boundary problems within the M/G/1 paradigm is as follows:

1. From the problem statement, determine $\mathcal{F}_a(z), \mathcal{F}_{b,j}(z)$ for $j = 0, 1, \ldots, C-1$ and express these polynomials in right polynomial fraction form as in (5.70).

2. Compute ν by using (5.72).

3. Compute the coefficients of $d(z)$ and $n_j(z)$ by using (5.73).

4. Using the results of previous step, (5.76), and (5.78), determine the matrices D, N, E, and A.

5. Perform a generalized Schur decomposition of A with respect to E according to Theorem 5.4. The LAPACK routine *dgges()* may be used for this purpose. This decomposition yields directly Q, n_u, n_s, E_{ss}, and A_{ss}.

6. Partition Q and Q^T to obtain Q_s, Q_u, and H.

7. Formulate the linear system of equations (5.85) and solve to obtain $\hat{x}_0$ and $\hat{y}_0$.

8. Solve for k_0 using (5.86).

9. Find x_0 and y_0 using (5.84).

10. Partition x_0 to find $\pi_0, \pi_1, \ldots, \pi_{C-1}$.

11. Compute all remaining desired π_j using (5.82).

We now discuss a few examples that illustrate the use of these techniques for examining the behavior of queueing systems in general.

EXAMPLE 5.6 (Continuation of Example 5.4). From Example5.4, we have $w(z) = (\lambda + \mu - \lambda z)(\lambda + \mu_e - \lambda z)$, $u(z) = \mu(\lambda + \mu_e - \lambda z)$, and $v(z) = \mu_e(\lambda + \mu - \lambda z)$. Thus, with $d(z) = zw(z) - u(z)$ and $n(z) = v(z) - w(z)$, we find $\nu_d = 3$ $\nu_n = 2$, so $\nu = \min\{3, 3\} = 3$. We readily find d_i and n_i for $i = 0, 1, 2, 3$ from (5.74). The results are shown in the Table 5.3.

From the table and (5.71) and (5.76), we have

$$N = [\,-\lambda(\lambda+\mu) \quad \lambda(2\lambda+\mu) \quad -\lambda^2\,],$$

$$D = \begin{bmatrix} -\mu(\lambda+\mu_e) & [\lambda\mu + (\lambda+\mu)(\lambda+\mu_e)] & -\lambda(2\lambda+\mu+\mu_e) \\ 0 & -\mu(\lambda+\mu_e) & [\lambda\mu + (\lambda+\mu)(\lambda+\mu_e)] \\ 0 & 0 & -\mu(\lambda+\mu_e) \end{bmatrix},$$

$$E = \begin{bmatrix} 1 & 0 & 0 \\ 0 & 1 & 0 \\ 0 & 0 & -\mu(\lambda+\mu_e) \end{bmatrix}, \text{ and } A = \begin{bmatrix} 0 & 0 & -\lambda^2 \\ 1 & 0 & \lambda(2\lambda+\mu+\mu_e) \\ 0 & 1 & -[\lambda\mu + (\lambda+\mu)(\lambda+\mu_e)] \end{bmatrix}.$$

Figure 5.3 shows a graph of survivor function that illustrates the effect of the service rate, μ_e, for the exceptional first service. From these graphs, we see that as the service rate decreases, the survivor functions, and therefore the moments of the occupancy distributions, increase.

Table 5.3. Formulae to compute parameter values for Example 5.6.

i	d_i	n_i
0	$-\mu(\lambda+\mu_e)$	$-\lambda(\lambda+\mu)$
1	$\lambda\mu + (\lambda+\mu)(\lambda+\mu_e)$	$\lambda(2\lambda+\mu)$
2	$-\lambda(2\lambda+\mu+\mu_e)$	$-\lambda^2$
3	λ^2	-

EXAMPLE 5.7 (Binomial Approximation of Poisson). In the M/D/1 system with unit service time, the number of arrivals that occur during a service period

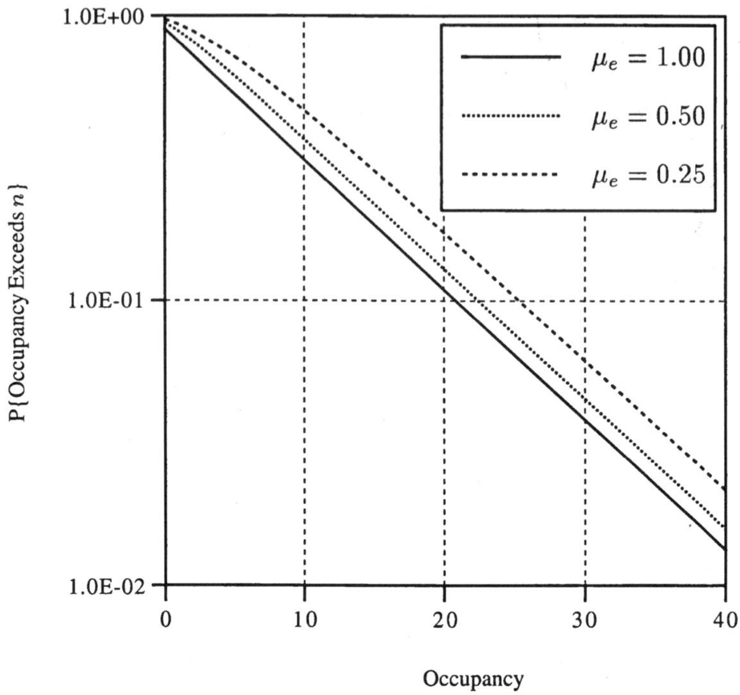

Figure 5.3. Survivor functions for system having exponential ordinary and exceptional first service $\mu = 1.0$, $\rho = 0.9$, and μ_e as a parameter.

is Poisson with parameter λ. It is well known that the binomial distribution with parameters N and p with Np fixed converges to the Poisson distribution with $\lambda = Np$ as $N \to \infty$. Suppose the number of arrivals during the service period has the binomial distribution with parameter N and $Np = 0.9$. We wish to plot the survivor functions of the occupancy distribution with N as a parameter in order to see whether or not the survivor function converges to that for the M/D/1, and if so, how fast.

In this example, $C = 1$, and $\mathcal{F}_a(z) = \mathcal{F}_b(z) = (pz + 1 - p)^N$. Figure 5.4 shows graphs of the survivor function of the occupancy distribution for $N = 4, 16, 64, \text{and}, 256$. We note that the graph for $N = 256$ and the graph for the M/D/1 system shown in Figure 5.1 are visually the same. In addition, we note that the graphs for $N = 64$ and $N = 256$ are virtually indistinguishable. On the other hand, there is a noticeable difference between the graphs for $N = 16$ and $N = 64$, especially at higher occupancy levels.

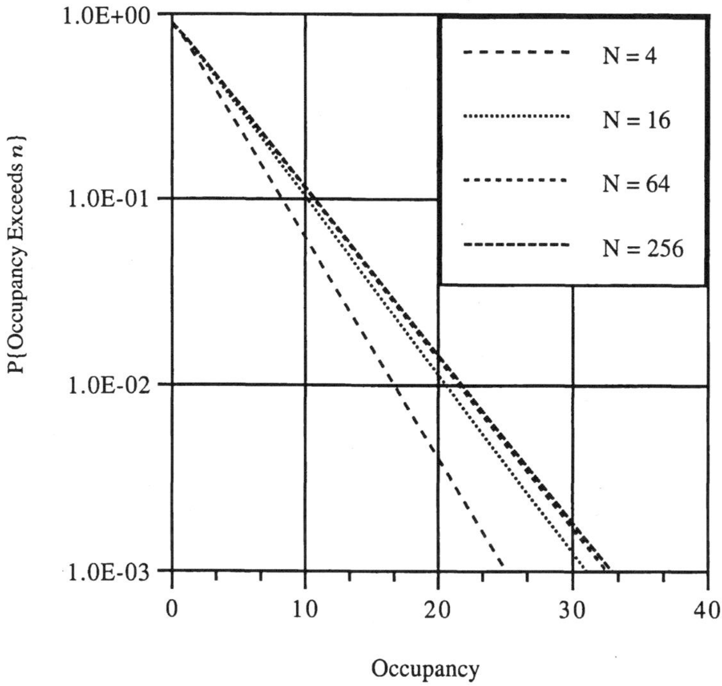

Figure 5.4. Survivor functions with unit deterministic service and binomially distributed arrivals with N as a parameter at $\rho = 0.9$.

EXAMPLE 5.8 (Batch size and queue length). Suppose a system's service time distribution is Erlang-K with unit mean and that arrivals to the system are Poisson with a rate that results in a traffic load of 90%. During each service interval, the server may serve a batch of up to C units. We wish to examine the form of the survivor function as the batch size is increased. Determine the functions $u(z)$, $v_j(z)$ $j = 0, 1, \ldots, C-1$, and $w(z)$ as a function of C and then determine ν. Find $d(z)$ and $n_j(z)$ for $K = 4$ and $C = 3$. Obtain plots of the survivor functions of the occupancy distributions for the cases of $K = 10$, $C = 1, 2, 4,$ and 8.

Solution: Since service on the boundaries is not exceptional, $\mathcal{F}_{b,j}(z) = \mathcal{F}_a(z)$ $\forall$ j. In general, since the Erlang-K random variable is the sum of K individual exponential random variables of equal rate, μ, the Laplace-Stieltjes transform of the service time distribution is

$$F_{\tilde{x}}^*(s) = \left(\frac{\mu}{\mu + s}\right)^K.$$

Table 5.4. Parameter values for Example 5.8.

i	d_i	$n_{0,i}$	$n_{1,i}$	$n_{2,i}$
0	-1.000	-6.872	1.000	1.000
1	0.000	12.688	-7.872	-0.000
2	0.000	-7.670	12.688	-7.872
3	7.872	2.061	-7.670	12.688
4	-12.688	-0.208	2.061	-7.670
5	7.669	-0.046	-0.208	2.061
6	-2.060	0.000	0.000	-0.208
7	0.208	-	-	-

Thus, the PGF of the number of arrivals during the service time is

$$\mathcal{F}_{\tilde{v}}(z) = \left(\frac{\mu}{\mu + \lambda(1-z)}\right)^K = \left(\frac{1}{1+\rho-\rho z}\right)^K,$$

where $\rho = \lambda/\mu$. In order to have unit service time, we must have $\mu = K$. Also, since the system serves batches of size C, the traffic load will be λ/C. Hence, for a load of 0.9, we need $\lambda = 0.9C$ so that $\rho = 0.9C/K$. Thus, we find

$$\mathcal{F}_{\tilde{v}}(z) = \left(\frac{1}{1+0.9C/K - [0.9C/K]\,z}\right)^K.$$

Therefore, $u(z) = v_j(z) = 1$, and

$$w(z) = (1 + 0.9C/K - [0.9C/K]\,z)^K.$$

Since the degree of v_j is zero for all j, the degree of $u(z)$ is also zero, and the degree of $w(z)$ is K, $\nu = C + K$. For $K = 4$ and $C = 3$, $f = 7$. Coefficients of $D(z)$ and $N_j(z)$ are given in Table 5.4.

Figure 5.5 shows the required graph of survivor functions. We note that the survivor functions are increasing functions of the batch size. At the end of each service period, the server removes up to C units from the queue. If there are less than C units, the server clears the queue. The curves demonstrate that backlog increases as batch size increases. This system is closely related to, but not identical to, a system of C parallel servers. In the case of C parallel servers, if we hold the service rate constant while increasing C, then we know the backlog increases. Thus, it is better to have one fast server than C slow

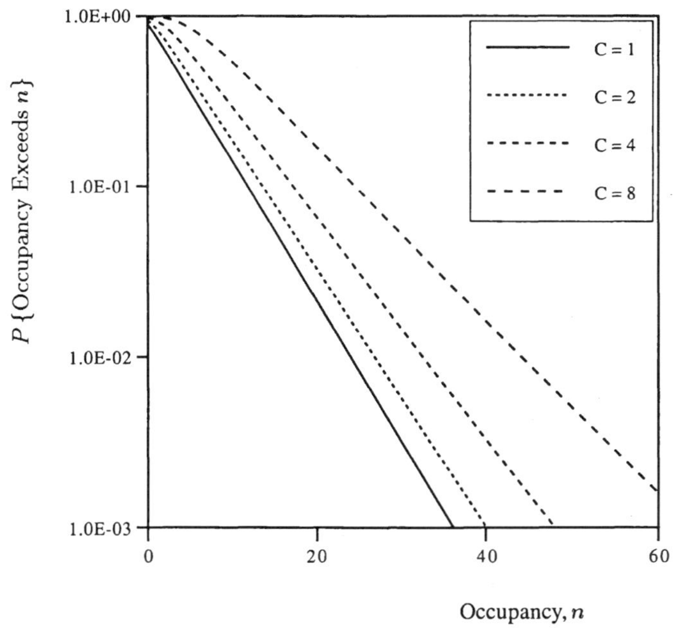

Figure 5.5. Survivor functions with unit-mean Erlang-10 service and Poisson arrivals with C as a parameter at a traffic load of 0.9.

ones. Similarly, it would be better to serve items one-at-a-time with a faster server than in batches with a slower server.

EXERCISE 5.31 Table 5.5 gives numerical values for the survivor function of the occupancy distributions shown in Figure 5.5. From this table, determine the probability masses for the first few elements of the distributions and then compute the mean number of units served during a service time for $C = 1, 2, 4,$ and 8. Analyze the results of your calculations.

EXAMPLE 5.9 (Erlang-K approximation of deterministic). The deterministic service time is often approximated by the Erlang-K distribution. Consider the M/D/1 system. We wish to investigate the effect of the choice of K on the queue length distribution at $\rho = 0.9$.

From the previous example, we have

$$\mathcal{F}_{\tilde{v}}(z) = \left(\frac{1}{1 + 0.91/K - [0.9/K]\,z}\right)^K .$$

Table 5.5. Occupancy values as a function of the number of units served during a service period for the system analyzed in Example 5.8.

Occupancy	$C = 1$	$C = 2$	$C = 4$	$C = 8$
0	1.00000	1.00000	1.00000	1.00000
1	0.9000	0.9493	0.9861	0.9985
2	0.7633	0.8507	0.9453	0.9920
3	0.6350	0.7347	0.8772	0.9762
4	0.5261	0.6233	0.7914	0.9481
5	0.4356	0.5254	0.6997	0.9071
6	0.3606	0.4419	0.6107	0.8547
7	0.2985	0.3715	0.5292	0.7942
8	0.2472	0.3123	0.4569	0.7292

Therefore, $u(z) = v(z) = 1$, and

$$w(z) = (1 + 0.9/K - [0.9/K]\, z)^K \,.$$

Since the degree of v is zero for all j, the degree of $u(z)$ is also zero, and the degree of $w(z)$ is K, $\nu = K$. Figure 5.6 shows the resulting survivor functions for $K = 2^i$ with $i \in \{0, 1, 2, 3, 4, 5, 6, 10\}$. To avoid clutter and because the survivor functions decrease with increasing K, the graphs are not individually labeled. From the graph, we see that the survivor function decreases rapidly as K is increased, with very little difference between $K = 2^6$ and $K = 2^{10}$. It is clear that the choice of K has at least some effect upon the results even for very large K.

EXAMPLE 5.10 (Pade approximation of deterministic). Consider again the M/D/1 system. We wish to consider use of the Pade approximation to the deterministic service time. We will do this by comparing the survivor function obtained using the Erlang 2^{10} distribution and survivor function obtained from the Pade approximation.

The idea of the Pade approximation is to approximate the LST of a given distribution by a ratio of polynomials of degrees n and ℓ such that the first $n + \ell$ moments of the actual service time distribution match with those of the approximation. In this case, $F^*_{\tilde{x}}(s) = e^{-s}$, and the Pade approximation has the

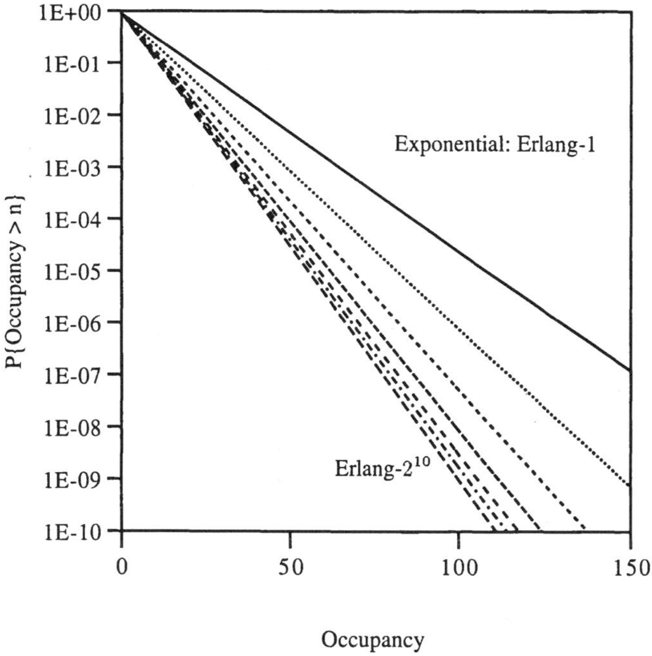

Figure 5.6. Survivor functions with unit-mean Erlang-K service and Poisson arrivals with K as a parameter at a traffic load of 0.9.

following closed-form solution [Akar and Arikan [1996]]:

$$\hat{F}^*_{n,\ell}(s) = \frac{\sum_{i=0}^{n}(n+\ell-i)!\binom{n}{i}(-1)^i(s)^i}{\sum_{i=0}^{\ell}(n+\ell-i)!\binom{\ell}{i}(s)^i}.$$

We note in passing that in order to avoid dealing with large numbers, we rearrange the above formula by first dividing the numerator and denominator by $n\,\ell$.

We then find

$$\hat{\mathcal{F}}_{n,\ell}(z) = \hat{F}^*_{n,\ell}(s)|_{s=\lambda(1-z)}.$$

We then have $\nu = \max\{n+1, \ell+1\}$.

In terms of $u(z)$ and $w(z)$, we find

$$u(z) = \frac{1}{n\,\ell}\sum_{i=0}^{n}(n+\ell-i)!\binom{n}{i}(-1)^i(s)^i\Big|_{s=\lambda(1-z)},$$

and

$$w(z) = \frac{1}{n\,\ell} \sum_{i=0}^{\ell} (n+\ell-i)! \binom{\ell}{i} (s)^i \Big|_{s=\lambda(1-z)}.$$

Figure 5.7 shows graphs of survivor functions obtained by using an $\hat{F}^*_{2,2}(s)$ approximation to e^{-s} and an Erlang 2^{10} service time distribution at $\rho = \lambda = 0.9$. We note that the curves are indistinguishable on the graph.

In order to investigate more closely, we present some selected values of the survivor function in the Table 5.6. Note that, except for the Pade$(1,1)$ case, the survivor function values for the Pade(n,ℓ) are always less than the Erlang-2^{10} values. Given that the Erlang-K values decrease with K, it appears that the Pade$(2,2)$ would be a very good approximation to the deterministic service. Note in addition the very small differences between the Pade$(2,2)$ and Pade$(3,3)$ values. We examined a number of addition Pade(n,ℓ) approximations and found similar results.

Table 5.6. Comparison of values of survivor function computed using various Pade approximations for service time in Example 5.10.

Occupancy	*Pade (1,1)*	*Pade (2,2)*	*Pade (3,3)*	*Erlang-2^{10}*
0	9.0000E-01	9.0000E-01	9.0000E-01	9.0000E-01
1	7.3636E-01	7.5425E-01	7.5404E-01	7.5423E-01
2	6.0248E-01	6.1617E-01	6.1640E-01	6.1679E-01
3	4.9294E-01	5.0129E-01	5.0135E-01	5.0187E-01
4	4.0331E-01	4.0756E-01	4.0755E-01	4.0813E-01
146	1.6995E-13	6.8424E-14	6.8468E-14	7.2473E-14
147	1.3905E-13	5.5621E-14	5.5657E-14	5.8936E-14
148	1.1377E-13	4.5214E-14	4.5244E-14	4.7928E-14
149	9.3082E-14	3.6755E-14	3.6779E-14	3.8976E-14
150	7.6158E-14	2.9878E-14	2.9898E-14	3.1696E-14

EXAMPLE 5.11 (Pade approximation vs Erlang approximation to deterministic). Consider the M/D/16 system. We wish to compare the results that would be obtained by using Pade (n,ℓ) and Erlang-K approximations to the deterministic service time for various values of n, ℓ, and K.

We do this by comparing the survivor functions obtained using the Erlang-K distribution and survivor function obtained from the Pade approximation with $n=\ell$.

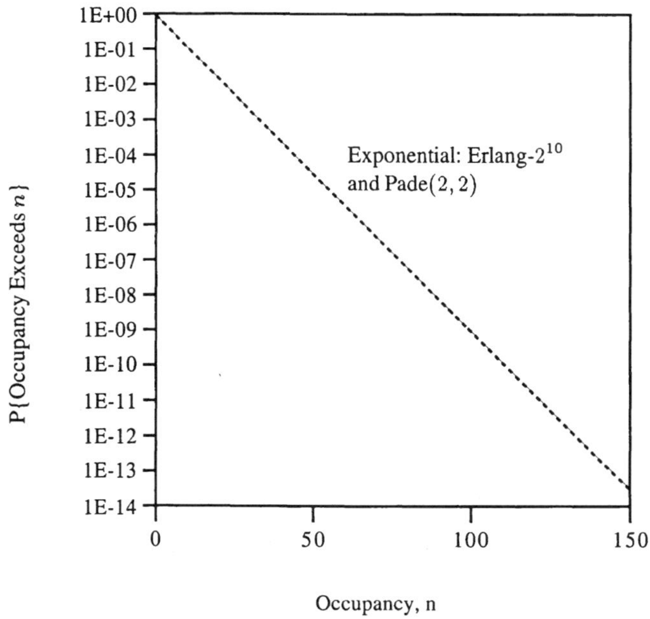

Figure 5.7. Survivor functions with Erlang-2^{10} and Pade$(2,2)$ service, Poisson arrivals, and a traffic load of 0.9.

Figures 5.8 and 5.9 show the survivor functions for the queue length distributions obtained for these approximations. Note that the results are not all actually survivor functions; some of the probabilities computed are actually negative for larger values of K. In fact, the Erlang approximation breaks down at values below $K = 64$ for this particular computational technique, but in any event, the approximation to deterministic is not very good over any range of K where the computational procedure appears to be stable. On the other hand, the results obtained for the Pade approximations are virtually unaffected by choice of n for values of n in the range of 16 to 128.

EXAMPLE 5.12 (Bulk service with random bulk size). Suppose that a system has poisson arrivals and deterministic service time, but that during each service period, the server serves a random number of units, say $\tilde{c}$, which has support on the integers in $[0, C]$. We wish to compare the queue lengths for the special case where $\tilde{c}$ has the binomial distribution with parameters 64 and 0.25 to the case where $\tilde{c}$ is deterministic with $C = 16$ with a load of $\rho = 0.9$.

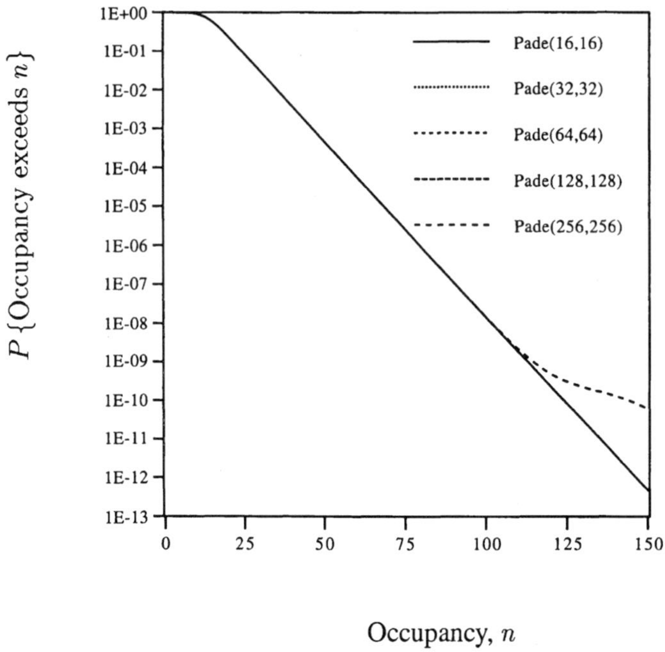

Figure 5.8. Survivor functions for deterministic (16) batch sizes with Pade(n, ℓ)-approximated deterministic service and Poisson arrivals at a traffic load of 0.9 for various choices of $n = \ell$.

Solution: For the case of general service times, it can then be shown that

$$\mathcal{F}_{\tilde{b},j}(z) = \mathcal{F}_{\tilde{v}}(z) \left[\sum_{i=0}^{j} z^i P\{\tilde{c} = j - i\} + P\{\tilde{c} > j\} \right],$$

and

$$\mathcal{F}_{\tilde{a}}(z) = \mathcal{F}_{\tilde{v}}(z) z^C \mathcal{F}_{\tilde{c}}(z^{-1}),$$

where, as usual, $\mathcal{F}_{\tilde{v}}(z)$ denotes the probability generating function for the number of arrivals that occur during a service time and $\mathcal{F}_{\tilde{c}}(z)$ denotes the probability generating function for the number of units served during a service interval.

As always, $\mathcal{F}_{\tilde{v}}(z) = F_{\tilde{x}}^*(\lambda[1 - z])$. Since $\rho = 0.9$, we need $\lambda = 0.9 \times 16 = 14.4$. Figure 5.10 shows the survivor function for the two cases, where a Pade $(32, 32)$ approximation to the deterministic service time distribution has been taken. From this figure, it is quite obvious that there is a significant difference between the queueing behavior of the two systems. For example, for the random batch size case, the probability that there will be more than

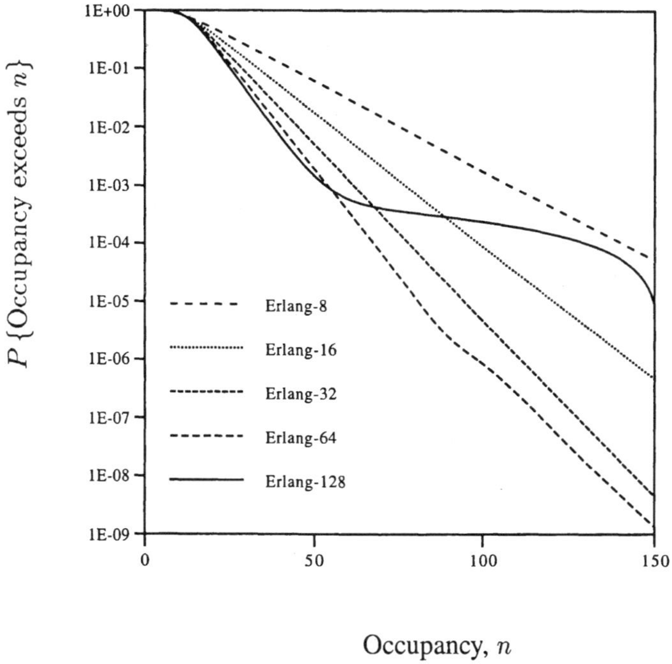

Figure 5.9. Survivor functions for deterministic (16) batch sizes with Erlang-K-approximated deterministic service and Poisson arrivals at a traffic load of 0.9 for various choices of K.

75 packets in the queue is approximately 5×10^{-4} while the probability that the queue size exceeds 75 in the deterministic batch-size case is only about 2.5×10^{-6}. Thus, it is about 200 times more likely to find a queue length exceeding 75 in the case of binomially distributed batch sizes with a mean of 16 than it is in the case of a system serving fixed batches of size 16.

We now briefly discuss computational strategy. There are two major parts of the computational procedure: specification of the D, N, A, and E matrices, and solving the remaining equations. The elements of the matrices D, N, A, and E are expressible in terms of the coefficients of $d(z)$ and $n_j(z)$ as follows: $E = \text{diag }(1, 1, \ldots, 1, d_0)$,

$$D = \begin{bmatrix} d_0 & d_1 & d_2 & \cdots & d_{\nu-1} \\ 0 & d_0 & d_1 & \cdots & d_{\nu-2} \\ 0 & 0 & d_0 & \cdots & d_{\nu-3} \\ \vdots & \vdots & \ddots & \ddots & \cdots \\ 0 & 0 & 0 & \cdots & d_0 \end{bmatrix}, \quad A = \begin{bmatrix} 0 & 0 & \cdots & \cdots & -d_\nu \\ 1 & 0 & \cdots & \cdots & -d_{\nu-1} \\ 0 & 1 & \cdots & \cdots & -d_{\nu-2} \\ \vdots & \ddots & \ddots & \ddots & \vdots \\ 0 & 0 & \cdots & 1 & -d_1 \end{bmatrix},$$

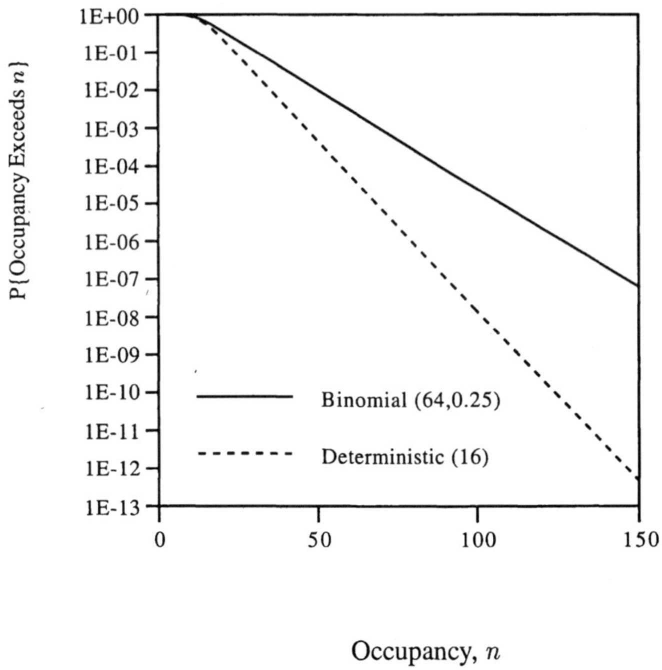

Figure 5.10. Survivor functions for binomial (64,0.25) and deterministic (16) batch sizes with deterministic service approximated by a Pade$(32, 32)$ approximation and Poisson arrivals at a traffic load of 0.9.

and

$$N = \begin{bmatrix} n_{01} & n_{02} & \cdots & n_{0,\nu-1} \\ n_{11} & n_{12} & \cdots & n_{1,\nu-1} \\ \vdots & \vdots & \cdots & \vdots \\ n_{C-1,1} & n_{C-1,2} & \cdots & n_{C-1,\nu 1} \end{bmatrix},$$

where $\nu = \max\{\nu_d, \nu_n + 1\}$, and

$$\begin{aligned} d_i &= w_{i-C} - u_i, \text{ for } i = 0, 1, \ldots, \nu, \\ n_{j,i} &= v_{j,i} - w_{i-j}, \text{ for } i = 0, 1, \ldots, \nu - 1, \text{ and } j = 0, 1, \ldots, C - 1. \end{aligned}$$

Because ν, the d_i, and the $n_{j,i}$ are always determined by the same formulae, the computational procedures are identical in all cases once C, $w(z)$, $u(z)$, and $v_j(z)$, $j = 0, 1, \ldots, C - 1$ are known. Therefore, a sensible approach is to write a procedure that delivers the solution given C, $w(z)$, $u(z)$, and $v_j(z)$, $j = 0, 1, \ldots, C - 1$ and then invoke this procedure from a program that generates C, $w(z)$, $u(z)$, and $v_j(z)$, $j = 0, 1, \ldots, C - 1$, passing this information to the solution procedure via a data structure and receiving the solution via a returned data structure.

As an example, a sensible data structure for the input, in pseudocode, would appear to be as shown in Table 5.7.

Table 5.7. Possible data structure for representing the input parameters in a program to implement the scalar case of the generalized state space approach.

data type	*variable*	*comment*
int	C;	Number of boundaries
int	Nv[0..C-1];	Degrees of the polynomials $v_j(z)$
double	v[0..C-1][0..Nv]	Array of coefficients of $v_j(z)$
int	Nu	Degree of the polynomial $u(z)$
double	u[0..Nu]	Coefficients of $u(z)$
int	Nw	Degree of the polynomial $w(z)$
double	w[0..Nw]	Coefficients of $w(z)$

Outputs from the program can also be specified in a uniform way. The solution is completely specified by C, x_0, ν, g, F, and H. Thus, a data structure that captures these characteristics would be sufficient to return all of the results. A sensible data structure would be as shown in Table 5.8.

Table 5.8. Possible data structure for representing the output parameters in a program to implement the scalar case of the generalized state space approach.

data type	*variable*	*comment*
int	C;	Number of boundaries
double	x0[0..C-1]	Boundary probabilities
int	Ns;	Number stable eigenvalues; dimension of F, g, and H
double	g[0..Ns-1]	Row vector
double	F[0..Ns-1][0..Ns-1]	Geometric rate matrix
double	H[0..Ns-1]	Column vector

The results desired by the user could then be computed by postprocessing the output data structure.

5.3 Expected Values For M/G/1 Via Renewal Theory

In this section, we present methodology for direct computation of expected waiting and sojourn times as well as busy-period lengths. We begin our presentation by reviewing our approach to computing the expected waiting time for the M/M/1 system and showing where this approach fails when applied to the M/G/1 system. At that point, we introduce renewal processes and present a few elementary but useful results from renewal theory. These results are then used to complete the derivation of the expected waiting time for the M/G/1 system. Next, we turn to the direct computation of the expected length of the busy period. At this point, we introduce alternating renewal processes and state a major result from the theory of alternating renewal processes in the form of a theorem. The theorem is then used to compute the expected length of the busy period directly.

5.3.1 Expected Waiting Times and Renewal Theory

We indicated earlier via an exercise that the expected waiting and sojourn times for the M/M/1 queueing system can be computed directly by applying Little's result in combination with the memoryless property of the exponential distribution. In particular, we suggested that the waiting time is the sum of the waiting time due to the customers in the queue, $\tilde{w}_q$, and the waiting time due to the customers in service, $\tilde{w}_s$, if any. That is,

$$E[\tilde{w}] = E[\tilde{w}_q] + E[\tilde{w}_s]. \tag{5.87}$$

Now, because the service times of the customers in the queue are independent of the number of customers in the queue,

$$E[\tilde{w}_q] = E[\tilde{n}_q]E[\tilde{x}], \tag{5.88}$$

where $\tilde{n}_q$ denotes the expected number of customers in the queue. Also,

$$E[\tilde{w}_s] = \rho E[\tilde{x}_s], \tag{5.89}$$

where ρ is the probability that there is a customer in service at an arbitrary point in time and $\tilde{x}_s$ denotes the remaining service time for the customer in service, if any. Using $E[\tilde{n}_q] = \lambda E[\tilde{w}_q]$ in (5.88) and then substituting the result together with (5.89) into (5.87) leads to

$$E[\tilde{w}] = \frac{\rho E[\tilde{x}_s]}{1-\rho}. \tag{5.90}$$

Because of the memoryless property of the exponential distribution, $E[\tilde{x}_s] = E[\tilde{x}] = 1/\mu$. Thus, for the M/M/1 system, we find

$$E[\tilde{w}] = \frac{\rho/\mu}{1-\rho}. \tag{5.91}$$

EXERCISE 5.32 Use a busy period argument to establish the validity of (5.90). [*Hint*: Consider the M/G/1 system under the nonpreemptive LCFS service discipline.]

A little thought reveals that (5.87) through (5.90) are still valid for the M/G/1 queueing system, but that (5.91) is no longer valid. That is, the service-time distribution is not memoryless so that the distribution of remaining time for the customer in service, if any, is not equal to the ordinary service-time distribution.

Since Poisson arrivals see the system in stochastic equilibrium, it is natural to conjecture that the expected length of time until the customer in service completes service would be one-half of the expected length of an ordinary service interval. But, we know that this quantity is equal to the entire expected length of a service interval if the service times are exponentially distributed. Thus there seems to be paradox here. The paradox, called the inspection paradox, is resolved by noting that the probability that random observers are more likely to observe longer intervals is higher than the probability that these longer intervals occur. The following example illustrates the paradox.

Suppose that $P\{\tilde{x} = 1\} = \frac{3}{4}$ and $P\{\tilde{x} = 2\} = \frac{1}{4}$. Now suppose we draw four customers at random without replacement from a group of thousands of customers and that the lengths of the service times corresponding to these customers are $x_{292} = 1$, $x_{2085} = 1$, $x_{1605} = 2$, and $x_{3176} = 1$. So, we have been lucky: We have drawn the customers such that the proportions of service times of each length in the sample are in exact proportion to their probabilities of occurrence. We now string these four intervals out in time and pick an arbitrary point in time in the interval covered by all four intervals as shown in Figure 5.11. If the point falls in an interval of length 2, we say that we "observe an interval of length 2." Because the total proportion of the line covered by intervals of length 2 is $\frac{2}{5}$, the probability that the interval observed has length 2 is $\frac{2}{5}$. If we let $\tilde{x}_o$ denote the length of the observed intervals, we find that $P\{\tilde{x}_o = 1\} = \frac{3}{5}$ and $P\{\tilde{x}_o = 2\} = \frac{2}{5}$. Thus, $E[\tilde{x}_o] = \frac{7}{5}$ while $E[\tilde{x}] = \frac{5}{4}$. We see that, in this particular example, $E[\tilde{x}_o] > E[\tilde{x}]$; that is, the expected length of the service times of the *observed* customers is greater than the expected service time taken over *all* customers.

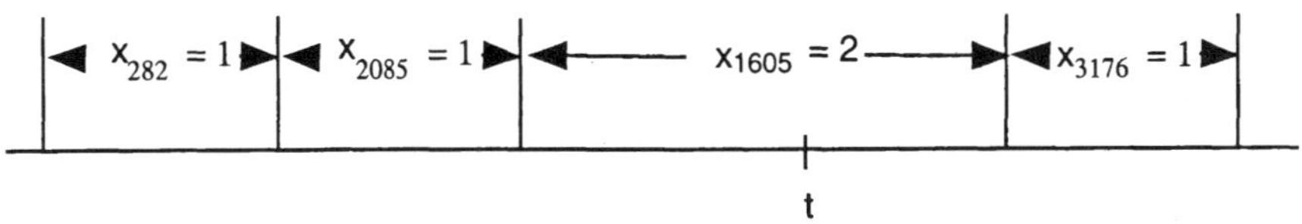

Figure 5.11. A sample of service times.

It turns out that, in the general case, the expected length of the observed intervals is greater than or equal to the expected length taken over all intervals with equality holding if and only if the intervals have fixed length.

DEFINITION 5.3 **Renewal process.** Let $\{\tilde{z}_i, i \geq 1\}$ denote a sequence of independent, identically distributed nonnegative random variables with $P\{\tilde{z} > 0\} > 0$ so that $E[\tilde{z}] > 0$, where $\tilde{z}$ denotes a generic $\tilde{z}_i$. Let $\tilde{s}_n = \sum_{i=1}^{n} \tilde{z}_i$ and $\tilde{s}_0 = 0$ with probability 1, and let $\{\tilde{n}(t), t \geq 0\} = \sup\{n|\tilde{s}_n \leq t\}$. Then the counting process $\{\tilde{n}(t), t \geq 0\}$ is called a renewal process, $\tilde{z}$ is called the renewal interval, and $\tilde{s}_n$ is called the time of the nth renewal.

For a general renewal process, it is intuitive that the probability that an interval of a particular length is observed is proportional to both the length in question and the probability of occurrence of an interval of the given length. That is,

$$P\{z \leq \tilde{z}_o \leq z + dz\} = KzP\{z \leq \tilde{z} \leq z + dz\}, \tag{5.92}$$

where $\tilde{z}_o$ denotes the length of the observed interval. Upon integrating both sides of (5.92), we find that $K = E[\tilde{z}]$ so that

$$\frac{d}{dz}F_{\tilde{z}_o}(z) = \frac{1}{E[\tilde{z}]}z\frac{d}{dz}F_{\tilde{z}}(z) \tag{5.93}$$

Now, from (5.93), we find that

$$E[\tilde{z}_o] = \frac{E[\tilde{z}^2]}{E[\tilde{z}]}. \tag{5.94}$$

Since $\text{Var}(\tilde{z}) = E[\tilde{z}^2] - E^2[\tilde{z}]$ and $\text{Var}(\tilde{z}) \geq 0$, we see that in general,

$$E[\tilde{z}_o] \geq E[\tilde{z}] \tag{5.95}$$

as we had observed earlier in our special case.

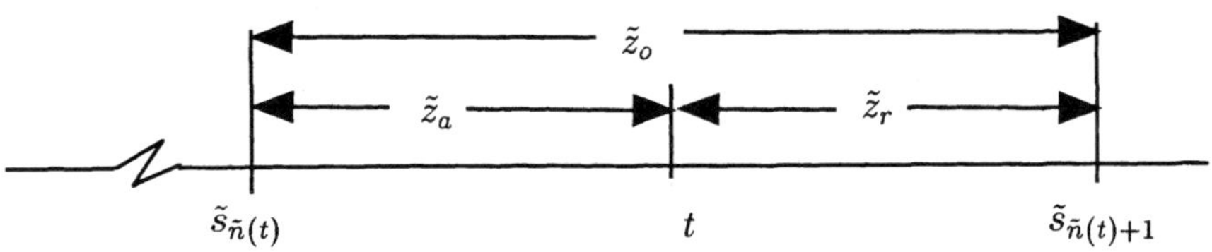

Figure 5.12. An observed interval of a renewal process.

Based on the above definition of the renewal process, we see that $\tilde{n}(t)$ is the number of renewals up to time t. This means that the time of the last renewal up to time t is given by $\tilde{s}_{\tilde{n}(t)}$, and the time of the first renewal after time t is given by $\tilde{s}_{\tilde{n}(t)+1}$. The interval $(\tilde{s}_{\tilde{n}(t)}, \tilde{s}_{\tilde{n}(t)+1}]$ is the observed interval as shown in Figure 5.12. Returning to the computation of the expected waiting time for the M/G/1 system, we note that if we condition on the system being busy, then the sequence of service times for the M/G/1 queueing system have the properties needed to form a renewal process. A little thought reveals that since the Poisson arrival observes the state of the system in exactly the same way as a random observer, then $E[\tilde{x}_s] = E\left[\text{length of } (t, \tilde{s}_{\tilde{n}(t)+1}]\right]$, and

$$E[\tilde{w}_s] = \rho\, E\left[\text{length of } (t, \tilde{s}_{\tilde{n}(t)+1}]\right].$$

That is, given that there is a customer in service upon an arbitrary customer's arrival, the expected waiting time due to the customer in service is given by the expected amount of time from the observance time until the customer in service completes service. One would guess that the expected length of this interval would be one-half the expected length of the observed interval. After the following definitions, we will specify the distribution of the length of this interval, and we will see that this is indeed the case.

DEFINITION 5.4 **Forward recurrence time (residual life).** The forward recurrence time or residual life for the renewal process $\{\tilde{n}(t), t \geq 0\}$ is defined as the interval $(t, \tilde{s}_{\tilde{n}(t)+1}]$. If $\tilde{z}$ denotes the length of a renewal interval, then the forward recurrence time for the renewal process will be denoted by $\tilde{z}_r$.

Based upon the above discussion, we see that $\tilde{x}_s$ of (5.90) is equivalent to the forward recurrence time of the renewal process whose underlying renewal interval is $\tilde{x}$. Thus,

$$E[\tilde{w}] = \frac{\rho E[\tilde{x}_r]}{1-\rho}. \tag{5.96}$$

DEFINITION 5.5 **Backward recurrence time (age).** The interval $(\tilde{s}_{\tilde{n}(t)}, t]$ is called the backward recurrence time or age for the renewal process $\{\tilde{n}(t), t \geq 0\}$. If $\tilde{z}$ denotes the length of a renewal interval, then the backward recurrence time for the renewal process will be denoted by $\tilde{z}_a$.

Now, $P\{\tilde{z}_r \leq r \mid \tilde{z}_o = z\} = r/z$ for $z \leq r$ because the point t is selected at random during the interval $\tilde{z}_o$. Thus, for $r \leq z$, we find

$$f_{\tilde{z}_r|\tilde{z}_o}(r \mid \tilde{z}_o = z) = \frac{d}{dr}P\{\tilde{z}_r \leq r \mid \tilde{z}_o = z\} = \frac{1}{z}.$$

The density function for $\tilde{z}_r$ is then given by

$$f_{\tilde{z}_r}(r) \quad = \quad \int_r^\infty \frac{d}{dr}P\{\tilde{z}_r \leq r \mid \tilde{z}_o = z\}dF_{\tilde{z}_o}(z)$$

$$
\begin{aligned}
&= \int_r^\infty \frac{1}{z} \frac{z f_{\tilde{z}}(z)}{E[\tilde{z}]} dz \qquad (5.97) \\
f_{\tilde{z}_r}(z) &= \frac{[1 - F_{\tilde{z}}(z)]}{E[\tilde{z}]}.
\end{aligned}
$$

EXERCISE 5.33 Show that the Laplace-Stieltjes transform for the distribution of the residual life for the renewal process having renewal intervals of length $\tilde{z}$ is given by

$$
F^*_{\tilde{z}_r}(s) = \frac{1 - F^*_{\tilde{z}}(s)}{s\, E[\tilde{z}]}. \qquad (5.98)
$$

We note $E[\tilde{z}] = \int_0^\infty [1 - F_{\tilde{z}}(z)]dz$ for $\tilde{z}$ a nonnegative random variable so that $f_{\tilde{z}_r}(z)$ is indeed a density function. Turning to the computation of the expected value of the residual service life, we find

$$
\begin{aligned}
E[\tilde{z}_r] &= \int_0^\infty r f_{\tilde{z}_r}(r) dr \\
&= \int_0^\infty r \left[\int_r^\infty \frac{1}{z} \frac{z f_{\tilde{z}}(z)}{E[\tilde{z}]} dz \right] dr \\
&= \frac{1}{E[\tilde{z}]} \int_0^\infty r \left[\int_r^\infty f_{\tilde{z}}(z) dz \right] dr \\
&= \frac{1}{E[\tilde{z}]} \int_0^\infty \left[\int_0^z r dr \right] f_{\tilde{z}}(z) dz \\
&= \frac{1}{E[\tilde{z}]} \int_0^\infty \frac{z^2}{2} f_{\tilde{z}}(z) dz.
\end{aligned}
$$

Thus,

$$
E[\tilde{z}_r] = \frac{E[\tilde{z}^2]}{2E[\tilde{z}]}. \qquad (5.99)
$$

Comparing (5.96) and (5.99), we see that

$$
E[\tilde{z}_r] = \frac{1}{2} E[\tilde{z}_o]
$$

as we would expect intuitively.

It is sometimes useful to express $E[\tilde{z}_r]$ in terms of the coefficient of variation of $F_{\tilde{z}}(z)$ which we will denote by $C_{\tilde{z}}$, where

$$
C_{\tilde{z}} \triangleq \frac{\sqrt{\mathrm{Var}(\tilde{z})}}{E(\tilde{z})}.
$$

Since $E[\tilde{z}^2] = \text{Var}(\tilde{z}) + E^2[\tilde{z}]$, we find that

$$E[\tilde{z}_r] = \frac{E[\tilde{z}](1 + C_{\tilde{z}}^2)}{2}. \tag{5.100}$$

Then, substituting (5.100) into (5.96) with $\tilde{z} = \tilde{x}$, we find for the M/G/1 system that

$$E[\tilde{w}] = \frac{\rho\, E[\tilde{z}]}{1-\rho}\left[\frac{1 + C_{\tilde{z}}^2}{2}\right]. \tag{5.101}$$

It is reasonable to question the general applicability of the theorems and definitions stated above. For example, the renewal intervals for a renewal process must all be drawn independently from the same distribution. One might ask, "What about processes for which the first renewal interval or the first several renewal intervals are drawn from a distribution other than the common distribution from which the length of all remaining intervals are drawn? Does this affect the stochastic equilibrium distribution of the length of observed intervals and backward and forward recurrence times?" The answer is that processes that are defective in this way behave the same as nondefective processes once they reach stochastic equilibrium. In this book, we take these results for granted, but we proceed carefully. These results may be found in any good text on stochastic processes (see, for example, Ross[1983]).

EXERCISE 5.34 For an arbitrary nonnegative random variable, $\tilde{x}$, show that

$$E\left[\tilde{x}_r^n\right] = \frac{E\left[\tilde{x}^{n+1}\right]}{(n+1)E[\tilde{x}]}. \tag{5.102}$$

EXERCISE 5.35 For the M/G/1 system, suppose that $\tilde{x} = \tilde{x}_1 + \tilde{x}_2$ where $\tilde{x}_1$ and $\tilde{x}_2$ are independent, exponentially distributed random variables with parameters μ_1 and μ_2, respectively. Show that $C_{\tilde{x}}^2 \leq 1$ for all μ_1, μ_2 such that $E[\tilde{x}] = 1$.

EXERCISE 5.36 Compute the expected waiting time for the M/G/1 system with unit mean deterministic service times and for the M/G/1 system with service times drawn from the unit mean Erlang-2 distribution. Plot on the same graph $E[\tilde{w}]$ as a function of ρ for these two distributions and for the M/M/1 queueing system with $\mu = 1$ on the same graph. Compare the results.

EXERCISE 5.37 For the M/G/1 system, suppose that $\tilde{x}$ is drawn from the distribution $F_{\tilde{x}_1}(x)$ with probability p and from $F_{\tilde{x}_2}(x)$ otherwise, where $\tilde{x}_1$ and $\tilde{x}_2$ are independent, exponentially distributed random variables with parameters μ_1 and μ_2, respectively. Let $E[\tilde{x}] = 1$. Show that $C_{\tilde{x}}^2 \geq 1$ for all $p \in [0, 1]$.

EXERCISE 5.38 With $\tilde{x}$ and p defined as in Exercise 5.37, let $p = \frac{1}{2}$. Find μ_1 and μ_2 such that $C_{\tilde{x}}^2 = 2$. Would it be possible to determine p, μ_1, and μ_2 uniquely for a given value of $C_{\tilde{x}}^2$? Explain.

5.3.2 Busy Periods and Alternating Renewal Theory

We now turn to the direct computation of the expected length of the busy period for the M/G/1 queueing system. We introduce alternating renewal processes and a major result from the theory of alternating renewal processes as a tool for approaching this computation. It will be seen that formulating problems in terms of alternating renewal processes provides a very powerful conceptual framework for dealing with important aspects of the behavior of complicated stochastic processes.

Alternating renewal processes are special types of renewal processes. In particular, an alternating renewal process is a renewal process in which the renewal interval comprises two subintervals. For example, consider the ordinary M/G/1 queueing system. Periods of time alternate between idle periods and busy periods. If we define a *cycle* to be the period of time between successive entries into idle periods, then the process that counts the number of cycles completed up to time t is an alternating renewal process.

Alternating renewal theory provides a very useful framework through which to conceptualize the functioning of more advanced queueing systems. In this subsection, we will present a formal definition of an alternating renewal process, state a basic theorem from the theory of alternating renewal processes, and compute the average length of a busy period by using the basic theorem. The theorem will not be proved, but an intuitive explanation of why it is true will be provided.

DEFINITION 5.6 **Alternating renewal process.** Let $\{\tilde{x}_i, i \geq 1\}$ and $\{\tilde{y}_i, i \geq 1\}$ denote sequences of independent and identically distributed nonnegative random variables, but with $\tilde{x}_i$ and $\tilde{y}_i$ not necessarily independent. Let $\tilde{x}$ and $\tilde{y}$ denote generic random variables for $\tilde{x}_i$ and $\tilde{y}_i$, respectively; and let $P\{\tilde{x} > 0\} > 0$ and $P\{\tilde{y} > 0\} > 0$ so that $E[\tilde{x}] > 0$ and $E[\tilde{y}] > 0$. Define $\tilde{z}_i = \tilde{x}_i + \tilde{y}_i$, and let $\tilde{z}$ denote a generic $\tilde{z}_i$. Further, define $\tilde{s}_n = \sum_{i=1}^{n} \tilde{z}_i$ and $\tilde{s}_0 = 0$ with probability 1, and let $\{\tilde{n}(t), t \geq 0\} = \sup\{n | \tilde{s}_n \leq t\}$. Then the counting process $\{\tilde{n}(t), t \geq 0\}$ is called an alternating renewal process, $\tilde{z}$ is called the renewal interval, and $\tilde{s}_n$ is called the time of the nth renewal.

We envision the alternating renewal process as alternating between x-periods and y-periods, and we think of each completion of an *x-period* followed by a *y-period* as the completion of a cycle. That is, time evolves as a succession of intervals as follows: $\tilde{x}_1, \tilde{y}_1, \tilde{x}_2, \tilde{y}_2, \ldots$. The following is a useful theorem from the theory of alternating renewal processes.

THEOREM 5.5 *Let $\{\tilde{n}(t), t \geq 0\}$ be an alternating renewal process that alternates between x-periods and y-periods as defined above. Then, the probability that the process is in an x-period at an arbitrary point in time is given by the ratio of the expected length of the x-period to the expected cycle length. That is,*

$$\begin{aligned} P_x &= \frac{E[\tilde{x}]}{E[\tilde{z}]} \\ &= \frac{E[\tilde{x}]}{E[\tilde{x}] + E[\tilde{y}]}, \end{aligned}$$

where

$$P_x = \lim_{t \to \infty} P\{\text{system is in an } x\text{–}period \text{ at time } t\}.$$

Similarly,

$$P_y = \frac{E[\tilde{y}]}{E[\tilde{x}] + E[\tilde{y}]},$$

where

$$P_y = \lim_{t \to \infty} P\{\text{system is in an } y\text{–}period \text{ at time } t\}.$$

Proof The truth of this theorem follows directly from the theory of Markov chains. We can define the process to be in state 0 whenever it is in an x-period and state 1 otherwise. Then, because the system alternates between x-periods and y-periods , it is clear that the proportion of transitions into each state is one-half. Then, from the theory of Markov chains, the proportion of time spent in state 0 is simply

$$\frac{(1/2)E[\tilde{x}]}{(1/2)E[\tilde{x}] + (1/2)E[\tilde{y}]} = \frac{E[\tilde{x}]}{E[\tilde{x}] + E[\tilde{y}]},$$

and the proportion of time spent in state 1 is

$$\frac{E[\tilde{y}]}{E[\tilde{x}] + E[\tilde{y}]}.$$

□

We will make use of the above theorem extensively in our study of priority queueing systems. Here, we use this theorem to determine the expected length of the busy period for the ordinary M/G/1 queueing system. As pointed out above, if we consider the idle periods as x-periods and the busy periods as y-periods, then the counting process that counts the number of cycles completed by time t is an alternating renewal process. Thus the probability that the process is in a busy period is given simply as the ratio of the expected length of the busy period to the expected length of the cycle. But the expected length of the cycle is simply the sum of the expected lengths of the idle and busy periods. Therefore,

$$P\{\text{busy}\} = \frac{E[\tilde{y}]}{E[\tilde{\imath}] + E[\tilde{y}]}.$$

But, the expected length of the idle period is $1/\lambda$, and the probability that the server is busy is found to be $\rho = \lambda E[\tilde{x}]$. Thus

$$\rho = \frac{E[\tilde{y}]}{1/\lambda + E[\tilde{y}]}.$$

Upon solving the above equation for $E[\tilde{y}]$, we readily find that

$$E[\tilde{y}] = \frac{1/\mu}{1-\rho},$$

as we previously determined in at least two other ways.

One can begin to appreciate the power of the seemingly trivial Theorem 5.5 by working a slightly more complicated example. The following exercise provides such an example.

EXERCISE 5.39 (Ross[1989]) Consider an ordinary renewal process with renewal interval $\tilde{z}$. Choose a real number c arbitrarily. Now suppose the renewal process is observed at a random point in time, t_0. If the age of the observed interval is less than c, define the system to be in an x-period, else define the system to be in a y-period. Thus, the expected cycle length is $E[\tilde{z}]$, and the expected length of the x-period is $E[\min\{c, \tilde{z}\}]$. Show that

$$E[\min\{c, \tilde{z}\}] = \int_0^c [1 - F_{\tilde{z}}(z)]dz$$

so that

$$\frac{d}{dz} F_{\tilde{z}_a}(z) = \frac{1 - F_{\tilde{z}}(z)}{E[\tilde{z}]},$$

as was shown in the previous subsection.

From the above example, we see that parameters of interest can sometimes be computed very simply by application of Theorem 5.5. In addition, the proof of the theorem explains why the expected length of the busy period is not affected by the form of the service-time distribution. The basic reason for this is simply that the probability that the server is busy is a time-averaged probability. For the time period prior to time t, the proportion of time spent in the busy period is simply the sum of the amount of time spent in the busy state divided by t. So long as there is at least one visit to the busy state, both numerator and denominator can be divided by the number of visits by time t, which is a random variable. An application of the strong law of large numbers and the elementary renewal theorem (see Wolff [1989]) then produces the desire result.

EXERCISE 5.40 Formalize the informal discussion of the previous paragraph.

It should be noted again that the truth of Theorem 5.5 does not depend upon independence of the length of x-periods and y-periods of the same cycle. The lengths of the x-periods must be drawn independently of each other from a common distribution, and the lengths of the y-periods must be drawn independently of each other from a common distribution, but there is no requirement that the lengths of the x-period and y-period of the same cycle be drawn independently of each other. For this particular example, however, the busy and idle periods are independent of each other.

As in the case of ordinary renewal process and defective renewal processes, if the lengths of the initial cycles have a distribution other than that of the common distribution from which all remaining intervals are chosen, the above theorem is still valid.

If the lengths of the x-periods and y-periods are drawn independently of each other, then it does not matter how one thinks of grouping the intervals of a cycle so long as a cycle consists of an x-period and a y-period. For example, one can think of a cycle as being y_i then x_i; x_i then y_i, or y_i then x_{i+1}. We will see in the next chapter that this property makes the theory even more useful in conceptualizing, from a mathematical point of view, the behavior of complicated systems.

5.4 Supplementary Problems

5-1 Consider a communication system in which messages are transmitted over a communication line having a capacity of C octets/sec. Suppose the messages have length $\tilde{m}$ (in octets), and the lengths are drawn from a geometric distribution having a mean of $E[\tilde{m}]$ octets, but truncated at a and b characters on the lower and upper ends of the distribution, respectively. That is, message lengths are drawn from a distribution characterized as follows:

$$P\{\tilde{m} = m\} = k\theta(1-\theta)^{m-1} \quad \text{for} \quad a \le m \le b,$$

where $\tilde{m}$ is the number of characters in a message and k is a normalizing constant.

(a) Given that

$$P\{\tilde{m} = m\} = k\theta(1-\theta)^{m-1} \quad \text{for} \quad a \leq m \leq b,$$

show that

$$k = \left[(1-\theta)^{a-1} - (1-\theta)^{b}\right]^{-1},$$

$$E[z^{\tilde{m}}] = z^{(a-1)} \frac{\theta z}{1-(1-\theta)z} \frac{1-[(1-\theta)z]^{(b-[a-1])}}{1-(1-\theta)^{(b-[a-1])}},$$

and

$$E[\tilde{m}] = a - 1 + \frac{1}{\theta} - \frac{(b-[a-1])(1-\theta)^{(b-[a-1])}}{1-(1-\theta)^{(b-[a-1])}}.$$

(b) Rearrange the expression for $E[\tilde{m}]$ given above by solving for θ^{-1} to obtain an equation of the form

$$\frac{1}{\theta} = f(E[\tilde{m}], a, b, \theta),$$

and use this expression to obtain a recursive expression for θ of the form

$$\frac{1}{\theta_{i+1}} = f(E[\tilde{m}], a, b, \theta_i).$$

(c) Write a simple program to implement the recursive relationship defined in part (b) to solve for θ in the special case of $a = 10$, $b = 80$, and $E[\tilde{m}] = 30$. Use $\theta_0 = E^{-1}[\tilde{m}]$ as the starting value for the recursion.

(d) Argue that $F_{\tilde{x}}^*(s) = F_{\tilde{m}}^*(s/C)$, where C is the transmission capacity in octets/sec.

(e) Use a computer program to obtain the complementary occupancy distribution for the transmission system under its actual message length distribution at a traffic utilization of 95%, assuming a transmission capacity of 30 characters/sec.

(f) Compare this complementary distribution to one obtained under the assumption that the message lengths are drawn from an ordinary geometric distribution. Comment on the suitability of making the geometric assumption.

5-2 Using the properties of the probability generating function, determine a formula for $E[\tilde{n}^2]$, the second moment of the occupancy distribution for the ordinary M/G/1 system, in terms of the first three moments of $F_{\tilde{x}}(x)$, the service time distribution. Verify the formula for $E[\tilde{n}]$ along the way. [*Hint*: The algebra will be greatly simplified if (5.8) is first rewritten as

$$\mathcal{F}_{\tilde{n}}(z) = \alpha(z)/\beta(z),$$

where

$$\alpha(z) = (1-\rho)F^*_{\tilde{x}}(\lambda[1-z]),$$
$$\beta(z) = 1 - \rho F^*_{\tilde{x}_r}(\lambda[1-z]),$$

and $F_{\tilde{x}_r}(x)$ is the distribution for the forward recurrence time of the service time. Then, in order to find

$$\lim_{z\to 1} \frac{d^2}{dz^2}\mathcal{F}_{\tilde{n}}(z),$$

first find the limits as $z \to 1$ of $\alpha(z)$, $\beta(z)$, $d\alpha(z)/dz$, $d\beta(z)/dz$, $d^2\alpha(z)/dz^2$, $d^2\beta(z)/dz^2$, and then substitute these limits into the formula for the second derivative of the ratio.]

5-3 Jobs arrive to a single server system at a Poisson rate λ. Each job consists of a random number of tasks, $\tilde{m}$, drawn from a general distribution $F_{\tilde{m}}(m)$, independent of everything. Each task requires a service time drawn from a common distribution, $F_{\tilde{x}_t}(x)$, independent of everything.

(a) Determine the Laplace-Stieltjes transform of the job service-time distribution.

(b) Determine the mean forward recurrence time of the service-time distribution using the result of part (a) and transform properties.

(c) Determine the stochastic equilibrium mean sojourn time for jobs in this system.

(d) Determine the mean number of tasks remaining for a job in service at an arbitrary point in time, if any.

5-4 Consider a queueing system in which ordinary customers have service times drawn from a general distribution with mean $1/\mu$. There is a special customer who receives immediate service whenever she enters the system, her service time being drawn, independently on each entry, from a general distribution, $F_{\tilde{x}_s}(x)$, which has mean $1/\alpha$. Upon completion of service, the special customer departs the system and then returns after an exponential, rate β, length of time. Let $\tilde{x}_{si}$ denote the length of the ith interruption of an ordinary customer by the special customer, and let $\tilde{n}$ denote the number of interruptions. Also, let $\tilde{c}$ denote the time that elapses from the instant an ordinary customer enters service until the instant the ordinary customer departs.

(a) Suppose that service time for the ordinary customer is chosen once. Following an interruption, the ordinary customer's service resumes from the point of interruption. Determine $P\{\tilde{n} = n|\tilde{x} = x\}$, the conditional probability that the number of interruptions is n, and $\mathcal{F}_{\tilde{n}}(z)$, the probability generating function for the number of interruptions suffered by the ordinary customer.

(b) Determine $F_{\tilde{c}}^{*}(s)$, the Laplace-Stieltjes transform for $\tilde{c}$ under this policy. [*Hint*: Condition on the the length of the service time of the ordinary customer and the number of service interruptions that occur.]

(c) Compare the results of part (b) with the Laplace-Stieltjes transform for the length of the M/G/1 busy period. Explain the relationship between these two results.

(d) Determine $E[\tilde{c}]$ and $E[\tilde{c}^2]$.

(e) Determine the probability that the server will be busy at an arbitrary point in time in stochastic equilibrium, and the stability condition for this system.

5-5 Consider a queueing system that services customers from a finite population of K identical customers. Each customer, while not being served or waiting, *thinks* for an exponentially distributed length of time with parameter λ and then joins a FCFS queue to wait for service. Service times are drawn independently from a general service time distribution $F_{\tilde{x}}(x)$.

(a) Given the expected length of the busy period for this system, describe a procedure through which you could obtain the expected waiting time. [*Hint*: Use alternating renewal theory.]

(b) Given the expected length of the busy period with $K = 2$, describe a procedure for obtaining the expected length of the busy period for the case of $K = 3$.

5-6 For the M/G/∞ queueing system, it is well known that the stochastic equilibrium distribution for the number of busy servers is Poisson with parameter $\lambda E[\tilde{x}]$, where λ is the arrival rate and $\tilde{x}$ is the holding time.

(a) Suppose that $\tilde{x} = i$ with probability p_i for $i = 1, 2, 3$ with $p_1 + p_2 + p_3 = 1$. Determine the equilibrium distribution of the number of servers that are busy serving jobs of length i for $i = 1, 2, 3$, and the distribution of the number of servers that are busy serving all jobs.

(b) Determine the probability that a job selected at random from all of the jobs in service at an arbitrary point in time in stochastic equilibrium will have service time i, $i = 1, 2, 3$.

(c) Calculate the mean length of an arbitrary job that is in service at an arbitrary point in time in stochastic equilibrium.

(d) Suppose that job service times are drawn from an arbitrary distribution $F_{\tilde{x}}(x)$. Repeat part (c).

(e) What can be concluded about the distribution of remaining service time of a customer in service at an arbitrary point in time in stochastic equilibrium for the M/G/∞ system?

Chapter 6

THE M/G/1 QUEUEING SYSTEM WITH PRIORITY

In Chapter 5, we developed basic tools to analyze non-Markovian systems, and we used those tools to analyze the basic M/G/1 system. Towards the end of Chapter 5, we introduced renewal theory and discussed some elementary properties of M/G/1 systems in the context of renewal theory. In this chapter, we turn our attention to more complicated systems, namely those having priority.

By using the results of Section 5.3, we will see that the Pollaczek-Khintchine transform equations for the waiting and sojourn times can be expressed as geometrically weighted sums of random variables. This characteristic had long eluded logical explanation, but was finally explained in terms of the unfinished work for the M/G/1 system under the last-come-first-serve (LCFS) service discipline. This explanation, due to Kelly [1979] and Cooper and Niu [1986], is provided in Section 6.1. The material illustrates the analytical advantage of substituting a seemingly difficult queueing-system question for a relatively easy one that has the same solution.

In Section 6.2, we analyze the M/G/1 queueing system with exceptional first service; that is, the service times of all customers except the first customer of each busy period are chosen independently from a common distribution $F_{\tilde{x}}$ whereas the service time of the first customer of each busy period is chosen independently from the distribution $F_{\tilde{x}_e}$. We begin our development by deriving the Pollaczek-Khintchine transform equation of the occupancy distribution using the same argument by which Fuhrmann-Cooper decomposition was derived (Fuhrmann [1984] and Fuhrmann and Cooper [1985]); this approach avoids the difficulties of writing and solving difference equations. We then derive the probability generating function of the occupancy distribution for the M/G/1 queueing system with exceptional first service, again using the ideas of Fuhrmann-Cooper decomposition. Finally, we derive the probability generating function of the occupancy distribution for the M/G/1 queueing system with set-up as a variant of the M/G/1 queueing system with exceptional first service.

The techniques explored in the study of the M/G/1 queueing system with exceptional first service are used in Section 6.3 to study the M/G/1 queueing system with externally assigned priorities and head-of-the-line service. That is, the customers arriving belong to a certain priority group, where the arrival processes of the various classes are Poisson with their parameter dependent upon their class. There are K classes, and the service times for the class i customers are drawn independently of everything from the distribution $F_{\tilde{x}_i}$, $1 \leq i \leq K$. Transform equations are developed for the occupancy, waiting-time and sojourn-time distributions.

In Section 6.5, we develop expressions for the average waiting and sojourn times for the M/G/1 queueing system under both preemptive and nonpreemptive priority disciplines. This section is basically an extension of Section 5.3, and it concludes this chapter.

6.1 M/G/1 Under Last-Come-First-Served, Preemptive-Resume Discipline

Under the last-come-first-served, preemptive-resume (LCFS-PR) service discipline, newly arriving customers immediately enter into service. If there is currently a customer in service, that customer's service is suspended until service for the newly arrived customer and his descendants is completed. Then, service for the suspended customer is resumed. Clearly, the sojourn time for a customer has the same distribution as the length of the busy period in an ordinary (FCFS) M/G/1 system. That is,

$$F^*_{\tilde{s}_{LCFS-PR}}(s) = F^*_{\tilde{y}}(s), \tag{6.1}$$

where $\tilde{s}_{LCFS-PR}$ is the sojourn time for the M/G/1 under the LCFS-PR discipline and $F^*_{\tilde{y}}(s) = F^*_{\tilde{x}}[s + \lambda - \lambda F^*_{\tilde{y}}(s)]$.

EXERCISE 6.1 Argue the validity of (6.1).

EXERCISE 6.2 Derive an expression for the Laplace-Stieltjes transform of the sojourn-time distribution for the M/G/1 system under the LCFS-PR discipline conditional on the customer's service time requirement. [*Hint*: See Exercise 5.13].

EXERCISE 6.3 Compare the means and variances of the sojourn times for the ordinary M/G/1 system and the M/G/1 system under the LCFS-PR discipline.

Two other quantities of interest in this system are the unfinished work, $\tilde{u}$, and the occupancy distribution. Clearly the unfinished work for the LCFS-PR discipline is equivalent to the waiting time for the ordinary M/G/1 system. That is, the unfinished work in any work-conserving system is independent of order of service (Wolff [1970]), and because the Poisson arrival sees the system in stochastic equilibrium, the waiting time for the ordinary M/G/1 system is the same as the unfinished work for that system. Thus, for the M/G/1 system under the LCFS-PR discipline, $\tilde{u} = \tilde{w}$, and from (5.41), we find,

$$F_{\tilde{u}}^*(s) = F_{\tilde{w}}^*(s) = (1-\rho)\sum_{i=0}^{\infty}[\rho F_{\tilde{x}_r}^*(s)]^i. \tag{6.2}$$

The observation represented by (6.2) is attributed to Kelly [1979] in Cooper and Niu [1986].

We now turn to the computation of the occupancy distribution, for which purpose we follow Cooper and Niu [1986]. Clearly, the customers left in the system by any customer are exactly the same as the ones found in the system by that customer. Thus the distribution of the number of customers seen by a departing customer is certainly the same as the distribution of the number of customers found by an arriving customer. Consequently, the distribution of the number of customers left in the system by an arbitrary departing customer is the same as the stochastic equilibrium occupancy distribution.

Now, the customers in the queue at an arbitrary point in time have all been preempted at least once. Indeed, they have been preempted only while in service and then only by customers arriving from a Poisson process. Clearly, the remaining service time for all customers in the queue is independent and identically distributed. One would suspect that the distribution of the remaining service time for the customers in the queue is the same as the distribution of the residual service-life variables because the interrupting (observing) process is Poisson. Thus, we will denote the remaining service time for the customers in the queue by $\tilde{x}_r$.

Let $P_j = P\{j$ customers in the system at an arbitrary point in time$\}$. Then, clearly, from Littles's result, $P_0 = 1-\rho$. Also, for $j > 1$, an arbitrary customer who arrives at time t_0 (call this customer "tagged") will find j customers in the system if and only if one of the following two conditions holds:

1. the most recent epoch prior to t_0 was an arrival that found $j-1$ customers in the system, or

2. the most recent epoch prior to t_0 was a departure that left j customers is the system.

Since we have argued that the arrival, departure, and stochastic equilibrium occupancy distributions are identical, we find

$$P_j = P_{j-1}P\{\tilde{t} \le \tilde{x}\} + P_j P\{\tilde{t} \le \tilde{x}_r\} \tag{6.3}$$

where $\tilde{t}$ is the interarrival time. Thus

$$P_j = P_{j-1}\int_0^\infty (1 - e^{-\lambda x})dF_{\tilde{x}}(x) + P_j\int_0^\infty (1 - e^{-\lambda x})dF_{\tilde{x}_r}(x) \tag{6.4}$$

or

$$P_j = P_{j-1}[1 - F_{\tilde{x}}^*(\lambda)] + P_j[1 - F_{\tilde{x}_r}^*(\lambda)]. \tag{6.5}$$

After collecting terms, we find

$$P_j = \frac{1 - F_{\tilde{x}}^*(\lambda)}{F_{\tilde{x}_r}^*(\lambda)}P_{j-1} \tag{6.6}$$

or

$$P_j = \left(\frac{1 - F_{\tilde{x}}^*(\lambda)}{F_{\tilde{x}_r}^*(\lambda)}\right)^j P_0. \tag{6.7}$$

From the requirement that the probabilities sum to unity, we find that

$$P_0 = 1 - \frac{1 - F_{\tilde{x}}^*(\lambda)}{F_{\tilde{x}_r}^*(\lambda)}. \tag{6.8}$$

But, since $P_0 = 1 - \rho$, (6.8) implies

$$\rho = \frac{1 - F_{\tilde{x}}^*(\lambda)}{F_{\tilde{x}_r}^*(\lambda)}. \tag{6.9}$$

Substitution of (6.9) into (6.8) and the result into (6.7) yields the occupancy distribution

$$P_j = (1 - \rho)\rho^j. \tag{6.10}$$

From (6.10), we see that the occupancy distribution for the M/G/1 under the LCFS-PR discipline is independent of the service-time distribution and identical to that of the ordinary M/M/1 system. In addition, we find from (6.9) that

$$F_{\tilde{x}_r}^*(\lambda) = \frac{1 - F_{\tilde{x}}^*(\lambda)}{\lambda E[\tilde{x}]} \tag{6.11}$$

where $\rho = \lambda E[\tilde{x}] < 1$. Substituting s for λ in (6.11) yields

$$F_{\tilde{x}_r}^*(s) = \frac{1 - F_{\tilde{x}}^*(s)}{sE[\tilde{x}]}. \tag{6.12}$$

This expression holds for for all s for which $F^*_{\tilde{z}}(s)$ is defined by the analytic continuity property of regular functions. Since (6.12) and (5.98) are identical, the remaining service times for the customers in the queue for the M/G/1 having the LCFS-PR discipline are indeed given by the residual service time as conjectured; that is, $\tilde{z}_r = \tilde{x}_r$.

Having determined the occupancy distribution and the distribution of the remaining service time for the customer in the system, we can readily determine the distribution for the unfinished work at an arbitrary point in time. We find

$$F_{\tilde{u}}(u) = (1-p)\sum_{i=0}^{\infty} \rho^i [F_{\tilde{x}_r}(u)^{(i)}]. \tag{6.13}$$

where $[F_{\tilde{x}_r}(u)^{(j)}]$ denotes the j-fold convolution of $F_{\tilde{x}_r}(u)$ with itself, and $[F_{\tilde{x}_r}(u)^{(0)}] = 1$. Upon taking the Laplace-Stieltjes transform of both sides of (6.13) and comparing to (6.2), we readily find that $F^*_{\tilde{u}}(s) = F^*_{\tilde{w}}(s)$ as expected. We note in passing that the result given in (6.13) is valid for any work-conserving system.

Thus the behavior of the M/G/1 system under the LCFS-PR discipline provides an intuitive explanation for (6.2) as follows. The waiting-time distribution for the ordinary M/G/1 system is the same as the unfinished work in the M/G/1 system under the LCFS-PR discipline. The occupancy distribution under the LCFS-PR discipline is independent of the service-time distribution and identical to the occupancy distribution for the ordinary M/M/1 system. The remaining service times for the customers in the system are independent and identically distributed, and their distribution is the same as that of the residual service time. The unfinished work is then just the geometrically weighted sum of the j-fold convolutions of the residual service-time distribution.

6.2 M/G/1 System with Exceptional First Service

In this section, we develop transform equations for the M/G/1 queueing system with exceptional first service, that is, an M/G/1 system in which the first customer served in each busy period has a special service time, $\tilde{x}_e$, and all other customers have service time $\tilde{x}$. The M/G/1 queueing system with exceptional first service is interesting in its own right, but it is also a very useful tool in understanding priority queueing systems which are in turn extremely valuable in examining the behavior of many practical systems.

This section discusses four basic models. First, we present a very simple method for specifying the PGF for the number of customers left by an arbitrary departing customer in an ordinary M/G/1 queueing system. Our development is based upon a decomposition principle similar to that used by Fuhrmann and Cooper [1985] to examine the M/G/1 queueing system with vacations (which we will discuss later). This approach is more direct than the method presented earlier in this text because it does not require the specification and manipulation of recursive equations. Next, we use the decomposition principle to develop the PGF for the occupancy distribution for the M/G/1 queueing system with exceptional first service; the LSTs for the waiting- and sojourn-time distributions are left as exercises. We then consider the M/G/1 queueing system with set-up times, which is a special case of exceptional first service. Finally, we consider the M/G/1 queueing system with multiple vacations.

We begin our development for the M/G/1 system with exceptional first service by presenting an alternate derivation for the PGF of the occupancy distribution of the ordinary M/G/1 system. The development is more direct than that previously presented in that it does not involve solving recursive equations. In addition, the alternate development has the advantage of introducing extremely powerful analysis techniques in a simple setting.

The idea exploited is very simple; namely, the number of customers left in the queue by an arbitrary departing customer-which we will again refer to as the *tagged customer*-is independent of the order of service so long as the order of service is not based upon the service-time requirement. Thus we arrange the order of service in a conceptually simple way, and the Pollaczek-Khintchine transform equation appears as a result.

In a manner parallel to Fuhrmann and Cooper [1985], we organize servicing in the following way. First, we classify customers as belonging to one of the following two types:

1. Type 2 customers are those who arrive during the busy period but after the expiration of the first service of the busy period.

2. Type 1 customers are all customers who are not type 2, including those who arrive when the server is idle and therefore start busy periods.

Let $\tilde{x}_1$ denote the service time of the first service of the busy period. Then upon expiration of $\tilde{x}_1$, the system contains 0 or more type 1 customers and no type 2 customers. If there are 0 type 1 customers, the busy period ends. Otherwise, service of customers during the remainder of the busy period is organized as a sequence of type 2 *sub-busy periods* each of which is initiated by a type 1 customer. That is, following $\tilde{x}_1$, for the remainder of the busy period we begin service for a type 1 customer only when there are no type 2 customers waiting. Type 2 customers may then arrive and generate a sub-busy period. Upon expiration of the first sub-busy period we select another type 1 customer, if any are left, and we initiate another sub-busy period, which has the same statistics as the first sub-busy period. We continue this process until all type 1 customers have been served. The busy period ends upon expiration of the sub-busy period of the last type 1 customer.

We note that the dynamics of the system during each of these type 2 sub-busy periods is identical to that of an ordinary M/G/1 busy period. Thus, the

PGF for the number of type 2 customers left in the system by an arbitrary departing customer during the sub-busy period, $\tilde{n}_2$, is the same as that for the ordinary M/G/1 system;

$$\mathcal{F}_{\tilde{n}_2}(z) = \mathcal{F}_{\tilde{n}}(z), \tag{6.14}$$

which we assume to be unknown.

Now, the tagged customer may arrive either during a busy period or not. Since arrivals are Poisson, the probability that the tagged customer arrives during a busy period is given by the probability that the server is busy at an arbitrary point in time. By Little's result, this quantity is readily computed to be $\rho = \lambda E[\tilde{x}]$.

If the tagged customer arrives during the idle period, then the number of type 2 customers left in the system by the tagged customer is identically zero. Also, the number of type 1 customers left in the system under this condition is equal to the number of customers that arrive during $\tilde{x}_1$, the PGF for which is found to be $F^*_{\tilde{x}}(\lambda[1 - z])$ by Theorem 5.2, because $\tilde{x}_1$ is drawn from $F_{\tilde{x}}$.

If the tagged customer arrives during the busy period, then the tagged customer may leave both type 1 and type 2 customers in the system upon departure. Because the statistics of the sub-busy period are identical to those of the ordinary M/G/1 system, the PGF for $\tilde{n}_2$–the number of type 2 customers left in the system by the tagged customer given that the tagged customer arrived during a busy period–is given by (6.14).

We now consider the number of type 1 customers, $\tilde{n}_1$, left by the tagged customer given that customer arrived during a busy period. Because only one type 1 customer is served in any sub-busy period, these customers are exactly the type 1 customers left behind by the departing type 1 customer who started the sub-busy period in which the tagged customer is serviced. In turn, these type 1 customers are exactly the ones who arrived during $\tilde{x}_1$ but after the type 1 customer who started the sub-busy period in which the tagged customer is serviced.

Now, the sequence of first service times of the busy periods constitutes a sequence of renewal intervals in a renewal process. Thus the distribution of the remaining service time that an arbitrary type 1 customer sees is simply the distribution of the residual life of $\tilde{x}_1$. This quantity is given by $F_{\tilde{x}_r}(x)$, and the PGF for the number of type 1 customers who arrive during this period is thus given by $\mathcal{F}_{\tilde{n}_1}(z) = F^*_{\tilde{x}_r}(\lambda[1 - z])$, where $\tilde{n}_1$ denotes the number of type 1 customers left in the system by the tagged customer given that customer arrived during a busy period.

Since $\tilde{n}_1$ is independent of $\tilde{n}_2$, the PGF for the total number of customers left in the system by the tagged customer given that the tagged customer arrived during the busy period is given by

$$F^*_{\tilde{x}_r}(\lambda[1 - z])\mathcal{F}_{\tilde{n}}(z).$$

Therefore, upon conditioning on whether or not the tagged customer arrived during a busy period, we find

$$\mathcal{F}_{\tilde{n}}(z) = (1-\rho)F_{\tilde{x}}^*(\lambda[1-z]) + \rho F_{\tilde{x}_r}^*(\lambda[1-z])\mathcal{F}_{\tilde{n}}(z). \tag{6.15}$$

Upon solving this last equation for $\mathcal{F}_{\tilde{n}}(z)$, we find

$$\mathcal{F}_{\tilde{n}}(z) = \frac{(1-\rho)F_{\tilde{x}}^*(\lambda[1-z])}{1-\rho F_{\tilde{x}_r}^*(\lambda[1-z])}. \tag{6.16}$$

Of course, (6.16) is also the PGF for the distribution of the number of customers found in the system by an arbitrary arriving customer and the stochastic equilibrium occupancy distribution.

We now turn to the determination of the PGF for the stochastic equilibrium occupancy distribution for the M/G/1 system with exceptional first service. Our reasoning is almost identical to that used to obtain (6.15), but there are two differences. First, the probability that an arriving customer finds the system busy in stochastic equilibrium is no longer given by our previously defined ρ; and second, the initial service time is drawn from the distribution $F_{\tilde{x}_e}$ rather than from $F_{\tilde{x}}$.

Let P_{busy} denote the probability that the tagged customer arrives during the busy period. We then find that the PGF for the total number of customers left in the system in the case of exceptional first service is given by

$$\mathcal{F}_{\tilde{n}_e}(z) = (1-P_{\text{busy}})F_{\tilde{x}_e}^*(\lambda[1-z]) + P_{\text{busy}}F_{\tilde{x}_{er}}^*(\lambda[1-z])\mathcal{F}_{\tilde{n}}(z). \tag{6.17}$$

We now develop an expression for P_{busy} by using results from alternating renewal theory. From Takàcs [1962], and also from Exercise 5.12, we know that the expected length of the busy period in which the expected initial backlog is $E[\tilde{x}_e]$ is given by

$$\frac{E[\tilde{x}_e]}{1-\rho}, \tag{6.18}$$

where $\rho = \lambda E[\tilde{x}]$, and the expected length of the idle period is given by $1/\lambda$. Because the busy and idle periods form an alternating renewal process and the Poisson arrivals observe the system in stochastic equilibrium, we find that the probability that an arbitrary arriving customer finds the system busy is given by the ratio of the expected length of the busy period to the expected length of the renewal cycle; that is,

$$\begin{aligned} P_{\text{busy}} &= \frac{E[\tilde{x}_e]/(1-\rho)}{1/\lambda + E[\tilde{x}_e]/(1-\rho)} \\ &= \frac{\rho_e}{1-\rho+\rho_e} \end{aligned} \tag{6.19}$$

where ρ_e is defined to be $\lambda E[\tilde{x}_e]$.

Substituting (6.18) into (6.17), we find

$$\begin{aligned}\mathcal{F}_{\tilde{n}_e}(z) &= \frac{1-\rho}{1-\rho+\rho_e} F^*_{\tilde{x}_e}(\lambda[1-z]) \\ &\quad + \frac{\rho_e}{1-\rho+\rho_e} F^*_{\tilde{x}_{er}}(\lambda[1-z])\mathcal{F}_{\tilde{n}}(z). \qquad (6.20)\end{aligned}$$

A special case of exceptional first service is the M/G/1 queueing system with set-up times, which was studied by Levy and Kleinrock [1986]. For this system, we assume $\tilde{x}_e = \tilde{x}_s + \tilde{x}$, where $\tilde{x}_s$ represents a *set-up* time independent of $\tilde{x}$. For this special case, $\rho_e = \rho + \rho_s$, where $\rho_s = \lambda \tilde{x}_s$, and (6.19) reduces to

$$P_{\text{busy}} = \frac{\rho_e}{1+\rho_s}, \qquad (6.21)$$

where ρ_e is defined to be $\lambda E[\tilde{x}_e]$. Because $\tilde{x}_s$ and $\tilde{x}$ are independent, we find that $F^*_{\tilde{x}_e}(s) = F^*_{\tilde{x}}(s)F^*_{\tilde{x}_s}(s)$ and

$$\begin{aligned}F^*_{\tilde{x}_{er}}(s) &= \frac{\rho}{\rho_e} F^*_{\tilde{x}_r}(s)F^*_{\tilde{x}_s}(s) + \frac{\rho_s}{\rho_e} F^*_{\tilde{x}_{sr}}(s) \\ &= \frac{\rho}{\rho_e} F^*_{\tilde{x}_r}(s) + \frac{\rho_s}{\rho_e} F^*_{\tilde{x}_{sr}}(s)F^*_{\tilde{x}}(s). \qquad (6.22)\end{aligned}$$

The expression (6.22) can be obtained as follows. We consider an alternating renewal process for which the renewal interval is $\tilde{x} + \tilde{x}_s$. The forward recurrence time for the process can then be obtained by conditioning on whether a random observer observes the system during an $\tilde{x}$ period or during an $\tilde{x}_s$ period, the probabilities of which are $E[\tilde{x}]/\{E[\tilde{x}] + E[\tilde{x}_s]\}$ and $E[\tilde{x}_s]/\{E[\tilde{x}] + E[\tilde{x}_s]\}$, respectively. These probabilities can be rewritten as ρ/ρ_e and ρ_s/ρ_e by multiplying numerator and denominator by λ. Now, if the observer observes an $\tilde{x}$ period in progress, then the time until the end of the cycle is $\tilde{x}_r + \tilde{x}_s$, the LST of the distribution of which is given by $F^*_{\tilde{x}_r}(s)F^*_{\tilde{x}_s}(s)$. On the other hand, if the observer finds the system in an $\tilde{x}_s$ interval, then the time until the end of the cycle is $\tilde{x}_{sr}$, the residual life of $\tilde{x}_s$, the LST of the distribution of which is $F^*_{\tilde{x}_{sr}}(s)$. We note that it is natural to think of $\tilde{x}_s$ as preceding $\tilde{x}$, but the distribution of the remaining time in the cycle is the same as that with $\tilde{x}$ preceding $\tilde{x}_s$.

The PGF for the equilibrium occupancy distribution for the M/G/1 queueing system with set-up times can now be obtained by substituting (6.22) into (6.20). We find

$$\begin{aligned}\mathcal{F}_{\tilde{n}_s}(z) &= \frac{1-\rho}{1+\rho_s} F^*_{\tilde{x}}(\lambda[1-z])F^*_{\tilde{x}_s}(\lambda[1-z]) + \frac{\rho_e}{1+\rho_s} \\ &\quad \Big(\frac{\rho}{\rho_e} F^*_{\tilde{x}_r}(\lambda[1-z])F^*_{\tilde{x}_s}(\lambda[1-z])\end{aligned}$$

$$+\frac{\rho_s}{\rho_e} F^*_{\tilde{x}_{sr}}(\lambda[1-z])\Big) \mathcal{F}_{\tilde{n}}(z), \tag{6.23}$$

which, with a minimum of algebra, can be reduced to

$$\mathcal{F}_{\tilde{n}_s}(z) = \left(\frac{1}{1+\rho_s} F^*_{\tilde{x}_s}(\lambda[1-z]) + \frac{\rho_s}{1+\rho_s} F^*_{\tilde{x}_{sr}}(\lambda[1-z]) \right) \mathcal{F}_{\tilde{n}}(z). \tag{6.24}$$

The form of (6.24) suggests that the number of customers left in the system by an arbitrary departing customer can be obtained as the sum of two independent random variables as pointed out by Fuhrmann and Cooper [1985]. We modify our definition of type 1 and type 2 customers to facilitate the explanation: type 1 customers are those who arrive before the first service of the busy period begins; all other customers are of type 2. The $\mathcal{F}_{\tilde{n}}(z)$ part of the expression then corresponds to the number of type 2 customers left in the system. The grouped term of (6.24) corresponds to the number of type 1 customers and has a simple interpretation.

Interpretation of the grouped term of (6.24) is as follows. Note that the distribution of the number of type 1 customers left in the system by an arbitrary departing customer is the same as the distribution of the number of type 1 customers left by an arbitrary departing type 1 customer, there being one type 1 customer for each sub-busy period. Now, if a type 1 customer arrives to find the system empty, then the number of type 1 customers left by that customer is exactly the number which arrive during the set-up time. If, on the other hand, the type 1 customer arrives while the set-up is in progress, because the type 1 customer arrivals are Poisson, the number of type 1 customers left is the number who arrive during the forward recurrence time of the renewal process in which the underlying renewal interval is $\tilde{x}_s$. The relative probabilities of these two events are readily obtained by setting up an alternating renewal process over the intervals of time during which type 1 customers arrive. The underlying renewal interval is then the length of an idle period plus the length of the set-up interval. Hence, the probability that a random observer finds the system in an idle interval is the ratio of the expected length of a idle period to the expected length of the cycle,

$$\frac{1/\lambda}{1/\lambda + E[\tilde{x}_s]} = \frac{1}{1+\rho_s}.$$

Also, the probability that a random observer finds the system in a set-up interval is the ratio of the expected length of a set-up period to the expected length of the cycle, which is

$$\frac{\rho_s}{1+\rho_s}.$$

Note that the proportion of type 1 customers that arrive while the system is in a set-up period is also given by this ratio. Thus the PGF for the number of type 1 customers left in the system by an arbitrary customer is

$$\frac{1}{1+\rho_s}F^*_{\tilde{x}_s}(\lambda[1-z]) + \frac{\rho_s}{1+\rho_s}F^*_{\tilde{x}_{sr}}(\lambda[1-z]).$$

Alternate forms for the expressions (6.20) and (6.24) that use only the ordinary service-time distributions rather than both ordinary distributions and residual life distributions are now given. Recall that for any nonnegative random variable, $\tilde{x}$,

$$F^*_{\tilde{x}_r}(s) = \frac{1 - F^*_{\tilde{x}}(s)}{sE[\tilde{x}]}. \tag{6.25}$$

In particular,

$$F^*_{\tilde{x}_{er}}(s) = \frac{1 - F^*_{\tilde{x}_e}(s)}{sE[\tilde{x}_e]}, \tag{6.26}$$

and

$$F^*_{\tilde{x}_{sr}}(s) = \frac{1 - F^*_{\tilde{x}_s}(s)}{sE[\tilde{x}_s]}. \tag{6.27}$$

If we substitute (6.26) and (6.27) into (6.20) and (6.24), respectively, the following alternate expressions can be obtained from simple algebra:

$$\mathcal{F}_{\tilde{n}_e}(z) = \frac{1-\rho}{1-\rho+\rho_e} \frac{zF^*_{\tilde{x}_e}(\lambda[1-z]) - F^*_{\tilde{x}}(\lambda[1-z])}{z - F^*_{\tilde{x}}(\lambda[1-z])}, \tag{6.28}$$

and

$$\begin{aligned}\mathcal{F}_{\tilde{n}_s}(z) &= \frac{\rho_s}{1+\rho_s} \frac{1 - zF^*_{\tilde{x}_s}(\lambda[1-z])}{\lambda[1-z]E[\tilde{x}_s]} \\ &\quad \frac{(1-\rho)(z-1)F^*_{\tilde{x}}(\lambda[1-z])}{z - F^*_{\tilde{x}}(\lambda[1-z])}.\end{aligned} \tag{6.29}$$

As we have previously pointed out in our discussions of (6.24), the product of the first two fractions of (6.29) is the probability generating function for the number of type 1 customers left in the system by an arbitrary departing customer. The third fraction is the familiar Pollaczek-Kintchine transform equation for the occupancy of the ordinary M/G/1 system, which is in turn, the probability generating function for the number of type 2 customers left in the system by an arbitrary departing customer.

An interesting and useful variation of the M/G/1 system with set-up times is the M/G/1 system with server vacations (Cooper [1972], [1981]). In the simplest version of this system, the server takes a *vacation* each time the queue becomes empty. Upon return from each vacation, the server begins a busy period if any customers are waiting; otherwise, the server takes another vacation. This model is called the *multiple vacation model* as opposed to the *single vacation model*, in which the server remains idle once having returned from vacation if there are no customers present. The duration of each vacation is a random variable $\tilde{x}_v$ drawn from the distribution $F_{\tilde{x}_v}$.

If we now define the type 1 customers as those who arrive during vacation periods, we can readily see that the distribution of the number of type 1 customers left by an arbitrary departing customer is the same as the distribution of the number of customers who arrive from the Poisson process during the residual life of the vacation period. The primary distinction between this multiple vacation model and the set-up time model is that *all* of the type 1 customers in the vacation model arrive during the server vacation while all type 1 customers except the first arrive during the server set-up time with the first customer arriving during the idle period. From (6.24) and (6.25), we find that the probability generating function for the number of customers left in the system by an arbitrary departing customer is, therefore,

$$\mathcal{F}_{\tilde{n}_v}(z) = \frac{1 - F^*_{\tilde{x}_v}(\lambda[1-z])}{\lambda[1-z]E[\tilde{x}_v]} \frac{(1-\rho)(z-1)F^*_{\tilde{x}}(\lambda[1-z])}{z - F^*_{\tilde{x}}(\lambda[1-z])}. \tag{6.30}$$

Vacation models have a number of interesting applications. For example, consider a variation of the M/G/1 queue in which the server works in the following way (Wortman and Disney [1990]).

> The server works on an auxiliary task for a period of time $\tilde{v}$ and then checks the system occupancy. If there are at least K customers waiting, the server serves a batch of K customers and then returns to the auxiliary task, regardless of the queue length. If there are less than K customers waiting, the server immediately returns to the auxiliary task.

Note that in the general analysis of vacation systems, the successive vacation periods are not required to be mutually independent. Some interesting applications of server vacation models include the study of server breakdowns and polling systems. The reader is referred to Doshi [1986, 1990], Levy and Sidi [1990], and Takagi [1986a, 1986b, 1990] for articles that survey the application of vacation models.

This concludes our discussion of the M/G/1 queueing system with exceptional first service and its variants. We will use the ideas presented here later in the development of the transform equations for priority queueing systems. It should be noted that $\mathcal{F}_{\tilde{n}_e}(z)$ and $\mathcal{F}_{\tilde{n}_s}(z)$ can be readily inverted using the methods based on discrete Fourier transforms, which were presented earlier in this chapter to obtain the distributions of $\tilde{n}_e$ and $\tilde{n}_s$, respectively.

6.3 M/G/1 under Head-of-the-Line Priority

We now turn our attention to the analysis of queueing systems having externally assigned priorities, that is, priorities that are assigned prior to or upon entry into the system. We assume that there is an integer number, I, of customer classes. Class i, $1 \leq i \leq I$, customers arrive to the system according to a Poisson process with rate λ_i, and their service times are drawn independently from the distribution $F_{\tilde{x}_i}(x)$. Class i customers have priority over class j customers if $i < j$. Upon arrival, a customer joins the queue ahead of all customers whose priority is lower than that of the arriving customer and behind all customers whose priority is at least as high as the arriving customer. The service discipline is illustrated in Figure 6.1.

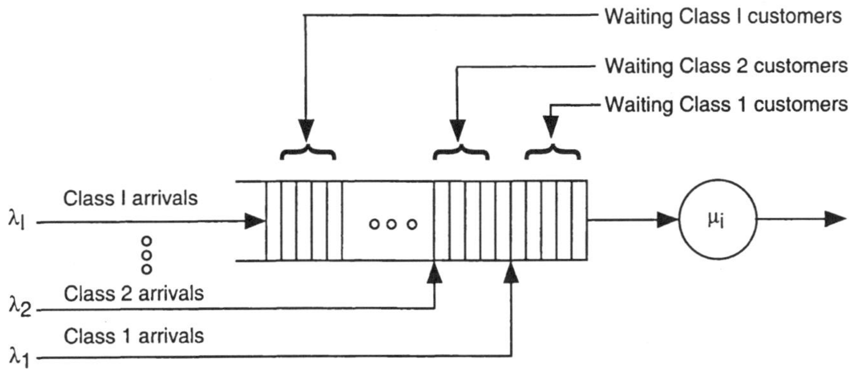

Figure 6.1. HOL service discipline.

There are two primary variations of this service discipline: nonpreemptive and preemptive resume. In the nonpreemptive version, denoted by HOL, servicing of a customer is never interrupted. That is, once servicing of a given customer begins, the server serves the customer to completion independent of the arrival process. Upon service completions, the server begins service on behalf of the customer who is currently at the head of the line.

In the case of preemptive resume, denoted by HOL-PR, an entering customer whose priority is higher than the customer currently in service immediately gains access to the server. Servicing of the preempted customer resumes from the point at which it was preempted as soon as there are no longer any customers present in the system whose priority exceeds that of the interrupted customer. Thus, under the HOL-PR discipline, customers of a given priority never receive service while any higher priority customers are in the system. There may be no more than one customer of a given priority preempted at any given time, and the maximum possible number of preempted customers is $I - 1$.

We analyze the ordinary HOL system for the special case of $I = 2$, and leave extension to the case of arbitrary I and the HOL-PR discipline to the exercises. We derive probability generating functions for the occupancy distribution of each customer class, and then use these distributions to specify the Laplace-Stieltjes transforms for the waiting- and sojourn-time distributions. Our derivation is based upon a variation of the Fuhrmann-Cooper decomposition principle discussed earlier in combination with results on alternating renewal processes from Section 5.3 and exceptional first service from Section 6.2. Customers with higher and lower priorities are examined in separate subsections in hopes of avoiding confusion about definitions of types and classes for the different points of view we will take.

6.3.1 Customers with Higher Priority

As we did in the case of the ordinary M/G/1 analysis using the principles of Fuhrmann-Cooper decomposition, we will obtain the probability generating function for the ergodic occupancy distribution by computing the probability generating function for the distribution of the number of customers left in the system by an arbitrary departing customer in stochastic equilibrium, because these two distributions are equal. We will designate an arbitrary customer as the tagged customer. Because arrivals are Poisson, the tagged customer sees the system in stochastic equilibrium as we stated earlier.

To begin our development, we note that at any given time, the server is in one of three possible states: idle, busy serving a class 1 customer, or busy serving a class 2 customer-the respective ergodic probabilities being $1-\rho_1-\rho_2$, ρ_1, and ρ_2, where $\rho_i = \lambda_i E[\tilde{x}_i]$. We separate the class 1 customers into two groups as follows:

1. Type 1 customers are those class 1 customers who arrive to the system either during an idle period or during a period during which a class 2 customer is being serviced, and

2. Type 2 customers are those class 1 customers who arrive to the system while a class 1 customer is being serviced.

As in the case of the ordinary M/G/1 queueing system, we envision all servicing as being organized as a series of sub-busy periods, all of which are started by type 1 customers, generated by type 2 customers, and are statistically identical to ordinary M/G/1 busy periods in which the arrival rate is λ_1 and the service times are drawn from the distribution $F_{\tilde{x}_2}$.

Let $\tilde{n}_{11}$ and $\tilde{n}_{12}$ denote the number of type 1 and type 2 customers, respectively, left in the system by an arbitrary departing class 1 customer, and let $\tilde{n}_1 = \tilde{n}_{11} + \tilde{n}_{12}$. Because all class 1 customers depart the system during one of the sub-busy periods just defined and all of the sub-busy periods are statistically identical, the distributions of $\tilde{n}_{11}$ and $\tilde{n}_{12}$ are the same as the distributions of the numbers of type 1 and type 2 customers left in the system by an arbitrary customer departing during an arbitrary sub-busy period. Thus we can study the system by studying an arbitrary sub-busy period.

Since type 1 and type 2 customers arrive according to a Poisson process over nonoverlapping intervals of time, $\tilde{n}_{11}$ and $\tilde{n}_{12}$ are statistically independent. Thus by Theorem 5.1 we see that

$$\mathcal{F}_{\tilde{n}_1}(z) = \mathcal{F}_{\tilde{n}_{11}}(z)\mathcal{F}_{\tilde{n}_{12}}(z). \tag{6.31}$$

We will compute $\mathcal{F}_{\tilde{n}_{11}}(z)$ and $\mathcal{F}_{\tilde{n}_{12}}(z)$ separately in the following paragraphs and then combine the result to obtain $\mathcal{F}_{\tilde{n}_1}(z)$.

From the definitions of type 1 and type 2 customers and the remark following their definition, it is clear that the distribution of the number of type 2 customers left in the system by an arbitrary departing customer is identically the same as that of an ordinary M/G/1 system in which the arrival rate is λ_1 and the service times are drawn from the distribution $F_{\tilde{x}_1}(x)$. Thus we find from the Pollaczek-Khintchine transform equation for the occupancy distribution,

$$\mathcal{F}_{\tilde{n}}(z) = \frac{(1-\rho)F^*_{\tilde{x}}(\lambda[1-z])}{1-\rho F^*_{\tilde{x}_r}(\lambda[1-z])},$$

that

$$\mathcal{F}_{\tilde{n}_{12}}(z) = \frac{(1-\rho_1)F^*_{\tilde{x}_1}(\lambda_1[1-z])}{1-\rho_1 F^*_{\tilde{x}_{1r}}(\lambda_1[1-z])}. \tag{6.32}$$

It remains to specify $\mathcal{F}_{\tilde{n}_{11}}(z)$. We note that the type 1 customers left by an arbitrary departing customer during a sub-busy period are identically those type 1 customers left by the first departing customer from the sub-busy period. But the first departing customer from an arbitrary sub-busy period is simply an arbitrary type 1 customer. Hence we define the tagged type 1 customer to be the type 1 customer who started the busy period during which the tagged customer is served. Then, the distribution of the number of type 1 customers left by the tagged class 1 customer is the same as the distribution of the number of type 1 customers left by the tagged type 1 customer. These are, in turn, the same as the type 1 customers who arrive while an arbitrary type 1 customer is in the system.

If the tagged type 1 customer arrives while the server is idle, the number of type 1 customers left by the tagged type 1 customer upon departure is 0 with probability 1, and the probability generating function for this distribution is identically 1. This is because the tagged type 1 customer immediately begins service so that all class 1 customers who arrive during the time the customer is in the system are of type 2.

If, on the other hand, the tagged type 1 customer arrives while the server is serving a class 2 customer, then the period of time over which type 1 customers arrive following the arrival of the tagged type 1 customer is the same as the distribution of the forward recurrence time of the renewal process in which the lengths of the renewal intervals are drawn from the distribution $F_{\tilde{x}_2}(x)$. This is because the tagged type 1 customer is drawn arbitrarily from a Poisson stream[1]. Therefore, by Theorem 5.1, the probability generating function for the number of type 1 customers left by the tagged type 1 customer given that the tagged type 1 customer arrived during a class 2 service period is given by $F^*_{\tilde{x}_{2r}}(\lambda_1[1-z])$, where $\tilde{x}_{2r}$ denotes the residual life of $\tilde{x}_2$.

Combining the results of the previous two paragraphs, we find that

$$\mathcal{F}_{\tilde{n}_{11}}(z) = \frac{1-\rho_1-\rho_2}{1-\rho_1} + \frac{\rho_2}{1-\rho_1} F^*_{\tilde{x}_{2r}}(\lambda_1[1-z]), \tag{6.33}$$

where it is easily seen that

$$\frac{1-\rho_1-\rho_2}{1-\rho_1}$$

is the conditional probability that the server is idle given that the server is either idle or serving class 2 customers, and

$$\frac{\rho_2}{1-\rho_1}$$

is the conditional probability that the server is serving class 2 customers given that the server is either idle or serving class 2 customers. Upon combining (6.31)-(6.33), we find

$$\begin{aligned} \mathcal{F}_{\tilde{n}_1}(z) &= \left(\frac{1-\rho_1-\rho_2}{1-\rho_1} + \frac{\rho_2 F^*_{\tilde{x}_{2r}}(\lambda_1[1-z])}{1-\rho_1} \right) \\ &\quad \frac{(1-\rho_1)F^*_{\tilde{x}_1}(\lambda_1[1-z])}{1-\rho_1 F^*_{\tilde{x}_{1r}}(\lambda_1[1-z])}. \end{aligned} \tag{6.34}$$

EXERCISE 6.4 Compare the probability generating function for the class 1 occupancy distributions for the HOL system to that of the M/G/1 system with set-up times discussed in Section 6.2. Do they have exactly the same form? Explain why or why not intuitively.

Because the class 1 customers left in the HOL system by the departing class 1 customers are identically the customers which arrive during the class 1 sojourn time, we find directly from Theorem 5.1 that

$$F^*_{\tilde{s}_1}(s) = \left(\frac{1-\rho_1-\rho_2}{1-\rho_1} + \frac{\rho_2 F^*_{\tilde{x}_{2r}}(s)}{1-\rho_1} \right) \frac{(1-\rho_1)F^*_{\tilde{x}_1}(s)}{1-\rho_1 F^*_{\tilde{x}_{1r}}(s)}, \tag{6.35}$$

[1]The Class 1 arrival process is Poisson and thus has stationary and independent increments. Thus, if we observe the arrival process over only those intervals during which class 2 customers are being serviced, the observed arrival process will still have stationary and independent increments, and the probability of an arrival over an interval of length h will still be $\lambda_1 h$; hence the arrival process will still be Poisson with rate λ_1.

where $\tilde{s}_1$ denotes the Laplace-Stieltjes transform of the class 1 sojourn time. Similarly,

$$F^*_{\tilde{w}_1}(s) = \left(\frac{1 - \rho_1 - \rho_2}{1 - \rho_1} + \frac{\rho_2 F^*_{\tilde{x}_{2r}}(s)}{1 - \rho_1} \right) \frac{(1 - \rho_1)}{1 - \rho_1 F^*_{\tilde{x}_{1r}}(s)}. \tag{6.36}$$

where $\tilde{w}_1$ denotes the Laplace-Stieltjes transform of the class 1 waiting time.

6.3.2 Customers with Lower Priority

We now turn our attention to the determination of the PGF for the class 2 occupancy distribution and the corresponding class 2 sojourn time. The mathematics for obtaining the desired probability generating function is quite simple, but the logic behind it is a little complicated due to the number of concepts which have to be juggled simultaneously. We will therefore approach the problem in a roundabout way. First, we will examine the queueing behavior of the system only during those busy periods which are started by a class 2 customer. We are then in a position to examine the queueing behavior during all busy periods during which class 2 customers are served.

Consider the evolution of the service system during busy periods started by class 2 customers. For reasons which will become clear later, we will refer to this type of busy period as a *type 2 busy period*, and we will denote its length by $\tilde{y}_{22}$. Initially, the server is idle. Upon arrival of a class 2 customer, service begins immediately. During servicing of the class 2 customer, class 1 and class 2 customers may arrive. But, in any event, the server will not be available to service the second class 2 customer until the service of the first class 2 customer is complete and there are no class 1 customers in the system. The distribution of the length of time between the start of service of the first class 2 customer of the busy period and the second class 2 customer of the same busy period, if any, is identical to the distribution of the length of a class 1 busy period started by a class 2 customer. This period of time is called the *class 2 completion time*, and we denote its length by $\tilde{x}_{2c}$. From the results of our analysis of M/G/1 queueing systems with exceptional first service, we find

$$F^*_{\tilde{x}_{2c}}(s) = F^*_{\tilde{x}_2}(s + \lambda_1 - \lambda_1 F^*_{\tilde{y}_{11}}(s)), \tag{6.37}$$

where $\tilde{y}_{11}$ denotes the length of a class 1 busy period started by a class 1 customer[2], which we will refer to as a "type 1 busy period", and $F^*_{\tilde{y}_{11}}(s)$ satisfies the functional equation

$$F^*_{\tilde{y}_{11}}(s) = F^*_{\tilde{x}_1}(s + \lambda_1 - \lambda_1 F^*_{\tilde{y}_{11}}(s)). \tag{6.38}$$

[2]That is, $\tilde{y}_{11}$ would be the length of the busy period if the system served only class 1 customers.

From this discussion, it is clear that the busy period started by a class 2 customer evolves as a sequence of class 2 completion times. The waiting time of a class two customer served in a type 2 busy period therefore has the same distribution as that of an ordinary M/G/1 queueing system in which the service times are drawn from the distribution $F_{\tilde{x}_{2c}}$. Consequently, the probability generating function for the distribution of the number of class 2 customers left in the system at the time an arbitrary class 2 service is completed is given by the Laplace-Stieltjes transform of the waiting time distribution evaluated at $\lambda_2(1-z)$. In addition, the number of class 2 customers left in the system by a departing class 2 customer is simply the sum of the number in the system at the time the customer entered service and the number who arrive while the class 2 customer is in service. The probability generating function for the distribution of the latter quantity is given by the Laplace-Stieltjes transform of the service-time distribution evaluated at $\lambda_2(1-z)$. Thus, if we denote by $\tilde{n}_{22}$ the number of class 2 customers left in the system by a departing class 2 customer during a class 2 busy period started by a class 2 customer, we find

$$\mathcal{F}_{\tilde{n}_{22}}(z) = \frac{(1-\lambda_2 E[\tilde{x}_{2c}])F^*_{\tilde{x}_2}(\lambda_2[1-z])}{1-\lambda_2 E[\tilde{x}_{2c}]F^*_{\tilde{x}_{2cr}}(\lambda_2[1-z])}, \tag{6.39}$$

where $\tilde{x}_{2cr}$ denotes the residual life of $\tilde{x}_{2c}$. Paralleling our definition of ρ, we define

$$\gamma_2 = \lambda_2 E[\tilde{x}_{2c}], \tag{6.40}$$

and upon substituting (6.40) into (6.39), we find

$$\mathcal{F}_{\tilde{n}_{22}}(z) = \frac{(1-\gamma_2)F^*_{\tilde{x}_2}(\lambda_2[1-z])}{1-\gamma_2 F^*_{\tilde{x}_{2cr}}(\lambda_2[1-z])}. \tag{6.41}$$

Now consider the evolution of a busy period for class 2 customers started by a class 1 customer. Initially, the server is idle. Then, upon arrival of a class 1 customer, the busy period starts. First, the server serves an initial busy period of class 1 customers, the length of which is $\tilde{y}_{11}$. Just as in the case of the ordinary M/G/1 system, we can think of the remainder of this busy period as evolving as a sequence of type 2 sub-busy periods, all of which have the same distribution as an ordinary type 2 busy period. Thus we see that for any busy period during which class 2 customers are served, the periods over which class 2 customers are served can be thought of as evolving as a sequence of type 2 busy periods.

Corresponding to our definition of type 1 and type 2 customers above, we classify type 1 customers as those class 2 customers who arrive prior to the start of service of the first class 2 customer served in a busy period, and we classify all other class 2 customers as type 2. Without affecting the distribution of the number of class 2 customers left in the system by an arbitrary departing class 2 customer, we may think of the order of service of the type 1 and type 2 customers as being the same as in Section 6.2. Then each sub-busy period behaves exactly the same as the type 2 busy period described above. It is then easy to see that the PGF for the number of type 2 customers left in the system by an arbitrary departing type 2 customer is given by $\mathcal{F}_{\tilde{n}_{22}}$.

The PGF for number of type 1 customers left by the tagged class 2 customer can be obtained in a manner analogous to the explanation of the bracketed term of (6.35). To begin with, we consider only that portion of the time line during which type 1 customers arrive. During that portion of the time line, the system behaves as though the class 2 customers never enter the system; they are merely observers. Periods of time under this condition alternate between class 1 idle periods, the lengths of which are drawn from an exponential distribution with parameter λ_1 and ordinary class 1 busy periods, the lengths of which are drawn from the distribution $F_{\tilde{y}_{11}}(y)$.[3]

A random observer who arrives during this portion of the time-line, and who consequently is an arbitrary type 1 arrival, finds the system idle with probability

$$\frac{1/\lambda_1}{1/\lambda_1 + E[\tilde{y}_{11}]} = \frac{1}{1+\lambda_1 E[\tilde{y}_{11}]} = \frac{1}{1+\gamma_1}$$

where we have defined $\gamma_1 = \lambda_1 E[\tilde{y}_{11}]$, and busy with probability one minus that quantity.

If the observer finds the server idle during these periods, then the number of type 1 customers left will be identically zero; otherwise, the number of type 1 class 2 customers left will be equal to the number which arrive during the residual life of $\tilde{y}_{11}$, which we denote by $\tilde{y}_{11r}$. Thus the PGF for the distribution of the number of type 1 customers left in the system by the tagged class 2 customer, $\tilde{n}_{21}$, is given by

$$\mathcal{F}_{\tilde{n}_{21}}(z) = \frac{1}{1+\gamma_1} + \frac{\gamma_1}{1+\gamma_1} F^*_{\tilde{y}_{11r}}(\lambda_2[1-z]). \tag{6.42}$$

[3]The validity of this assertion is not necessarily obvious, but it can be reasoned as follows. Due to the properties of the exponential distribution, the length of each idle period is drawn independently from an exponential distribution with parameter $\lambda_1 + \lambda_2$, which is the distribution of the minimum of two independent exponential distributions having parameters λ_1 and λ_2, respectively. In addition, each idle period is terminated, independently of its length, by a class 2 customer with probability $\lambda_2/(\lambda_1+\lambda_2)$. Thus the total amount of the idle time observed before an idle period is terminated by a class 1 customer is the geometric, parameter $\lambda_2/(\lambda_1 + \lambda_2)$, sum of idle periods whose lengths are drawn independently from an exponential distribution having parameter $\lambda_1 + \lambda_2$. The overall length of this idle period is then exponentially distributed with parameter λ_1.

Because $\mathcal{F}_{\tilde{n}_2}(z) = \mathcal{F}_{\tilde{n}_{21}}(z)\mathcal{F}_{\tilde{n}_{22}}(z)$, we find from (6.41) and (6.42) that

$$\mathcal{F}_{\tilde{n}_2}(z) = \left(\frac{1}{1+\gamma_1} + \frac{\gamma_1 F^*_{\tilde{x}_{11r}}(\lambda_2[1-z])}{1+\gamma_1}\right) \frac{(1-\gamma_2)F^*_{\tilde{x}_2}(\lambda_2[1-z])}{(1-\gamma_2)F^*_{\tilde{x}_{2cr}}(\lambda_2[1-z])}. \tag{6.43}$$

Using the general relationship between the LST of the distribution of random variable and that of its residual,

$$F^*_{\tilde{x}_r}(s) = \frac{1 - F^*_{\tilde{x}}(s)}{sE[\tilde{x}]}, \tag{6.44}$$

in (6.43), we find, after rearranging terms that

$$\mathcal{F}_{\tilde{n}_2}(z) = \frac{1-\gamma_2}{1+\gamma_1} \left(\frac{(1-z) + (\lambda_1/\lambda_2)\{1 - F^*_{\tilde{y}_{11}}(\lambda_2[1-z])\}}{F^*_{\tilde{x}_{2c}}(\lambda_2[1-z]) - z}\right) F^*_{\tilde{x}_2}(\lambda_2[1-z]). \tag{6.45}$$

Because the class 2 customers left in the system are precisely those who arrive during the sojourn time of the class 2 customer, it follows from Theorem 5.1 that $\mathcal{F}_{\tilde{n}_2}(z) = F^*_{\tilde{s}_2}(\lambda_2[1-z])$. Thus, from (6.45), we find

$$F^*_{\tilde{s}_2}(s) = \frac{1-\gamma_2}{1+\gamma_1} \left(\frac{s + \lambda_1\{1 - F^*_{\tilde{y}_{11}}(s)\}}{s - \lambda_2 + \lambda_2 F^*_{\tilde{x}_{2c}}(s)}\right) F^*_{\tilde{x}_2}(s). \tag{6.46}$$

EXERCISE 6.5 Derive the expression for $\mathcal{F}_{\tilde{n}_2}(z)$ for the case of the HOL-PR discipline with $I = 2$.

EXERCISE 6.6 Derive expressions for $\mathcal{F}_{\tilde{n}_1}(z)$, $\mathcal{F}_{\tilde{n}_2}(z)$, and $\mathcal{F}_{\tilde{n}_3}(z)$ for the ordinary HOL discipline with $I = 3$. Extend the analysis to the case of arbitrary I.

EXERCISE 6.7 Extend the analysis of the previous case to the case of HOL-PR.

6.4 Ergodic Occupancy Probabilities for Priority Queues

We now turn to the topic of inversion of (6.34) and (6.45) to obtain the equilibrium occupancy distributions. These transform equations can be inverted using the techniques based on discrete Fourier transforms that were discussed in Section 5.2. Recall that the techniques presented in Section 5.2 require evaluation of the probability generating function at points equally spaced around the unit circle of the complex plane.

In the case of (6.34), the results presented in Section 5.2 can be applied directly; no modifications whatsoever are required. However, in the case of (6.45), the right-hand side contains the terms $F^*_{\tilde{x}_{2c}}(\lambda_2[1-z])$ and $F^*_{\tilde{y}_{11}}(\lambda_2[1-z])$. These expressions are defined in 6.37 and 6.38, respectively, and repeated here for continuity:

$$F^*_{\tilde{x}_{2c}}(s) = F^*_{\tilde{x}_2}(s + \lambda_1 - \lambda_1 F^*_{\tilde{y}_{11}}(s)),$$

and

$$F^*_{\tilde{y}_{11}}(s) = F^*_{\tilde{x}_1}(s + \lambda_1 - \lambda_1 F^*_{\tilde{y}_{11}}(s)).$$

In order to apply the techniques of Section 5.2, we must first evaluate each of these expressions at points equally spaced around the unit circle of the complex plane. Upon evaluating each equation at $s = \lambda_2[1-z]$, we find

$$F^*_{\tilde{x}_{2c}}(\lambda_2[1-z]) = F^*_{\tilde{x}_2}(\lambda_2[1-z] + \lambda_1 - \lambda_1 F^*_{\tilde{y}_{11}}(\lambda_2[1-z])), \tag{6.47}$$

and

$$F^*_{\tilde{y}_{11}}(\lambda_2[1-z]) = F^*_{\tilde{x}_1}(\lambda_2[1-z] + \lambda_1 - \lambda_1 F^*_{\tilde{y}_{11}}(\lambda_2[1-z])). \tag{6.48}$$

Our technique is to first evaluate $F^*_{\tilde{y}_{11}}(\lambda_2[1-z])$ at points around the unit circle of the complex plane, then substitute the result into (6.47) to obtain $F^*_{\tilde{x}_{2c}}(\lambda_2[1-z])$, and finally to substitute back into (6.45) to obtain the required evaluations.

Define z^* to be any point on the unit circle of the complex plane and define $\nu = F^*_{\tilde{y}_{11}}(\lambda_2[1-z^*])$. Upon rewriting (6.48) in terms of ν we find

$$\nu = F^*_{\tilde{x}_1}(\lambda_2[1-z^*] + \lambda_1[1-\nu]). \tag{6.49}$$

We then have the following theorem, which is proved in Daigle and Roughan [1999].

THEOREM 6.1 *The value* $\nu = F^*_{\tilde{y}_{11}}(\lambda_2[1-z^*])$ *can always be determined from the expression* $\nu_{i+1} = F^*_{\tilde{x}_1}(\lambda_2[1-z^*] + \lambda_1[1-\nu_i])$ *by iteration on* i *starting with* $\nu_0 = z^*$. *That is, if the sequence* $\{\nu_0, \nu_1, \ldots\}$ *is defined by the recursion* $\nu_{i+1} = F^*_{\tilde{x}_1}(\lambda_2[1-z^*] + \lambda_1[1-\nu_i])$ *with* $\nu_0 = z^*$, *where* z^* *is any point on the unit circle, then* $\lim_{i\to\infty} \nu_i = F^*_{\tilde{y}_{11}}(\lambda_2[1-z^*])$. □

Remark. We point out that our experience is that the iterative procedure of the above theorem converges very fast. In fact the number of iterations required to converge to a high degree of accuracy is usually on the order of ten iterations.

The above theorem may be used to compute the values of $F^*_{\tilde{y}_{11}}(\lambda_2[1-z_k])$ for $z_k \in \{z_0, z_1, \ldots, z_K\}$, a set of $K+1$ points evenly spaced around the unit circle. By following this procedure, we can find $\mathcal{F}_{\tilde{n}_2}(z)$ for $z_k \in \{z_0, z_1, \ldots, z_K\}$, and then we can apply the techniques discussed in Section 5.2 to determine the ergodic occupancy probabilities.

A number of numerical examples are given in Daigle and Roughan [1999]. In addition, they present modifications to the methods presented in Section 5.2 for the case in which one or both of the service time distributions have long tails, and in which the tail probabilities of the occupancy distribution are not necessarily geometrically decreasing.

6.5 Expected Waiting and Sojourn Times for M/G/1 under HOL Priority

As in Section 6.3, we will compute the average waiting time and average sojourn time for each customer class under two service disciplines: head of the line (HOL) and HOL-preemptive resume (HOL-PR). Under the ordinary HOL discipline, class i customers arriving to the system join the service queue ahead of all customers of lower priority and behind all customers whose priorities are at least as high. The HOL-PR discipline is similar except that a newly arriving class i customer goes directly into service if the server is serving a customer of lower priority than the new customer.

We will again compute the delays by examining the system from the point of view of an arbitrary class i customer, whom we will refer to as the "tagged class i" customer. Owing to the Poisson arrivals, the tagged class i customer observes the system in stochastic equilibrium, so that the average delay observed by the tagged class i customer is the same as that for an arbitrary class i customer.

Suppose the tagged class i customer arrives to the system at time t_0. The waiting time in the queue that this customer experiences can then be thought of as resulting from two basic customer groups: early arrivals and late arrivals. The early arrivals are those customers who arrived to the system prior to t_0; the late arrivals are those customers who arrive to the system prior to t_0.

Let $\tilde{w}_{i,e_j}$ and $\tilde{w}_{i,\ell_j}$ denote the delay suffered by the tagged class i customer due to early and late class j customers, respectively. Then, we find that

$$\tilde{w}_i = \sum_{j=1}^{I} \tilde{w}_{i,e_j} + \sum_{j=1}^{I} \tilde{w}_{i,\ell_j}, \tag{6.50}$$

where $\tilde{w}_i$ denotes the waiting time for the class i customers. Now, the waiting time due to early arrivals can be further subdivided into the delay due to customers in the queue and the delay due to customers whose service has already begun at time t_0, which we will denote as $\tilde{w}_{i,q_j}$ and $\tilde{w}_{i,s_j}$, respectively. Thus,

(6.50) can be rewritten as

$$\tilde{w}_i = \sum_{j=1}^{I} \tilde{w}_{i,q_j} + \sum_{j=1}^{I} \tilde{w}_{i,s_j} + \sum_{j=1}^{I} \tilde{w}_{i,\ell_j}. \tag{6.51}$$

Clearly, under the HOL discipline, $\tilde{w}_{i,q_j} = 0$ if $j > i$, and $\tilde{w}_{i,\ell_j} = 0$ if $j \geq i$. But delays due to customers who may be in service at time t_0 depend on whether or not the service discipline is preemptive. Under HOL-PR, $\tilde{w}_{i,s_j} = 0$ for $j > i$, but under ordinary HOL, $\tilde{w}_{i,s_j}$ may be nonzero for all j. Based on the above discussions, we find

$$\tilde{w}_i = \sum_{j=1}^{i} \tilde{w}_{i,q_j} + \sum_{j=1}^{i-1} \tilde{w}_{i,\ell_j} + \sum_{j=1}^{I} \tilde{w}_{i,s_j} \tag{6.52}$$

so that

$$E[\tilde{w}_i] = \sum_{j=1}^{i} E[\tilde{w}_{i,q_j}] + \sum_{j=1}^{i-1} E[\tilde{w}_{i,\ell_j}] + \sum_{j=1}^{I} E[\tilde{w}_{i,s_j}]. \tag{6.53}$$

Clearly, $E[\tilde{w}_{i,q_j}] = E[N_{q_j}]E[\tilde{x}_j]$ for $j \leq i$ where N_{q_j} denotes the number of class j customers in the queue at time t_0. Due to Little's result, $E[N_{q_j}] = \lambda_j E[\tilde{w}_j]$. Thus,

$$E[\tilde{w}_{i,q_j}] = \rho_j E[\tilde{w}_j] \qquad \text{for} \qquad j \leq i. \tag{6.54}$$

As for the late arrivals, the tagged class i customer will be delayed in the queue by any class j late arrivals, $j < i$, who arrive while the tagged class i customer is still in the queue. That is, all class j, $j < i$, customers who arrive during $\tilde{w}_i$ will delay the tagged class i customer in the queue. Since class j arrivals are Poisson with rate λ_j and service times are drawn independently of everything, we find $E[\tilde{w}_{i,\ell_j}] = \lambda_j E[\tilde{w}_i]E[\tilde{x}_j]$, or equivalently,

$$E[\tilde{w}_{i,\ell_j}] = \rho_j E[\tilde{w}_i] \qquad \text{for} \qquad j < i. \tag{6.55}$$

Since $E[\tilde{w}_{i,s_j}]$ is dependent upon the service discipline, we will defer specifying a formula for its computation for the time being, and we solve (6.53) for the general case. Substitution of (6.54) and (6.55) into (6.53) yields

$$E[\tilde{w}_i] = \sum_{j=1}^{i} \rho_j E[\tilde{w}_j] + \sum_{j=1}^{i-1} \rho_j E[\tilde{w}_i] + \sum_{j=1}^{I} E[\tilde{w}_{i,s_j}]. \tag{6.56}$$

Now, define

$$\sigma_i = \sum_{j=1}^{i-1} \rho_j. \tag{6.57}$$

Then we can rewrite (6.56) in the following two versions:

$$(1-\sigma_{i-1})E[\tilde{w}_i] = \sum_{j=1}^{i} \rho_j E[\tilde{w}_j] + \sum_{j=1}^{I} E[\tilde{w}_{i,s_j}], \tag{6.58}$$

and

$$(1-\sigma_i)E[\tilde{w}_i] = \sum_{j=1}^{i-1} \rho_j E[\tilde{w}_j] + \sum_{j=1}^{I} E[\tilde{w}_{i,s_j}]. \tag{6.59}$$

Comparing (6.58) and (6.59), we find

$$(1-\sigma_i)E[\tilde{w}_i] = (1-\sigma_{i-2})E[\tilde{w}_{i-1}] + E[\tilde{w}_{i0}] - E[\tilde{w}_{i-1,0}], \tag{6.60}$$

where we have defined

$$\tilde{w}_{i0} = \sum_{j=1}^{I} \tilde{w}_{i,s_j} \tag{6.61}$$

and $\tilde{w}_{i0} = 0$ for $i \leq 0$. Thus, we find

$$E[\tilde{w}_i] = \frac{(1-\sigma_{i-2})E[\tilde{w}_{i-1}] + E[\tilde{w}_{i0}] - E[\tilde{w}_{i-1,0}]}{(1-\sigma_i)}. \tag{6.62}$$

6.5.1 HOL Discipline

Recall that $\tilde{w}_{i0}$ is the total delay suffered by the class i tagged customer due to customers whose service is in progress at time t_0. This quantity is clearly a function of the service discipline. Under the ordinary HOL discipline, $\tilde{w}_{i0} = \tilde{w}_{j0}$ for all i, $j \geq 1$, because customers who are in service remain in service regardless of the class to which the tagged customer belongs. Thus, for the HOL discipline, we find by solving (6.62) recursively that

$$E[\tilde{w}_i] = \frac{E[\tilde{w}_{i0}]}{(1-\sigma_i)(1-\sigma_{i-1})}. \tag{6.63}$$

It remains to specify $E[\tilde{w}_{i0}]$. From (6.61) we find

$$E[\tilde{w}_{i0}] = \sum_{j=1}^{I} E[\tilde{w}_{i,s_j}].$$

By conditioning on whether or not a class j customer is in service at time t_0, we find

$$E[\tilde{w}_{i0}] = \sum_{j=1}^{I} E[\tilde{w}_{i,s_j} | \text{ class } j \text{ in service at } t_0] \, P\{\text{class } j \text{ in service at } t_0\}. \tag{6.64}$$

Clearly, the delay a class i customer suffers due to a class j customer in service at time t_o is equal to the residual service time for a class j customer and the probability that a class j customer is in service is ρ_j. Thus, (6.64) reduces to

$$E[\tilde{w}_{i0}] = \sum_{j=1}^{I} \rho_j E[\tilde{x}_{r_j}], \tag{6.65}$$

or, equivalently,

$$E[\tilde{w}_{i0}] = \sum_{j=1}^{I} \rho_j \left(\frac{1 + C_{\tilde{x}_j}^2}{2}\right) E[\tilde{x}_j]. \tag{6.66}$$

Upon substituting (6.57) and (6.66) into (6.63), we find

$$E[\tilde{w}_i] = \frac{\sum_{j=1}^{I} \rho_j \left((1 + C_{\tilde{x}_j}^2)/2\right) E[\tilde{x}_j]}{(1 - \sigma_i)(1 - \sigma_{i-1})}. \tag{6.67}$$

6.5.2 HOL-PR Discipline

Under the HOL-PR discipline, $\tilde{w}_{i,0}$ is different for each i because class i customers are delayed by class j customers only if $j \leq i$. Thus, under this discipline,

$$\tilde{w}_{i0} = \sum_{j=1}^{i} \tilde{w}_{i,s_j}. \tag{6.68}$$

In addition class j customers are preempted by all higher priority customers who arrive while they are in service. As a result, the time required to complete service for a class j customer is the same as the length of a busy period started by a customer whose service time is $\tilde{x}_j$ and generated by all traffic having priority higher than class j. This period, called a "class j completion time", is denoted by $\tilde{x}_{c_j}$. Thus

$$E[\tilde{x}_{c_j}] = \frac{E[\tilde{x}_j]}{1 - \sigma_{j-1}}. \tag{6.69}$$

Now, at time t_0, a class j customer whose service has begun may be either preempted or actually in service. In either case, that customer's remaining service time is given by $\tilde{x}_{r_j}$. Also, the probability that there is a class j customer either in service or preempted at time t_0 is readily computed by applying Little's result. We find

$$P\{\text{class } j \text{ service in progress at } t_0\} = \frac{\lambda_j E[\tilde{x}_j]}{1 - \sigma_{j-1}}$$

or

$$P\{\text{class } j \text{ service in progress at } t_0\} = \frac{\rho_j}{1-\sigma_{j-1}}. \tag{6.70}$$

Applying (6.70) and our remaining service-time observation to (6.68) and conditioning, we find

$$E[\tilde{w}_{i0}] = \sum_{j=1}^{i} \frac{\rho_j E[\tilde{x}_{r_j}]}{1-\sigma_{j-1}}. \tag{6.71}$$

Substitution of (6.71) into (6.62) yields

$$E[\tilde{w}_i] = \frac{(1-\sigma_{i-1})E[\tilde{w}_{i-1}] + \{(\rho_i E[\tilde{x}_{ri}])/(1-\sigma_{i-1})\}}{(1-\sigma_i)}, \tag{6.72}$$

or, equivalently,

$$E[\tilde{w}_i] = \frac{(1-\sigma_{i-1})(1-\sigma_{i-2})E[\tilde{w}_{i-1}] + \rho_i E[\tilde{x}_{r_i}]}{(1-\sigma_i)(1-\sigma_{i-1})}. \tag{6.73}$$

Solving recursively, we find

$$E[\tilde{w}_i] = \frac{\sum_{j=1}^{i} \rho_j E[\tilde{x}_{r_j}]}{(1-\sigma_i)(1-\sigma_{i-1})}. \tag{6.74}$$

Thus we find that for the HOL-PR service discipline,

$$E[\tilde{w}_i] = \frac{\sum_{j=1}^{i} \rho_j[(1+C_{\tilde{x}_j}^2)/2]E[\tilde{x}_j]}{(1-\sigma_i)(1-\sigma_{i-1})}. \tag{6.75}$$

Comparison of (6.75) to (6.67) reveals that the only difference between the waiting times for HOL and HOL-PR is accounted for by the difference in perception of the delay due to the customer who may be in service at time t_0. Customers of all classes are relevant in the case of HOL whereas for HOL-PR, only customers having priority at least as high as that of the customer in question appear in the result. This is intuitively satisfying in that the only effect of preemption for customers who are serviced during the tagged customer's waiting time is to rearrange the order in which service is rendered. Additionally, from the tagged customer's point of view, the customer preempted at time t_0 has a lower priority and for all intents and purposes doesn't exist.

The sojourn time under the HOL discipline is obtained by simply adding the service time to the waiting time, whereas that for the HOL-PR system is obtained by adding the completion time to the waiting time. Thus we find for the HOL discipline,

$$E[\tilde{s}_i] = \frac{\sum_{j=1}^{I} \rho_j[(1+C_{\tilde{x}_j}^2)/2]E[\tilde{x}_j]}{(1-\sigma_i)(1-\sigma_{i-1})} + E[\tilde{x}_i], \tag{6.76}$$

and for the HOL-PR discipline,

$$E[\tilde{s}_i] = \frac{\sum_{j=1}^{i} \rho_j[(1 + C_{\tilde{x}_j}^2)/2]E[\tilde{x}_j]}{(1 - \sigma_i)(1 - \sigma_{i-1})} + \frac{E[\tilde{x}_i]}{1 - \sigma_{i-1}}. \tag{6.77}$$

EXERCISE 6.8 Suppose that the service time of the customers in an M/G/1 system are drawn from the distribution $F_{\tilde{x}_i}(x)$ with probability p_i such that $\sum_{i=1}^{I} p_i = 1$. Determine $E[\tilde{w}]$ for this system.

EXERCISE 6.9 **Conservation Law (Kleinrock [1976])** Under the conditions of Exercise 6.8, suppose the customers whose service times are drawn from the distribution $F_{\tilde{x}_i}(x)$ are assigned priority i and the service discipline is HOL. Show that $\sum_{i=1}^{I} \rho_i E[\tilde{w}_i] = \rho E[\tilde{w}]$ where $E[\tilde{w}]$ is as determined in Exercise 5.9. Explain the implications of this result. Does the result imply that the expected waiting time is independent of the priority assignment? Why or why not? If not, under what conditions would equality hold?

Chapter 7

VECTOR MARKOV CHAIN ANALYSIS: THE M/G/1 AND G/M/1 PARADIGMS

In the previous chapter of this book, we discussed the M/G/1 queueing system and some of its variants. We saw that the occupancy process $\{\tilde{n}(t), t \geq 0\}$ is a semi-Markov process, and we analyzed the system by first embedding a Markov chain, $\{\tilde{q}_n, n = 0, 1, 2, \ldots\}$, at instants of customer departure. Although we identified the process $\{\tilde{q}_n, n = 0, 1, 2, \ldots\}$ as a discrete-parameter Markov chain on the nonnegative integers, we did not explicitly present its transition probability matrix. In fact, such a specification was unnecessary because our analysis technique avoided this issue.

In this chapter, we provide a brief introduction to the G/M/1 and M/G/1 paradigms, which are useful in solving practical problems and are discussed at length in Neuts [1981a] and Neuts [1989], respectively. These paradigms are natural extensions of the ordinary M/G/1 and G/M/1 systems. In particular, the structure of the one-step transition probability matrices for the embedded Markov chains for these systems are simply matrix versions of the one-step transition probability matrices for the embedded Markov chains of the elementary systems.

In Section 7.1 we introduce the M/G/1 and G/M/1 paradigms. We first present a concise development of the one-step transition probability matrix for the embedded Markov chain of the M/G/1 system. Next, for the G/M/1 system, we define the embedded Markov chain $\{\tilde{q}'_n, n = 0, 1, 2, \ldots\}$, where $\tilde{q}'_n$ denotes the number of customers found in the system by the nth arriving customer, and we specify its one-step transition probability matrix. Markov chains of the M/G/1 and G/M/1 type are then defined.

The general solution procedure for models of the G/M/1 type is then discussed in Section 7.2. This presentation is very brief because of the extensive coverage of similar methodology presented in Chapter 3. In Section 7.3, we discuss matrix analytic solution procedures for solving models of the M/G/1 type with simple boundaries. In Section 7.4, application of M/G/1 paradigm ideas to statistical multiplexing is discussed by way of examples. In Section 7.5, we extend our earlier development of the generalized state space methods to the case of the Markov chains of the M/G/1 type with complex boundary conditions. The methodology presented in Section 7.5 is relatively new, and our experience with this methodology has been very positive. Finally, additional applications are discussed and conclusions are drawn in Section 7.6.

Entire books are devoted to solution procedures for models of the M/G/1 and G/M/1 types. There is no attempt here to provide complete coverage of the solution methodologies that have been developed over the last 30 years. Rather, we satisfy ourselves with a brief presentation of the main ideas and direct our readers to the appropriate references. On the other hand, generalized state-space procedures are relatively new, and we attempt to provide a thorough introduction.

7.1 The M/G/1 and G/M/1 Paradigms

Recall that for the M/G/1 system $\tilde{q}_n$, $n = 0, 1, 2, \ldots$, denotes the number of customers left in the system by the nth departing customer. From (5.1), we have

$$\tilde{q}_{n+1} = (\tilde{q}_n - 1)^+ + \tilde{v}_{n+1}, \tag{7.1}$$

where $(a)^+ = \max\{a, 0\}$ and $\tilde{v}_n$ is defined as the number of arrivals that occur during the nth customer's service.

As we have pointed out, the process $\{\tilde{q}_n, n = 0, 1, 2, \ldots\}$ is a discrete-parameter Markov chain on the nonnegative integers. Recall that for such Markov chains, the probability of being in state j after the $(n+1)$th state change given that the system was in state i after the nth state change is called the *one-step transition probability* from state i to state j. The matrix of these transition probabilities, $P_{ij} = P\{\tilde{q}_{n+1} = j | \tilde{q}_n = i\}$, is called the *one-step transition probability matrix.*

Upon conditioning on $\tilde{q}_n$, we find

$$P\{\tilde{q}_{n+1} = j\} = \sum_{i=0}^{\infty} P\{\tilde{q}_{n+1} = j | \tilde{q}_n = i\} P\{\tilde{q}_n = i\}. \tag{7.2}$$

Then, substitution of (7.1) into (7.2) yields

$$P\{\tilde{q}_{n+1} = j\} = \sum_{i=0}^{\infty} P\{(\tilde{q}_n - 1)^+ + \tilde{v}_{n+1} = j | \tilde{q}_n = i\} P\{\tilde{q}_n = i\}. \tag{7.3}$$

Because $\tilde{v}_{n+1}$ is independent of $\tilde{q}_n$, (7.3) is readily reduced to

$$P\{\tilde{q}_{n+1} = j\} = \sum_{i=0}^{\infty} P\{\tilde{v}_{n+1} = j - (i-1)^+\} P\{\tilde{q}_n = i\}. \tag{7.4}$$

Now, departures occur only one at a time. Therefore, the infinite summation of (7.4) can be replaced by the finite summation

$$P\{\tilde{q}_{n+1} = j\} = \sum_{i=0}^{j+1} P\{\tilde{v}_{n+1} = j - (i-1)^+\} P\{\tilde{q}_n = i\}. \tag{7.5}$$

Because the service times are a sequence of independent, identically distributed random variables with distribution $F_{\tilde{x}}(x)$, we see that the one-step transition probability of going from state i to state j is simply

$$P\{\tilde{q}_{n+1} = j | \tilde{q}_n = i\} = P\{\tilde{v} = j - (i-1)^+\},$$

where $\tilde{v}$ is the number of arrivals that occur during a service time, and these are nonzero for only $i+1$ entries. We define $a_k = P\{\tilde{v} = k\}$, and since arrivals occur according to a Poisson process with parameter λ, we find

$$a_k = \int_0^\infty \frac{(\lambda x)^k}{k!} e^{-\lambda x} dF_{\tilde{x}}(x). \tag{7.6}$$

The one-step transition probability matrix for the embedded Markov chain for the M/G/1 system is then

$$\mathcal{P}_{MG} = \begin{bmatrix} a_0 & a_1 & a_2 & a_3 & \cdots \\ a_0 & a_1 & a_2 & a_3 & \cdots \\ 0 & a_0 & a_1 & a_2 & \cdots \\ 0 & 0 & a_0 & a_1 & \cdots \\ \cdots & \cdots & \cdots & \cdots & \cdots \end{bmatrix}. \tag{7.7}$$

Where first service times are exceptional, the distribution of the number of arrivals that occur during the first service time of each busy period is different from the distribution of the number of arrivals that occur during service times other than the first. In this case, we define b_k as the probability that k arrivals occur during exceptional first service, and we find

$$b_k = \int_0^\infty \frac{(\lambda x)^k}{k!} e^{-\lambda x} dF_{\tilde{x}_e}(x), \tag{7.8}$$

where $\tilde{x}_e$ denotes the length of the exceptional first service. The one-step transition probability matrix is then

$$\mathcal{P}_{MG} = \begin{bmatrix} b_0 & b_1 & b_2 & b_3 & \cdots \\ a_0 & a_1 & a_2 & a_3 & \cdots \\ 0 & a_0 & a_1 & a_2 & \cdots \\ 0 & 0 & a_0 & a_1 & \cdots \\ \cdots & \cdots & \cdots & \cdots & \cdots \end{bmatrix}. \tag{7.9}$$

If we now define $\pi_j = \lim_{n\to\infty} P\{\tilde{q}_n = j\}$ and $\pi = [\pi_0 \quad \pi_1 \quad \cdots]$, then, at least in principle, we can obtain π by solving the matrix equations

$$\pi = \pi \mathcal{P}_{MG}, \quad \pi \mathrm{e} = 1$$

simultaneously. The limiting solution is known to exist so long as the service rate exceeds the arrival rate. But, as we have already seen in Chapter 5, the solution is not necessarily easily obtained. In addition, the entries of the matrices $\mathcal{P}_{MG}$ must also be calculated; this, in itself, is nontrivial for most distributions unless a special form such as a weighted sum of exponentials or a collection of point masses is assumed for the service time distribution.

Remark. We use the notation π rather than P for these probabilities to emphasize that the stationary probability vector is for a discrete-parameter Markov chain rather than for a continuous-time Markov chain. Therefore, the individual probabilities represent the proportion of entries to or exits from a given state rather than the time-averaged probability that the system is in the given state. For systems having Poisson arrivals, such as the ordinary M/G/1 system, these probabilities are equal, but they are not equal in the general case.

EXERCISE 7.1 Suppose that $P\{\tilde{x} = 1\} = 1$, that is, the service time is deterministic with mean 1. Determine $\{a_k, k = 0, 1, \ldots\}$ as defined by (7.6).

EXERCISE 7.2 Suppose that $\tilde{x}$ is a discrete valued random variable having support set $\mathcal{X} = \{x_0, x_1, \ldots, x_K\}$, where K is an integer. Define $\alpha_k = P\{\tilde{x} = x_k\}$ for $x_k \in \mathcal{X}$. Determine $\{a_k, k = 0, 1, \ldots\}$ as defined by (7.6). In order to get started, let $dF_{\tilde{x}}(x) = \sum_{x\in\mathcal{X}} \alpha_k \delta(x - x_k)$, where $\delta(x)$ is the Dirac delta function.

EXERCISE 7.3 Suppose that $\tilde{x}$ is an exponential random variable with parameter μ. Determine $\{a_k, k = 0, 1, \ldots\}$ as defined by (7.6).

EXERCISE 7.4 Suppose that $\tilde{x} = \tilde{x}_1 + \tilde{x}_2$, where $\tilde{x}_1$ and $\tilde{x}_2$ are exponential random variables with parameter μ_1 and μ_2, respectively. Determine $\{a_k, k = 0, 1, \ldots\}$ as defined by (7.6).

Similar to the case of the M/G/1 system, the G/M/1 system can be analyzed by embedding a Markov chain at points in time just prior to customer arrivals. As before, we denote the embedded Markov chain by $\{\tilde{q}'_n, n = 0, 1, 2, \ldots\}$. The state of this Markov chain is then defined as the number of customers found in the system by the nth arriving customer. It is easy to see that

$$\tilde{q}'_{n+1} = \tilde{q}'_n + 1 - \tilde{v}'_{n+1}, \tag{7.10}$$

where $\tilde{v}'_n$ denotes the number of service completions that occur during the nth interarrival interval. From (7.10), we can easily determine that

$$P\{\tilde{q}'_{n+1} = j\} = \sum_{i=0}^{\infty} P\{\tilde{v}' = i+1-j\} P\{\tilde{q}'_n = i\}.$$

Since arrivals occur only one at a time, this equation can be rewritten as

$$P\{\tilde{q}'_{n+1} = j\} = \sum_{i=(j-1)^+}^{\infty} P\{\tilde{v}' = i+1-j\} P\{\tilde{q}'_n = i\}. \tag{7.11}$$

From (7.11), we see that the one-step transition probability from state i to state j is given by $P\{\tilde{v}' = i+1-j\}$. Computation of these transition probabilities is slightly more involved than in the M/G/1 case because we have to distinguish between whether or not all customers present are served prior to the next arrival-that is, whether the system is left empty or not. If we let $\tilde{x}_k$ denote the service time of the ith customer served during the $(n+1)$th interarrival time, then we find

$$P\{\tilde{v}' = i+1-j\} = \begin{cases} P\left\{\sum_{k=1}^{i+1} \tilde{x}_k < \tilde{t}\right\} & \text{for } j = 0, \\ P\left\{\sum_{k=1}^{i+1-j} \tilde{x}_k < \tilde{t}, \sum_{k=1}^{i+1-j} \tilde{x}_k > \tilde{t}\right\} & \text{for } j > 0, \end{cases} \tag{7.12}$$

where $\tilde{t}$ denotes the interarrival time.

We therefore define b_k to be the probability that k customers are served during the $(n+1)$th interarrival time if the system is found empty by the $(n+1)$th arriving customer and a_k to be the corresponding probability otherwise. Because service times are exponential and the sum of $K+1$ independent exponentially distributed random variables has the gamma distribution with parameters K and μ, or equivalently, the Erlang-$(K+1)$ distribution, it is easy to see that

$$b_k = \int_0^{\infty} \frac{\mu(\mu x)^k}{k!} e^{-\mu x} F^c_{\tilde{t}}(x)dx \quad \text{for} \quad k = 0, 1, \ldots. \tag{7.13}$$

Owing to the exponential service times and the properties of the Poisson process, it is easy to see that a_k is simply the probability that exactly k arrivals from a Poisson process occur during the $(n+1)$th interarrival time. Thus,

$$a_k = \int_0^{\infty} \frac{(\mu x)^k}{k!} e^{-\mu x} dF_{\tilde{t}}(x) \quad \text{for} \quad k = 0, 1, \ldots. \tag{7.14}$$

The one-step transition probability matrix for the embedded Markov chain for the G/M/1 system is then

$$\mathcal{P}_{GM} = \begin{bmatrix} b_0 & a_0 & 0 & \cdots & \cdots \\ b_1 & a_1 & a_0 & 0 & \cdots \\ b_2 & a_2 & a_1 & a_0 & \cdots \\ b_3 & a_3 & a_2 & a_1 & \cdots \\ \cdots & \cdots & \cdots & \cdots & \cdots \end{bmatrix}. \tag{7.15}$$

As in the case of the M/G/1 system, if we define $\pi_j = \lim_{n\to\infty} P\{\tilde{q}'_n = j\}$ and $\pi = [\pi_0 \quad \pi_1 \quad \cdots]$, then we can obtain π by solving the matrix equations

$$\pi = \pi \mathcal{P}_{GM}, \quad \pi \mathbf{e} = 1$$

simultaneously. Again, the limiting solution is known to exist so long as the service rate exceeds the arrival rate. Unlike the case of the M/G/1 system, the solution to the embedded Markov chain for the G/M/1 system has a very simple form, namely,

$$\pi_j = (1-\omega)\omega^j \quad \text{for} \quad j \geq 0, \tag{7.16}$$

where ω is the unique (real) solution inside the unit circle to the functional equation

$$\omega = F_{\tilde{t}}^*(\mu[1-\omega]). \tag{7.17}$$

The value of ω can be obtained iteratively from the mapping

$$\omega_{i+1} = F_{\tilde{t}}^*(\mu[1-\omega_i]) \quad \text{with} \quad 0 \leq \omega_0 < 1, \tag{7.18}$$

as is shown in Tackàcs [1962].

The waiting time for an arbitrary arriving customer, given that there is at least one customer present, is then simply the geometric sum of independent, identically distributed exponential variables and is consequently exponential with parameter $(1-\omega)\mu$. Since the probability that an arriving customer finds at least one customer present is ω, we find

$$P\{\tilde{w} > t\} = \omega e^{-(1-\omega)\mu t}. \tag{7.19}$$

We note in passing that the expression on the right-hand side of (7.17) is, according to Theorem 5.2, the probability generating function for the number of arrivals that occur from a Poisson process having rate μ over a random period of time having distribution $F_{\tilde{t}}(t)$, where the arrival process is independent of $\tilde{t}$. By contrast, in the case of the M/G/1 system, the tail probabilities are approximately geometrically decreasing with the decay rate r_0 where r_0 is the inverse of the unique (real) solution outside the unit circle, z_0, to the functional equation

$$z = F_{\tilde{x}}^*(\lambda[1-z]). \tag{7.20}$$

The general form of the occupancy distribution for the M/G/1 system is much more complicated than that of the G/M/1 system seen in Chapter 5.

Markov chains whose one-step transition probability matrices have the structures of (7.9) and (7.15) are said to be *Markov chains of the M/G/1 type* and *Markov chains of the G/M/1 type*, respectively. The idea has been generalized by Neuts [1981a] to the cases in which these one-step transition matrices have block-partitioned structures of similar forms, as follows:

$$\mathcal{P}_{MG} = \begin{bmatrix} B_0 & B_1 & B_2 & B_3 & \cdots \\ A_0 & A_1 & A_2 & A_3 & \cdots \\ 0 & A_0 & A_1 & A_2 & \cdots \\ 0 & 0 & A_0 & A_1 & \cdots \\ \cdots & \cdots & \cdots & \cdots & \cdots \end{bmatrix} \tag{7.21}$$

and

$$\mathcal{P}_{GM} = \begin{bmatrix} B_0 & A_0 & 0 & \cdots & \cdots \\ B_1 & A_1 & A_0 & 0 & \cdots \\ B_2 & A_2 & A_1 & A_0 & \cdots \\ B_3 & A_3 & A_2 & A_1 & \cdots \\ \cdots & \cdots & \cdots & \cdots & \cdots \end{bmatrix}. \tag{7.22}$$

In this case, as in the cases covered in Chapter 3, the states of the Markov chain are denoted by (i, j), $i \geq 0$, and $0 \leq j \leq K$, where i denotes the occupancy (or level) and j denotes an abstract auxiliary descriptor that we refer to as the *phase*. The states are ordered lexicographically; that is, we define $\pi_i = [\pi_{i0} \quad \pi_{i1} \quad \cdots \quad \pi_{iK}]$, and $\pi = [\pi_0 \quad \pi_1 \quad \pi_2 \quad \pi_3 \quad \cdots]$. The Markov chains $\{\tilde{q}_n, n = 0, 1, 2, \cdots\}$ and $\{\tilde{q}'_n, n = 0, 1, 2, \cdots\}$ are then interpreted as vector-valued Markov chains.

Remark. As has been our practice throughout this book, we avoid inventing special notation to distinguish between scalar and vector quantities unless there is a specific gain to be made in the particular problem under consideration. In this case, there does not appear to be any.

7.2 G/M/1 Solution Methodology

We saw in Chapter 3 that some queueing systems have matrix-geometric solutions. Interestingly, all positive-recurrent Markov chains of the G/M/1 type have matrix-geometric solutions. That is, the occupancy probabilities have the form

$$\pi_{i+1} = \pi_i R \quad \text{for} \quad i \geq 0,$$

or, equivalently

$$\pi_i = \pi_0 R^i \quad \text{for} \quad i \geq 0. \tag{7.23}$$

This result is given in the following theorem from Neuts [1981a, pp. 10-11]:

THEOREM 7.1 *(Neuts) If the Markov chain $\mathcal{P}_{GM}$ is positive recurrent, then*

1. *for $i \geq 0$, we have $\pi_{i+1} = \pi_i \mathcal{R}$,*
2. *the eigenvalues of $\mathcal{R}$ lie inside the unit disk of the complex plane,*
3. *the matrix*

$$B(\mathcal{R}) = \sum_{k=0}^{\infty} \mathcal{R}^k B_k$$

   is stochastic[1], *and*

4. *the vector π_0 is a positive, left invariant eigenvector of $B(\mathcal{R})$ normalized by $\pi_0(I - \mathcal{R})^{-1}e = 1$.* □

Neuts [1981a] focuses on the algorithmic solution of Markov chains of the G/M/1 type. Neuts shows that the matrix $\mathcal{R}$ can be obtained by solving the matrix equation

$$\mathcal{R} = \sum_{k=0}^{\infty} \mathcal{R}^k A_k \tag{7.24}$$

for its minimal nonnegative solution, and that the minimal nonnegative solution can be obtained by solving the equation

$$\mathcal{R} = \left(\sum_{\substack{k=0 \\ k \neq 1}}^{\infty} \mathcal{R}^k A_k \right) (I - A_1)^{-1}$$

by successive substitutions starting with $\mathcal{R} = 0$. Note that (7.24) is simply a matrix version of (7.18) because

$$F_{\tilde{x}}^*(\mu[1 - \omega]) = \sum_{k=0}^{\infty} w^k a_k. \tag{7.25}$$

Some researchers have suggested that it is easier to solve the Markov chain $\mathcal{P}_{GM}$ directly by truncating the state space and approximating the probabilities; such an approach certainly has some advantages if the goal of the analysis is to obtain occupancy distributions only. However, the rate matrix, $\mathcal{R}$, obtained via the matrix-geometric approach, as we have seen in Chapter 3, is a

[1]That is, the elements of the matrix are nonnegative and the elements of each row sum to unity.

fundamental parameter of a given Markov chain. As such, the matrix $\mathcal{R}$ can be used to obtain much more than the occupancy distribution. For example, just as the scalar ω allows one to determine the FCFS waiting time in the ordinary G/M/1 system, the matrix $\mathcal{R}$ allows us to determine the waiting-time distribution from many points of view for queues of the G/M/1 type as in Ramaswami and Lucantoni [1985] and Daigle and Lucantoni [1990]. For a recent book that provides a thorough treatment of matrix analytic solution techniques applied to Markov chains of the G/M/1 type, the reader is referred to Latouche and Ramaswami [1999].

7.3 M/G/1 Solution Methodology

We turn now to the M/G/1 system. Analogous to our discussion of the ordinary M/G/1 system, we define the vector generating function

$$\mathcal{F}_{\tilde{q}}(z) = \sum_{j=0}^{\infty} z^j \pi_j. \tag{7.26}$$

Then, based on (7.21), it is easy to show that

$$\mathcal{F}_{\tilde{q}}(z)[Iz - \mathcal{A}(z)] = \pi_0[z\mathcal{B}(z) - \mathcal{A}(z)], \tag{7.27}$$

where $\mathcal{A}(z)$ and $\mathcal{B}(z)$ are defined as

$$\mathcal{A}(z) = \sum_{j=0}^{\infty} A_j z^j \quad \text{and} \quad \mathcal{B}(z) = \sum_{j=0}^{\infty} B_j z^j. \tag{7.28}$$

The scalar version of (7.27) is given in (4.94) and is

$$\mathcal{F}_{\tilde{n}_e}(z) = \frac{1-\rho}{1-\rho+\rho_e} \frac{zF^*_{\tilde{x}_e}(\lambda[1-z]) - F^*_{\tilde{x}}(\lambda[1-z])}{z - F^*_{\tilde{x}}(\lambda[1-z])}, \tag{7.29}$$

where the correspondence between the terms of (7.27) and (7.29) are obvious.

Just as in the case of the scalar version of the M/G/1 system, there are two difficulties: solving for the unknown constant (vector) π_0, and inverting the transform. Both of these operations are somewhat more involved than in the scalar case. A brief sketch of the techniques developed by Lucantoni, Neuts, and Ramaswami is given below. The reader is referred to Neuts [1989] for a thorough pedagogical presentation and to Lucantoni [1993] for a thorough reformulation of the solution methodology and some additional results.

We first discuss computation of π_0 and then present Ramaswami's technique for determining π_j, for $j = 1, 2, \ldots$, for the general case of the M/G/1 paradigm. We then consider a special case in which simplified algorithms can be developed. The special case has application to statistical multiplexing systems and the results of this section are used as a starting point in the next section, which discusses an application. The issue of stability is deferred for the time being, but is discussed in Section 7.5.

Determination of the vector π_0 is accomplished through the clever application of elementary Markov chain theory. Towards this end, let $\wp_n$ denote the phase of the system at the instant of the nth return to level 0. Then, due to the memoryless property of the arrival process, the stochastic process $\{\wp_n, n = 0, 1, \ldots\}$ is a discrete-parameter Markov chain on the space $\{0, 1, \ldots, K\}$. Let $\mathcal{P}_\wp$ denote the one-step transition probability matrix for this embedded Markov chain, and let κ denote its stationary probability vector.

From elementary Markov chain theory, it is well known that the proportion of transitions into level 0 is simply the inverse of the expected number of transitions between entries to level 0. Let κ_i^* denote the expected number of transitions between entries to level 0 given that the system enters level 0 in phase i, and let κ^* denote the column vector $[\,\kappa_0^* \quad \kappa_1^*, \cdots, \kappa_K^*\,]^T$. Then, it is readily seen that

$$\pi_0 = (\kappa\kappa^*)^{-1}\kappa. \tag{7.30}$$

It remains to specify $\mathcal{P}_\wp$, κ and κ^*. Towards this end, we begin by considering the first passage time between successive entries to level 0 of the Markov chain $\mathcal{P}_{MG}$ in stochastic equilibrium. In the first transition from level 0, the Markov chain transitions to level j with probability B_j. Having entered level j, the system cannot return to level 0 without having first passed through each level between j and 0.

Observation of the matrix $\mathcal{P}_{MG}$ reveals that the number of transitions required to decrease the level from j to $j - 1$ is independent of j. Thus the number of transitions required in the first passage time from level j to level 0 is simply the sum of j independent, identically distributed discrete-valued random variables. Define $G(z)$ to be the (matrix) generating function for the first passage time from level 1 to level 0.[2] The (matrix) generating function for the first passage time from level j to level 0 is then $[G(z)]^j$.

Remark. We distinguish between generating functions and probability generating functions. The idea of a generating function is to represent an arbitrary sequence of numbers $\{s_0, s_1, \ldots\}$, finite or infinite, by a power series, $\sum_{i=0}^{\infty} s_i z^i$, which may or may not be expressible in closed form. In the case of a probability generating function, the sequence in question is a probabil-

[2]That is, the number of transitions of the Markov chain $\mathcal{P}_{MG}$ in the first passage time from phase i of level 1 to level 0 is a discrete random variable. If the probability masses for this discrete random variable are partitioned according to the phase entered upon entry to level 0, and then the generating function is taken on the set of partitioned probability masses, then the result is a matrix of generating functions. If the elements of any row of the resulting matrix are summed, the result is a *probability* generating function in the ordinary sense.

ity mass function; that is, the s_i are probability masses for a discrete random variable.

By conditioning on the outcome of the first transition from level 1, we readily find that

$$G(z) = \sum_{j=0}^{\infty} zA_j[G(z)]^j. \tag{7.31}$$

Also, let $\mathcal{K}(z)$ denote the (matrix) generating function for the number of transitions between successive entries to level 0. Then, by conditioning on the outcome of the first transition from level 0, we find that

$$\mathcal{K}(z) = \sum_{j=0}^{\infty} zB_j[G(z)]^j. \tag{7.32}$$

We then find that $\mathcal{K}(1)$ is the one-step transition probability matrix for the Markov chain $\{\wp_n, n = 0, 1, \ldots\}$ defined above; that is,

$$\mathcal{P}_\wp = \mathcal{K}(1) = \sum_{i=0}^{\infty} B_i G^i, \tag{7.33}$$

where $G = G(1)$ is the unique stochastic matrix solution to the equation

$$G = \sum_{i=0}^{\infty} A_i G^i, \tag{7.34}$$

which is obtained by substituting $z = 1$ into (7.31). For consistency with Neuts's notation, from now on we will refer to $\mathcal{P}_\wp$ as $\mathcal{K}(1)$. We emphasize that G is the unique stochastic solution to (7.34) because, in general, (7.34) does not have a unique solution. For example, recall that in the scalar case, (7.34) has exactly the form $z = F_{\tilde{x}}^*(\lambda[1 - z])$. We have already seen that this functional equation has a solution $z = 1$, which is a 1×1 stochastic matrix, and an additional real-valued solution, z_0, the inverse of which determines the rate at which the tail of the occupancy distribution decreases.

In addition, using standard properties of probability generating functions, we readily find that

$$\kappa^* = \lim_{z \to 1} \frac{d}{dz} K(z)\mathbf{e} = K'(1)\mathbf{e}. \tag{7.35}$$

The desired solution for G can be obtained by solving (7.34) by successive substitutions, starting with $G = 0$. The transition matrix $\mathcal{K}(1)$ can then be obtained from (7.32), and then κ can be obtained by any of a number of techniques such as those described in Chapter 3. A normalizing constant $\kappa\kappa^*$ is needed to compute π_0 in (7.30). The usual technique for obtaining this normalizing constant is to differentiate (7.31) and (7.32) directly and then take advantage of the special structure of the problem at hand to obtain a reasonable computational formula. This involves a certain level of creativity on the part of the analyst, as may be seen from some of the published literature and our discussion here.

Ramaswami [1988a, 1988b] has devoted substantial energy to developing workable algorithms to solve for the stationary probability vector P of the Markov chain $\mathcal{P}_{MG}$. A primary theorem resulting from his work is quoted below for continuity, but the interested reader is encouraged to consult Ramaswami [1988b].

THEOREM 7.2 *(Ramaswami) For* $i \geq 1$

$$\pi_i = \left[\pi_0 \bar{B}_i + \sum_{j=1}^{i-1} \pi_j \bar{A}_{i+1-j}\right] (I - \bar{A}_1)^{-1}, \tag{7.36}$$

where $\bar{B}_i = \sum_{j=1}^{\infty} B_j G^{j-i}$ *and* $\bar{A}_i = \sum_{j=i}^{\infty} B_j G^{j-i}$. □

The key to finding the unknown probabilities in this approach is to determine the matrix G, and then to find the unknown vector of probabilities π_0. Beginning with π_0, Ramaswami's algorithm can be applied to find the remaining probabilities. Although Ramaswami has shown that a computational algorithm based on Theorem 7.2 is numerically stable, Lucantoni [1991] has pointed out that in practice it has not been feasible to implement the algorithm in its full generality.

Lucantoni [1993] considered a special case of considerable interest called the BMAP/G/1 queueing system, where BMAP stands for *batch Markovian arrival process*. He has developed new algorithms that allow for general implementation in terms of canned computer programs. The BMAP includes all the models discussed in this section. A general discussion of Lucantoni's methodology is beyond the scope of the current text, but the reader is urged to consult that reference, despite its age, for an up-to-date treatment of queueing systems in this class.

There are basically two steps in computing the level probabilities for Markov chains of the M/G/1 type: resolving the boundary probabilities and finding the level probabilities given the boundary probabilities. An interesting approach to the second step is discussed in Meini [1997], where an algorithm based on FFTs is presented for performing the computations specified in 7.2.

The matrix analytical approach can be readily modified to solve problems having complex or multiple boundary conditions in which case the one-step transition probability matrix has the form

$$\mathcal{P}_{MG} = \begin{bmatrix} B_{00} & B_{01} & B_{02} & B_{03} & \cdots \\ B_{10} & B_{11} & B_{12} & B_{13} & \cdots \\ \cdots & \cdots & \cdots & \cdots & \cdots \\ B_{C-1,0} & B_{C-1,1} & B_{C-1,2} & B_{C-1,3} & \cdots \\ A_0 & A_1 & A_2 & A_3 & \cdots \\ 0 & A_0 & A_1 & A_2 & \ddots \\ 0 & 0 & A_0 & A_1 & \ddots \\ 0 & 0 & 0 & A_0 & \ddots \\ \vdots & \ddots & \ddots & \ddots & \ddots \end{bmatrix}.$$

The approach is to group the level probabilities into groups of C blocks and group the matrices into blocks of $C \times C$ submatrices such that the resulting form is that of the standard M/G/1:

$$\mathcal{P}_{MG} = \begin{bmatrix} \hat{B}_0 & \hat{B}_1 & \hat{B}_2 & \hat{B}_3 & \cdots \\ \hat{A}_0 & \hat{A}_1 & \hat{A}_2 & \hat{A}_3 & \cdots \\ 0 & \hat{A}_0 & \hat{A}_1 & \hat{A}_2 & \ddots \\ 0 & 0 & \hat{A}_0 & \hat{A}_1 & \ddots \\ 0 & 0 & 0 & \hat{A}_0 & \ddots \\ \vdots & \ddots & \ddots & \ddots & \ddots \end{bmatrix}.$$

This approach was first described in Neuts [1981b], but will not be commented upon further. Rather, an alternate approach to solving the complex boundary case will be discussed in a later section. Before moving to that discussion, however, we present an application of the techniques discussed in this section to statistical multiplexing.

7.4 An Application to Statistical Multiplexing

A significant issue in the design of computer communication systems is the analysis of the occupancy distribution in statistical multiplexing systems. In particular, we envision a collection of individual users whose traffic is multiplexed onto a single high-capacity trunk. Traffic from individual users arrives to the statistical multiplexer over access lines, which can have lower capacity than the trunk. The users tend to transmit messages, whereas the statistical multiplexers tend to transfer data as fixed length packets. This gives rise to a high degree of correlation in the packet-arrival process of individual users. This correlation of packet interarrival times among the packets of individual users and the arrival process as a whole can be captured to a very high degree through the introduction of a discrete-time Markov chain, called the phase process, that characterizes the state of the arrival process. The distribution of the number of packets that arrive during a time slot is then dependent solely on the phase of the arrival process and the distribution of the number of arrivals per slot given the phase.

Under these conditions, the matrices $\mathcal{B}(z)$ and $\mathcal{A}(z)$ of (7.27) are equal and have the form $\mathcal{P}\mathcal{F}_{\tilde{a}}(z)$, where $\mathcal{P}$ is the one-step transition probability matrix for the phase process and $\mathcal{F}_{\tilde{a}}(z)$ is a diagonal matrix in which the ith diagonal element is a probability generating function. For this special case, it is possible to obtain a simplified algorithm for obtaining results.

We begin by making the substitutions in (7.27) to obtain

$$\mathcal{F}_{\tilde{q}}(z)\left[Iz - \mathcal{P}\mathcal{F}_{\tilde{a}}(z)\right] = \pi_0\left[z-1\right]\mathcal{P}\mathcal{F}_{\tilde{a}}(z). \tag{7.37}$$

Then differentiating on both sides, premultiplying by e, and taking limits as $z \to 1$ leads to

$$\left[1 - \mathcal{F}_{\tilde{q}}(1)\mathcal{P}\mathcal{F}_{\tilde{a}}^{(1)}(1)\mathrm{e}\right] = \pi_0\mathrm{e}. \tag{7.38}$$

We then note that (7.31) and (7.32) are identical under the restricted case. This means that $G(1)$ and $\mathcal{K}(1)$ are equal. Thus we can solve for $\mathcal{K}(1)$ by solving the fixed-point matrix equation

$$\mathcal{K}(1) = \sum_{j=0}^{\infty} B_j \mathcal{K}(1)^j. \tag{7.39}$$

Once $\mathcal{K}(1)$ is determined, we can solve for its stationary probability vector, which is denoted by κ.

EXERCISE 7.5 Show that the quantity $\mathcal{F}_{\tilde{q}}(1)$ corresponds to the stationary probability vector for the phase process.

EXERCISE 7.6 Derive equation (7.38).

Now, the implication of (7.30) is that π_0 is proportional to κ. But we know the value of $\pi_0\mathrm{e}$ from (7.38), so that we can readily determine the constant of proportionality. That is, we have

$$\pi_0 = \gamma\kappa, \tag{7.40}$$

where γ is an unknown constant. Thus

$$\pi_0\mathrm{e} = \gamma\kappa\mathrm{e} = \gamma,$$

where the final equality of the previous equation follows because κ is a stationary probability vector. Therefore, we have

$$\pi_0 = \left[1 - \mathcal{F}_{\tilde{q}}(1)\mathcal{P}\mathcal{F}_{\tilde{a}}^{(1)}(1)\mathrm{e}\right]\kappa. \tag{7.41}$$

An interesting special case is the one in which there are N statistically independent users, each of whose arrival process is governed by a phase process having M phases. In this case, the specifications of $\mathcal{P}$ and $\mathcal{F}_{\tilde{a}}(z)$ are not unique. It is easy to show that if we take $\mathcal{P}$ and $\mathcal{F}_{\tilde{a}}(z)$ to be the n-fold Kronecker products of $\mathcal{P}_s$ and $\mathcal{F}_{\tilde{a}}(z)$, respectively, then the resulting arrival process correctly characterizes the system. However, the description would contain redundant phases. For a system having N independent sources, each of which has M phases, it is relatively straightforward to show that the minimum dimension of the combined phase process is $\binom{N+M-1}{N}$, whereas a straightforward Kronecker product formulation would lead to $\mathcal{A}(z)$ having a dimension of M^N.

As an example, if $M = 3$, then the minimum dimension for the phase process is $(N+2)(N+1)/2$, and the dimension of the phase process, not taking redundancies into account, is 3^N. To put this in perspective, if $N = 5$, then $3^N = 343$ and $(N+2)(N+1)/2 = 21$. If $N = 50$, then $3^N \approx 7.2 \times 10^{23}$ and $(N+2)(N+1)/2 = 1326$.

EXERCISE 7.7 Suppose there are N identical traffic sources, each of which has an arrival process that is governed by an M-state Markov chain. Suppose the state of the combined phase process is defined by an M-vector in which the ith element is the number of sources currently in phase i. First, argue that this state description completely characterizes the phase of the arrival process. Next, show that the number of states of the phase process is given by $\binom{N+M-1}{N}$.

To illustrate the power of this approach, we present the analysis of a problem that has received much attention, but for which exact results do not seem to have been presented as of this time. Specifically, we show how to obtain the matrices $\mathcal{P}$ and $\mathcal{F}_{\tilde{a}}(z)$ for the special case where there are N identical individual users, each of whom has an access line that has exactly one-half the capacity of the trunk line onto which it is being statistically multiplexed. Following Example 7.1, we discuss the solution of the resulting model.

EXAMPLE 7.1 Consider a collection of N identical and independently operating sources. Suppose that each source alternates between active and inactive periods. During an active period, the source generates a packet every second time slot, and the number of packets generated during an active period is geometrically distributed with mean $1/(1-\alpha)$. The lengths of the inactive periods, in time slots, are geometrically distributed with mean $1/(1-\beta)$. We wish to specify $\mathcal{P}$ and $\mathcal{F}_{\tilde{a}}(z)$.

Solution: We find that the Markov chain governing the phase process of each source is

$$\mathcal{P}_s = \begin{bmatrix} \beta & 1-\beta & 0 \\ 0 & 0 & 1 \\ 1-\alpha & \alpha & 0 \end{bmatrix},$$

and the probability generating function for the number of arrivals in each slot is

$$\mathcal{F}_{\tilde{a}_s} z = \begin{bmatrix} 1 & 0 & 0 \\ 0 & 1 & 0 \\ 0 & 0 & z \end{bmatrix}.$$

Clearly, the arrival process can be described in the form

$$\mathcal{A}(z) = \mathcal{P}\mathcal{F}_{\tilde{a}} z,$$

and the methods described above can be used to obtain the unknown probability vector and then the occupancy distribution.

As mentioned above, the total number of phases required to characterize the phase of the combined process is $\binom{N+M+1}{N}$. For the special case under discussion, $M = 3$ so that $(N+2)(N+1)/2$ phases are required. For example, if $N = 2$, we define the six required phases as shown in Table 7.1.

For the given phase definitions, $\mathcal{F}_{\tilde{z}}(z) = \text{diag}\ (1, 1, 1, z, z, z^2)$. This indicates that in the first three phases of the aggregate process, no packets would arrive; during the fourth and fifth phases, one packet would arrive; and in the sixth phase, two packets would arrive. It is straightforward to generate an algorithm to define the minimum set of phases for arbitrary M and N.

Given the state definitions as described in Table 7.1, the resulting one-step probability transition matrix for the phase process, $\mathcal{P}$, is given by

$$\begin{bmatrix} \beta^2 & 2\beta(1-\beta) & (1-\beta)^2 & 0 & 0 & 0 \\ 0 & 0 & 0 & \beta & 1-\beta & 0 \\ 0 & 0 & 0 & 0 & 0 & 1 \\ \beta(1-\alpha) & \beta\alpha + (1-\beta)(1-\alpha) & \alpha(1-\beta) & 0 & 0 & 0 \\ 0 & 0 & 0 & (1-\alpha) & \alpha & 0 \\ (1-\alpha)^2 & 2\alpha(1-\alpha) & \alpha^2 & 0 & 0 & 0 \end{bmatrix}.$$

This completes the basic description of the model.

EXERCISE 7.8 Define the matrices $\mathcal{P}$ and $\mathcal{F}_{\tilde{a}}(a)$ for the model defined in Example 7.1 for the special case of $N = 3$, where the states of the phase process have the interpretations shown in Table 7.2. In the table, if the phase vector is ijk, then there are i sources in phase 0, j sources in phase 1, and k sources in phase 2.

In consideration of the form of $\mathcal{F}_{\tilde{a}}(z)$ and the process through which $\mathcal{K}(1)$ is obtained using (7.32), it is easy to see that column i of $\mathcal{K}(1)$ is nonzero if

Table 7.1. Definition of the phases for the problem solved in Example 7.1.

Aggregate Phase	*Phases of Individual Sources*
0	both sources in phase 0
1	one source in phase 0, one source in phase 1
2	both sources in phase 1
3	one source in phase 0, one source in phase 2
4	one source in phase 1, one source in phase 2
5	both sources in phase 2

Table 7.2. Definition of the phases for the system of Exercise 7.8.

Aggregate		*Phase Vector*	
0	3	0	0
1	2	1	0
2	1	2	0
3	0	3	0
4	2	0	1
5	1	1	1
6	0	2	1
7	1	0	2
8	0	1	2
9	0	0	3

and only if the ith diagonal element of $\mathcal{F}_{\tilde{a}}(z)$ is equal to 1. This fact is also made obvious by considering that $\mathcal{K}(1)$ is the one-step transition matrix for the Markov chain whose state is the phase of the arrival process upon entries of the queueing process to level 0. Level 0 can be entered only if there are no arrivals to the system during a slot. Because the condition for no arrivals during a slot is that the number of arrivals is equal to unity with probability 1, the generating function for the number of arrivals during the slot is equal to 1. Thus, an efficient algorithm can be devised to compute $\mathcal{K}(1)$. From this, κ can be determined and then the result normalized as in (7.41).

EXERCISE 7.9 In the previous example, we specified $\mathcal{F}_{\tilde{a}}(z)$ and $\mathcal{P}$. From these specifications, we have

$$\mathcal{P}\mathcal{F}_{\tilde{a}}(z) = \sum_{i=0}^{2} \mathcal{A}_i z^i.$$

Therefore,

$$\mathcal{K}(1) = \sum_{i=0}^{2} \mathcal{A}_i \left[\mathcal{K}(1)\right]^i,$$

with

$$\mathcal{A}_0 = \begin{bmatrix} \beta^2 & 2\beta(1-\beta) & (1-\beta)^2 & 0 & 0 & 0 \\ 0 & 0 & 0 & 0 & 0 & 0 \\ 0 & 0 & 0 & 0 & 0 & 0 \\ \beta(1-\alpha) & \beta\alpha + (1-\beta)(1-\alpha) & \alpha(1-\beta) & 0 & 0 & 0 \\ 0 & 0 & 0 & 0 & 0 & 0 \\ (1-\alpha)^2 & 2\alpha(1-\alpha) & \alpha^2 & 0 & 0 & 0 \end{bmatrix},$$

$$\mathcal{A}_1 = \begin{bmatrix} 0 & 0 & 0 & 0 & 0 & 0 \\ 0 & 0 & 0 & \beta & (1-\beta) & 0 \\ 0 & 0 & 0 & 0 & 0 & 0 \\ 0 & 0 & 0 & 0 & 0 & 0 \\ 0 & 0 & 0 & (1-\alpha) & \alpha & 0 \\ 0 & 0 & 0 & 0 & 0 & 0 \end{bmatrix},$$

and

$$\mathcal{A}_2 = \begin{bmatrix} 0 & 0 & 0 & 0 & 0 & 0 \\ 0 & 0 & 0 & 0 & 0 & 0 \\ 0 & 0 & 0 & 0 & 0 & 1 \\ 0 & 0 & 0 & 0 & 0 & 0 \\ 0 & 0 & 0 & 0 & 0 & 0 \\ 0 & 0 & 0 & 0 & 0 & 0 \end{bmatrix}$$

Now, suppose we compute $\mathcal{K}(1)$ iteratively; that is, we use the formula

$$\mathcal{K}_j(1) = \sum_{i=0}^{2} \mathcal{A}_i \left[\mathcal{K}_{j-1}(1)\right]^i \quad \text{for } j \geq 1, \tag{7.42}$$

with $\mathcal{K}_0(1) = 0$. Prove that the final three columns of $\mathcal{K}_j(1)$ are zero columns for all j.

EXERCISE 7.10 Suppose $\mathcal{K}(1)$ has the form

$$\mathcal{K}(1) = \begin{bmatrix} \mathcal{K}_{00} & 0 \\ \mathcal{K}_{10} & 0 \end{bmatrix}, \tag{7.43}$$

where $\mathcal{K}_{00}$ is a square matrix. Bearing in mind that $\mathcal{K}(1)$ is stochastic, prove that $\mathcal{K}_{00}$ is also stochastic and that κ has the form $[\,\kappa \quad 0\,]$, where κ_0 is the stationary probability vector for $\mathcal{K}_{00}$.

THEOREM 7.3 *Consider the algorithm*

$$\mathcal{K}_j(1) = \sum_{i=0}^{\infty} \mathcal{A}_i \left[\mathcal{K}_{j-1}(1)\right]^i \quad \text{for} \quad j \geq 1. \tag{7.44}$$

Suppose that any stochastic matrix is chosen for $\mathcal{K}_0(1)$. *Then* $\sum_{i=0}^{\infty} \mathcal{A}_i$ *is stochastic,* $[\mathcal{K}_0(1)]^i$ *is stochastic for every* $i \geq 0$, *and* $\mathcal{K}_j(1)$ *is stochastic for every* $j \geq 0$. □

EXERCISE 7.11 Prove Theorem 7.3.

The implication of Theorem 7.3 and Exercises 7.9, 7.10, and 7.11 is that the initial condition for the recursion

$$\mathcal{K}_j(1) = \sum_{i=0}^{\infty} \mathcal{A}_i \left[\mathcal{K}_{j-1}(1)\right]^i \quad \text{for} \quad j \geq 1$$

need not necessarily be chosen to be 0. In fact, experience has shown that convergence is especially slow when 0 is chosen as the initial condition. As an example, in working with the system discussed above, we defined $\mathcal{K}_0(1)$ in the following way:

1. each row of $\mathcal{K}_0(1)$ is exactly the same as the rows of $\mathcal{A}_0$ when the corresponding row of $\mathcal{A}_0$ is not zero, and
2. each row of $\mathcal{K}_0(1)$ has a 1 in its first column when the corresponding row of $\mathcal{A}_0$ is a row of zeros.

As an example, with $N = 2$, we chose $\mathcal{K}_0(1)$ as

$$\begin{bmatrix} \beta^2 & 2\beta(1-\beta) & (1-\beta)^2 & 0 & 0 & 0 \\ 1 & 0 & 0 & 0 & 0 & 0 \\ 1 & 0 & 0 & 0 & 0 & 0 \\ \beta(1-\alpha) & \beta\alpha + (1-\beta)(1-\alpha) & \alpha(1-\beta) & 0 & 0 & 0 \\ 1 & 0 & 0 & 0 & 0 & 0 \\ (1-\alpha)^2 & 2\alpha(1-\alpha) & \alpha^2 & 0 & 0 & 0 \end{bmatrix}. \tag{7.45}$$

For the case of $N = 4$, when $\mathcal{K}_0(1)$ was chosen according to the above procedure, the iterative algorithm converged to an acceptable level of accuracy in roughly 300 iterations; the same algorithm required more than 1700 iterations to converge when $\mathcal{K}_0(1)$ was chosen to be 0. The algorithm, of course, converged to the same values in either case. However, this illustrates that when multiple runs are to be made, it is worthwhile to experiment with the initial conditions.

An alternate approach to obtaining the level probabilities, $\pi_j\mathbf{e}$, is to first obtain π_0 as described above first, and then to use the discrete Fourier transform approach described in Chapter 5. Specifically, once π_0 is known, we can use (7.27) or (7.37) directly to solve for the value of $\mathcal{F}_{\tilde{q}}(z)\mathbf{e}$ at points around the unit circle of the complex plane and then use the IDFT to obtain the marginal or joint level probabilities. This procedure, of course, involves solving the complex linear system

$$\mathcal{F}_{\tilde{q}}(z_k)\left[Iz - \mathcal{P}\mathcal{F}_{\tilde{a}}(z_k)\right] = \pi_0\left[z_k - 1\right]\mathcal{A}(z_k). \tag{7.46}$$

where $z_k = e^{-j[2\pi k/(K+1)]}$ for $1 \le k \le K$, where $K = 2^n - 1$ for some integer-values n.

EXAMPLE 7.2 (Example 7.1 continued) Consider the statistical multiplexing system described in Example 7.1. Suppose it is known that the mean message length of the individual users is eight packets. We wish to determine the probability that the system occupancy will exceed a certain number of packets at a traffic load of $\rho = 0.9$ during an arbitrary slot in stochastic equilibrium as a function of the number of individual users served.

Solution: Each source operates according to an alternating renewal process, so the proportion of time each source spends in the active period is given by the ratio of the expected length of the active period to the expected length of the cycle. Because a source delivers a geometric number of packets with mean $1/(1-\alpha)$ during an active period, and because packets arrive only in alternate slots beginning with the second, the expected length of an active period, in slots, is $2/(1-\alpha)$. Similarly, the expected length of the idle period is $1/(1-\beta)$. Therefore the proportion of time each source spends in the active period is given by

$$\frac{2(1-\beta)}{(1-\alpha)+2(1-\beta)}.$$

Because packets are generated at rate 0.5 during active periods by each source and the service time of a packet is one slot, the traffic intensity is

$$\rho = \frac{N(1-\beta)}{(1-\alpha)+2(1-\beta)}.$$

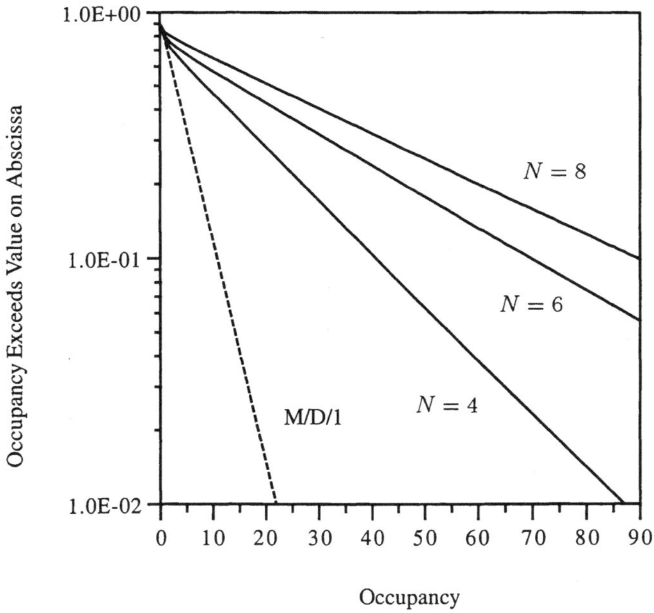

Figure 7.1. Survivor functions for occupancy distributions for statistical multiplexing system with 0.5 to 1.0 speed conversion at $\rho = 0.9$.

For fixed ρ and $(1 - \alpha)$, we can then solve for

$$\beta = 1 - \frac{(1-\alpha)\rho}{N - 2\rho}.$$

The average message length is eight packets, so $(1 - \alpha) = 0.125$. This, then, completely specifies the parameters for the model.

The recursion of (7.42) with $\mathcal{K}_0(1)$ as defined in (7.45) is then used to determine $\mathcal{K}(1)$ and its stationary probability vector, κ, is determined by solving the system $\kappa_0 = \kappa_0 \mathcal{K}_{00}$, $\kappa_0 \mathrm{e} = 1$. Then, $\mathcal{F}_{\tilde{a}}(1)$ is determined by solving the system $\mathcal{F}_{\tilde{q}}(1) = \mathcal{F}_{\tilde{q}}(1)\mathcal{K}_{00}$, $\mathcal{F}_{\tilde{q}}(1)\mathrm{e} = 1$. Next, (7.41) is used to obtain π_0. Finally, (7.46) and the IDFT procedure described in Chapter 5 are used to obtain the occupancy distribution. Numerical results are presented in Figure 7.1 and Table 7.3.

With reference to Figure 7.1, we see that the survivor function is an increasing function of the number of independent sources that generate traffic to the statistical multiplexer. This demonstrates that the often-cited claim that traffic arriving from a large number of sources can be treated as a Poisson arrival process must be carefully examined. In fact, often, as in the current case, exactly

Table 7.3. Mean and second moments of queue lengths for multiplexed lines with line speed conversion.

N	$E[\tilde{n}]$	% difference	$E[\tilde{n}^2]$	r_k
2	1.1025	2.9502E-17	1.50750	0.1296
4	15.9627	9.2715E-14	631.87702	0.9514
6	26.9358	1.0751E-13	1836.14439	0.9714
8	33.5743	1.1819E-13	2865.30967	0.9771

the opposite is true–a larger number of sources delivers traffic that is more bursty than that delivered by an ordinary Poisson process. This can be seen by observation of Figure 7.1, where the survivor function for the occupancy distribution for the M/D/1 system at a traffic intensity of 0.9 is also presented. Obviously, an M/D/1 approximation for the current system would miss the mark considerably.

Table 7.3 presents the mean and second moment of the occupancy distribution for several values of N. As a check on the accuracy of the PGF inversion routine, the mean values computed on the basis of the occupancy distribution were compared to the mean values computed analytically. The percent difference between the results of the two calculations is presented in the table. The table shows that the PGF inversion process works quite well. The general formula for the mean occupancy, derived in Daigle, Lee, and Magalhães [1990] following the parallel development in Lucantoni, Meier-Hellstern, and Neuts [1990], is

$$\begin{aligned} E[\tilde{q}] &= \mathcal{F}_{\tilde{q}}(1)\mathbf{e} \\ &= \frac{1}{1-\rho}\Big\{\frac{1}{2}\mathcal{F}_{\tilde{q}}(1)\mathcal{F}_{\tilde{a}}^{(2)}(1)\mathbf{e} + \pi_0\mathcal{F}_{\tilde{a}}^{(1)}(1)\mathbf{e} \\ &\qquad + \left(\pi_0\mathcal{P} - \mathcal{F}_{\tilde{q}}(1)\left[I - \mathcal{F}_{\tilde{a}}^{(1)}\right]\right) \\ &\qquad \left[I - \mathcal{P} + \mathbf{e}\mathcal{F}_{\tilde{q}}(1)\right]^{-1}\mathcal{P}\mathcal{F}_{\tilde{a}}^{(1)}\mathbf{e}\Big\} \end{aligned} \tag{7.47}$$

The development is rather involved and is deferred to the Supplemental Problems. We consider here only one aspect of the formula shown in (7.47). Note that $[I - \mathcal{P} + \mathbf{e}\mathcal{F}_{\tilde{q}}(1)]^{-1}$ is contained in the right-hand side of (7.47). This expression is called the *fundamental equation* for the Markov chain whose one-step transition probability matrix is $\mathcal{P}$ (see Hunter [1983]). For irreducible Markov chains, the fundamental matrix is always nonsingular. In addition,

$\mathbf{e}\mathcal{F}_{\tilde{q}}(1)\,[I - \mathcal{P} + \mathbf{e}\mathcal{F}_{\tilde{q}}(1)] = \mathbf{e}\mathcal{F}_{\tilde{q}}(1)$, the proof of which is deferred to Exercise 7.12.

EXERCISE 7.12 Suppose that $\mathcal{P}$ is the one-step transition probability matrix for an irreducible discrete-valued, discrete-parameter Markov chain. Define ϕ to be the stationary probability vector for the Markov chain. Prove that $\mathbf{e}\phi\,[I - \mathcal{P} + \mathbf{e}\phi] = \mathbf{e}\phi$ and that, therefore, $\mathbf{e}\phi = \mathbf{e}\phi\,[1 - \mathcal{P} + \mathbf{e}\phi]^{-1}$.

Further simplification of the computational technique for π_0 is readily obtained if the traffic arrival process for the multiplexing system under study has only one phase during which 0 packets can arrive. That is, the distribution of the number of packets that arrive during a slot is dependent on the phase of the arrival process, The phase is, in turn, governed by a discrete-time Markov chain. Now, suppose that the probability generating function for the number of arrivals during a slot, given the phase, is arbitrary except that the number of packets delivered during a time slot can be zero if and only if the phase of the arrival process is zero. Under that condition, the unknown probability vector, π_0, is trivially computed as described in the following paragraph.

In (7.38), the quantity $\mathcal{F}_{\tilde{q}}(1)$ corresponds to the stationary vector for the phase process, as is readily seen by taking limits on both sides of (7.37) as $z \to 1$. Now $\pi_0 = [\pi_{00} \quad \pi_{10} \quad \cdots \quad \pi_{N0}]$. But π_{j0} represents the probability that the phase process is in phase j and zero packets are present at the end of the slot. Hence $\pi_{j0} = 0$ except for $j = 0$. Thus, $\pi_{00} = [1 - \mathcal{F}_{\tilde{q}}(1)\mathcal{P}\mathcal{F}_{\tilde{a}}^{(1)}(1)\mathbf{e}]$.

As in Example 7.2, such an arrival process is useful in modeling the traffic arising from a collection of independently operating sources in an integrated services environment. This is shown in Example 7.3.

EXAMPLE 7.3 Consider a collection of N identical and independently operating sources. Suppose that each source alternates between inactive and active periods. During an active period, the source generates a packet every time slot, and the number of packets generated during an active period is geometrically distributed with mean $1/(1-\alpha)$. The lengths of the inactive periods, in time slots, are geometrically distributed with mean $1/(1-\beta)$. We wish to specify $\mathcal{P}$ and $\mathcal{F}_{\tilde{a}}(z)$, and to determine the occupancy distribution for the statistical multiplexing system as a function of N with $\rho = 0.9$ and $1/(1-\alpha) = 8$.

Solution: The solution is similar to that of Example 7.1. We find that the Markov chain governing the phase of each source and the probability generating function for the number of arrivals in each time slot are as follows:

$$\mathcal{P}_s = \begin{bmatrix} \beta & 1-\beta \\ 1-\alpha & \alpha \end{bmatrix} \quad \text{and} \quad \mathcal{F}_{\tilde{a}_s}(z) = \begin{bmatrix} 1 & 0 \\ 0 & z \end{bmatrix},$$

respectively. Clearly, the arrival process can be described in the form

$$\mathcal{A}(z) = \mathcal{P}\mathcal{F}_{\tilde{a}}(z),$$

and the methods described above can be used to obtain the unknown probability vector and then the occupancy distribution.

Packets arrive during every time slot when a source is active, so the state of the phase process can be defined simply as the number of sources in the active state. Thus the total number of phases required to characterize the phase of the combined arrival process is $N+1$.

It is relatively easy to show that the phase transition probabilities are obtained from the expression

$$\mathcal{P}_{mk} = \sum_{i=(k-[N-n])^+}^{\min\{k,n\}} \binom{n}{i}\binom{N-n}{k-i}\alpha^i(1-\alpha)^{n-i}(1-\beta)^{k-i}\beta^{(N-n-k+i)}.$$

For the given phase definitions, $\mathcal{F}_{\tilde{a}}(z) = \text{diag}\left(1, z, z^2, \ldots, z^N\right)$. This indicates that in the first phase of the aggregate process, no packets would arrive. This completes the basic description of the model.

By following the development of Example 7.2, we can readily find that the proportion of time each source spends in the active period is given by

$$\sigma = \frac{(1-\beta)}{(1-\alpha)+(1-\beta)}$$

and that the traffic intensity is given by

$$\rho = \frac{N(1-\beta)}{(1-\alpha)+(1-\beta)}.$$

For fixed ρ and $(1-\alpha)$, we can then solve for

$$\beta = 1 - \frac{(1-\alpha)\rho}{N-\rho}.$$

Because the average message length is eight packets, $(1-\alpha) = 0.125$. This, then, completely specifies the parameters for the model.

We could obtain $\mathcal{F}_{\tilde{q}}(1)$ by solving the system $\mathcal{F}_{\tilde{q}}(1) = \mathcal{F}_{\tilde{q}}(1)\mathcal{P}$, $\mathcal{F}_{\tilde{q}}(1)\mathbf{e} = 1$. However, because each of the sources operates according to an alternating renewal process, we can readily see that the equilibrium number of active sources has a binomial distribution with parameters N and σ. Thus, $\mathcal{F}_{\tilde{q}}(1)$ is simply the vector of equilibrium probabilities. Also, π_0 is trivially determined as $[\,\rho \quad 0 \quad \cdots \quad 0\,]$. Thus it remains only to use (7.46) and the IDFT procedure described in Chapter 5 to obtain the occupancy distribution. Numerical results are presented in Figure 7.2 and Table 7.4.

As in Example 7.2, we can see from Figure 7.2 that the survivor function is an increasing function of N. In fact, the departure from the survivor function

Table 7.4. Mean and second moments of queue lengths for multiplexed lines with no line speed conversion.

N	$E[\tilde{n}]$	*% difference*	$E[\tilde{n}^2]$	r_k
2	1.025	2.9502E-17	1.5075	0.0196
4	35.5275	9.3871E-14	3020.63366	0.9769
6	43.4250	2.3726E-13	4532.70825	0.9811
8	47.6775	2.6868E-13	5473.48123	0.9828
10	50.3262	2.2786E-13	6104.22453	0.9837
12	52.1325	2.2243E-13	6554.07506	0.9843
14	53.4426	2.4292E-13	6890.33668	0.9847

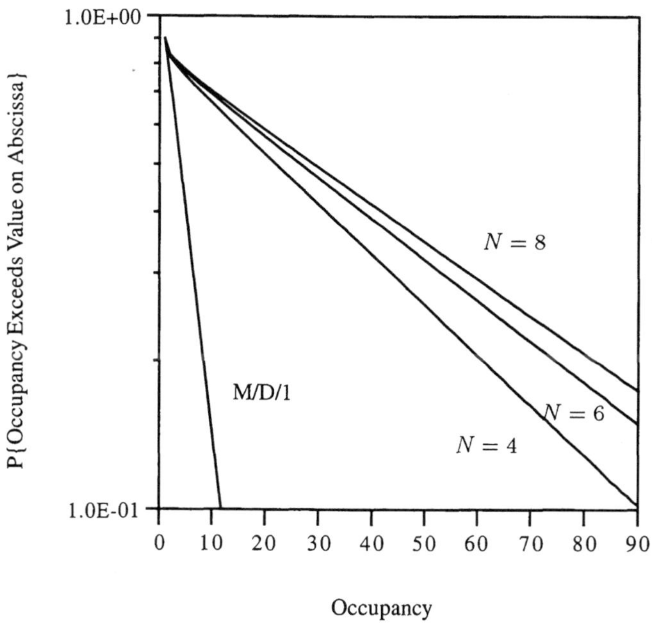

Figure 7.2. Survivor functions for occupancy distributions for statistical multiplexing system with equal line and trunk capacities at $\rho = 0.9$.

for the M/D/1 system having equal traffic intensity is even more pronounced than in Example 7.2. This is because the line speeds of the individual sources

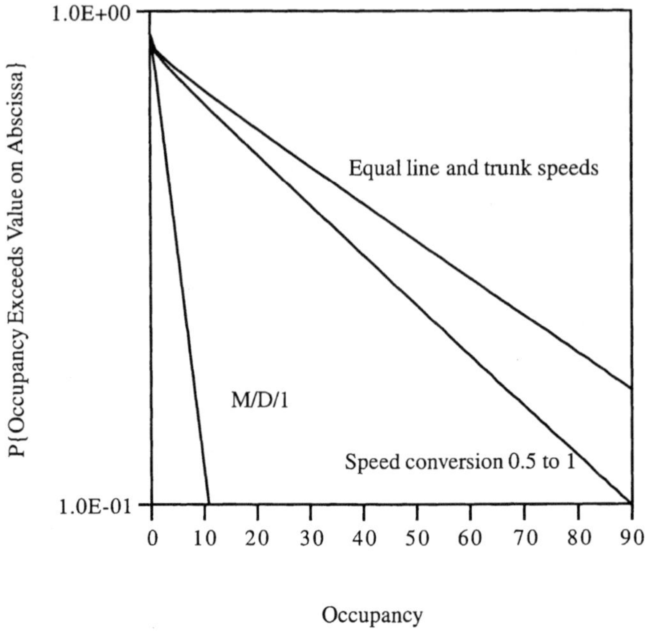

Figure 7.3. Survivor functions for occupancy distributions for statistical multiplexing system with and without line-speed conversion at $\rho = 0.9$.

are higher than in Example 7.2, thus giving rise to more bursty traffic. The differences among the M/D/1, the N-source model with transmission-speed conversion, and the N-source model without transmission-speed conversion are highlighted in Figure 7.3, where numerical results are presented for the special case of $N = 8$.

7.5 Generalized State Space Approach: Complex Boundaries

In a more general case, a queueing system may have C boundary conditions. A specific case of interest is a frame-based wireless transmission system in which the quality of the wireless link may vary from time slot to time slot. The number of units that can be served is dependent upon the wireless link quality. One possible approach is to model the link quality as a Markov chain in which the state of the chain determines the number of units that can be served during a frame. If we assume the maximum number of units that may be served during a frame is limited to say, C, then the system of state equations for the Markov chain in equilibrium would have multiple boundaries and the one-step state transition probability matrix, $\mathcal{P}$, would have the form

$$\mathcal{P}_{MG} = \begin{bmatrix} B_{00} & B_{01} & B_{02} & B_{03} & \cdots \\ B_{10} & B_{11} & B_{12} & B_{13} & \cdots \\ \cdots & \cdots & \cdots & \cdots & \cdots \\ B_{C-1,0} & B_{C-1,1} & B_{C-1,2} & B_{C-1,3} & \cdots \\ A_0 & A_1 & A_2 & A_3 & \cdots \\ 0 & A_0 & A_1 & A_2 & \ddots \\ 0 & 0 & A_0 & A_1 & \ddots \\ 0 & 0 & 0 & A_0 & \ddots \\ \vdots & \ddots & \ddots & \ddots & \ddots \end{bmatrix}. \tag{7.48}$$

If we assume that the number of units arriving to the service system during a frame is a sequence of independent, identically distributed random variables, then the dimension of the level probability vectors would be the same as the dimension of the Markov chain governing the service process, which we define to be $K + 1$. In that case, the matrices B_{ij} and A_j all have dimension $(K + 1) \times (K + 1)$.

In this section, we discuss solution of the system of equations $\pi = \pi\mathcal{P}_{MG}$ via the generalized state-space approach described in Akar, Oğuz, and Sohraby [1998]. Our formulation of the problem is more direct than theirs, but our solution approach follows more or less along the same lines.

As in Section 7.3, we define π to be the stationary vector of the Markov chain so that $\pi = \pi\mathcal{P}_{MG}$, π_j to be the level probability vector for level j, and the vector generating function as

$$\mathcal{F}_{\tilde{q}}(z) = \sum_{j=0}^{\infty} z^j \pi_j.$$

Further, the matrix generating functions of the sequences $\{A_i, i = 0, 1, \ldots\}$ and $\{B_{ji}, j = 0, 1, \ldots, C-1, i = 0, 1, \ldots\}$ are defined to be

$$\mathcal{A}(z) = \sum_{n=0}^{\infty} A_i z^i \quad \text{and} \quad \mathcal{B}_j(z) = \sum_{n=0}^{\infty} B_{ji} z^i.$$

It is then straightforward to show that

$$\mathcal{F}_{\tilde{q}}(z)\left[z^C I - \mathcal{A}(z)\right] = \sum_{j=0}^{C-1} \pi_j \left[z^C \mathcal{B}_j(z) - z^j \mathcal{A}(z)\right]. \tag{7.49}$$

By examination of (7.49) it is readily seen that in order to completely specify $\mathcal{F}_{\tilde{q}}(z)$, we must find the probability vector from level 0 to level $C-1$. Thus, we refer to the level probability vectors from level 0 to level $C-1$ as boundary probability vectors. Once the boundary probability vectors are known, in principle, the remaining level probability vectors can be determined. Alternative methods of determining the remaining level probability vectors include the FFT methods described earlier in the context of the ordinary and priority M/G/1 queues and a straightforward extension to the classical matrix analytic method.

In this section, we discuss an alternative method of finding both the boundary level probabilities and the remaining level probability vectors. The method presented here is a straightforward extension to the method presented earlier for the scalar M/G/1 queue and QBD processes. We show that the solution to (7.49) can always be expressed in terms of the boundary level probabilities, a vector g, and matrices F and H as follows:

$$\pi_{i+C} = gF^iH, \tag{7.50}$$

where the vector g is expressible in terms of the boundary probabilities.

As in the scalar case discussed in a Chapter 5, we begin our development with conversion of (7.49) to a form whose left-hand side does not involve the boundary level probabilities; that is, we wish to transform (7.49) such that the level probability vectors $\pi_0, \pi_1, \ldots, \pi_{C-1}$ appear only on the right-hand side. To this end, we define

$$\mathcal{F}_{\tilde{q}}^{(1)}(z) = \frac{1}{z}\left[\mathcal{F}_{\tilde{q}}(z) - \pi_0\right],$$

and form iteratively

$$\mathcal{F}_{\tilde{q}}^{(i)}(z) = \frac{1}{z}\left[\mathcal{F}_{\tilde{q}}^{(i-1)}(z) - \pi_{i-1}\right],$$

for $i = 2, 3, \ldots, C$. As in the scalar case, it is then straightforward to show that (7.49) reduces to

$$\mathcal{F}_{\tilde{q}}^{(C)}(z)\left[z^C I - \mathcal{A}(z)\right] = \sum_{j=0}^{C-1} \pi_j \left[\mathcal{B}j(z) - z^j I\right]. \tag{7.51}$$

EXERCISE 7.13 Define

$$\mathcal{F}_{\tilde{q}}^{(1)}(z) = \frac{1}{z}\left[\mathcal{F}_{\tilde{q}}(z) - \pi_0\right] \text{ and } \mathcal{F}_{\tilde{q}}^{(i+1)}(z) = \frac{1}{z}\left[\mathcal{F}_{\tilde{q}}^{(i)}(z) - \pi_i\right], i \geq 1.$$

Starting with (7.49), substitute a function of $\mathcal{F}_{\tilde{q}}^{(1)}(z)$ for $\mathcal{F}_{\tilde{q}}(z)$, then a function of $\mathcal{F}_{\tilde{q}}^{(2)}(z)$ for $\mathcal{F}_{\tilde{q}}^{(1)}(z)$, and continue step by step until a function of $\mathcal{F}_{\tilde{q}}^{(C)}(z)$ is substituted for $\mathcal{F}_{\tilde{q}}^{(C-1)}(z)$. Show that at each step, one element of

$$\sum_{j=0}^{C-1} \pi_j z^j \mathcal{A}(z)$$

is eliminated, resulting in (7.51).

Now, suppose that $\mathcal{A}(z)$ and $\mathcal{B}_j(z)$, $j = 0, 1, \ldots, C-1$, are all left multiples of the $(K+1)$-square matrix $\mathcal{W}^{-1}(z)$; that is, we can write $\mathcal{A}(z)$ and $\mathcal{B}_j(z)$ in right polynomial fraction form (Chen [1999]) as

$$\mathcal{A}(z) = \mathcal{U}(z)\mathcal{W}^{-1}(z) \quad \text{and} \quad \mathcal{B}_j(z) = \mathcal{V}_j(z)\mathcal{W}^{-1}(z). \tag{7.52}$$

Then, upon substituting (7.51) can be rewritten as follows:

$$\mathcal{F}_{\tilde{q}}^{(C)}(z)\left[z^C I - \mathcal{U}(z)\mathcal{W}^{-1}(z)\right] = \sum_{j=0}^{C-1} \pi_j \left[\mathcal{V}_j(z)\mathcal{W}^{-1}(z) - z^j\right].$$

After post multiplying both sides of the previous equation by $\mathcal{W}(z)$, we then find

$$\mathcal{F}_{\tilde{q}}^{(C)}(z)\left[z^C \mathcal{W}(z) - \mathcal{U}(z)\right] = \sum_{j=0}^{C-1} \pi_j \left[\mathcal{V}_j(z) - z^j \mathcal{W}(z)\right]. \tag{7.53}$$

Now define $\mathcal{D}(z) = z^C \mathcal{W}(z) - \mathcal{U}(z)$, ν_D as the degree of $\mathcal{D}(z)$, and D_i as the coefficient of z^i in $\mathcal{D}(z)$. Similarly, define $\mathcal{N}_j(z) = \mathcal{V}j(z) - z^j\mathcal{W}(z)$ for $j = 0, 1, \ldots, C-1$, ν_N as the maximal degree of $\mathcal{N}_j(z)$ over all j, and and $N_{j,i}$ as the coefficient of z^i in $\mathcal{N}_j(z)$. Note that some of the coefficients of $\mathcal{D}(z)$ and $\mathcal{N}_j(z)$ may be zero, but $D_{\nu_D} \neq 0$ and $N_{j,\nu_N} \neq 0$ for at least one value of j. Finally, define

$$\nu = \max\{\nu_D, \nu_N + 1\}. \tag{7.54}$$

Upon substituting the expressions defined in the previous paragraph into (7.53), the following equation results:

$$\mathcal{F}_{\tilde{q}}^{(C)}(z)\mathcal{D}(z) = \sum_{j=0}^{C-1} \pi_j \mathcal{N}_j(z). \tag{7.55}$$

The coefficients of $D(z)$ and $N_j(z)$ are given by

$$\begin{aligned} D_i &= W_{i-C} - U_i, \text{ for } i = 0, 1, \ldots, \nu, \\ N_{j,i} &= V_{j,i} - W_{i-j}, \text{ for } i = 0, 1, \ldots, \nu, j = 0, 1, \ldots, C-1. \end{aligned} \tag{7.56}$$

In review, we note the following:

- $\mathcal{F}_{\tilde{q}}^{(C)}(z)$ is a function of level vectors at or above level C only;

$$\mathcal{F}_{\tilde{q}}^{(C)}(z) = \sum_{i=0}^{\infty} \pi_{C+i} z^i.$$

- It is always true that $\nu \geq C$, $\nu \geq \nu_D$, and $\nu_N \leq \nu - 1$.

- The degrees of the polynomials in z within the matrix $z^C \mathcal{W}(z) - \mathcal{U}(z)$ are all less than or equal to ν_D and at least one of the coefficients has degree equal to ν_D. If $\nu > \nu_D$, we can augment the coefficients of $\mathcal{D}(z)$ with zero-valued coefficients so that we can write

$$\mathcal{D}(z) = \sum_{i=0}^{\nu} D_i z^i.$$

- The degrees of the polynomials in z within the matrices $\mathcal{V}_j(z) - z^j \mathcal{W}(z)$ for $j = 0, 1, \ldots, C-1$ are all less than or equal to ν_N and at least one of the coefficients has degree equal to ν_N. Therefore $\mathcal{N}_j(z)$ has degree ν_N at most, and $\nu_N \leq \nu - 1$. If $\nu_N < \nu - 1$, we can augment the coefficients of $\mathcal{N}_j(z)$ with zero-valued coefficients so that we can write

$$\mathcal{N}_j(z) = \sum_{i=0}^{\nu-1} N_{j,i} z^i.$$

Given the forms of $\mathcal{F}_{\tilde{q}}^{(C)}(z)$, $\mathcal{D}(z)$, and $\mathcal{N}_j(z)$ just listed, we can rewrite (7.55) in terms of powers of z so that (7.55) becomes

$$\left[\sum_{i=0}^{\infty} \pi_{C+i} z^i\right] \left[\sum_{k=0}^{\nu} D_k z^k\right] = \sum_{j=0}^{C-1} \pi_j \sum_{i=0}^{\nu-1} N_{j,i} z^i.$$

Upon reordering the summations of the previous equation, we find

$$\sum_{i=0}^{\infty} \left[\sum_{k=0}^{\min\{i,N\}} \pi_{C+i-k} D_k\right] z^i = \sum_{i=0}^{\nu-1} \left[\sum_{j=0}^{C-1} \pi_j N_{j,i}\right] z^i. \tag{7.57}$$

By matching coefficients of the powers of z in (7.57) for $i = 0, 1, \ldots, \nu - 1$, we obtain

$$y_0 D = x_0 N, \tag{7.58}$$

where

$$D = \begin{bmatrix} D_0 & D_1 & \cdots & D_{\nu-1} \\ 0 & D_0 & \ddots & D_{\nu-2} \\ \vdots & \ddots & \ddots & \vdots \\ 0 & 0 & \cdots & D_0 \end{bmatrix}, \quad N = \begin{bmatrix} N_{0,0} & N_{0,1} & \cdots & N_{0,\nu-1} \\ N_{1,0} & N_{1,1} & \cdots & N_{1,\nu-1} \\ \vdots & \vdots & \vdots & \vdots \\ N_{C-1,0} & N_{C-1,1} & \cdots & N_{C-1,\nu-1} \end{bmatrix},$$

$x_0 = [\,\pi_0 \quad \pi_1 \quad \cdots \quad \pi_{C-1}\,]$, and $y_0 = [\,\pi_C \quad \pi_{C+1} \quad \cdots \quad \pi_{C+\nu-1}\,]$.

The coefficient of z^i on the right-hand side of (7.57) is zero for $i > \nu - 1$. Thus, by setting the coefficient of z^i to zero for $i > \nu - 1$, we have from (7.57)

$$[\,\pi_{C+i} \quad \pi_{C+i+1} \quad \cdots \quad \pi_{C+i+\nu}\,] \begin{bmatrix} D_\nu \\ D_{\nu-1} \\ \vdots \\ D_0 \end{bmatrix} = 0 \text{ for } i > \nu,$$

or, equivalently,

$$\pi_{C+i+\nu} D_0 = [\,\pi_{C+i} \quad \pi_{C+i+1} \quad \cdots \quad \pi_{C+i+\nu-1}\,] \begin{bmatrix} -D_\nu \\ -D_{\nu-1} \\ \vdots \\ -D_1 \end{bmatrix} \text{ for } i \geq \nu.$$

If we now augment the previous equation with the simple equations $\pi_{C+i+j} = \pi_{C+i+j}$ for $j = 0, 1, \ldots, \nu - 1$, we obtain the system of equations

$$y_{i+1} E = y_i A, \tag{7.59}$$

where $y_i = [\,\pi_{C+i} \quad \pi_{C+i+1} \quad \cdots \quad \pi_{C+i+\nu-1}\,]$, $E = \text{diag } (I, I, \ldots, I, D_0)$, and

$$A = \begin{bmatrix} 0 & 0 & \cdots & 0 & -D_\nu \\ I & 0 & \ddots & \vdots & -D_{\nu-1} \\ 0 & I & \ddots & \vdots & \vdots \\ \vdots & \ddots & \ddots & \vdots & \vdots \\ 0 & 0 & \cdots & I & -D_1 \end{bmatrix}.$$

To solve (7.59), we now apply generalized Schur decomposition, as we did previously in solving for the equilibrium probabilities for QBD and scalar M/G/1 systems. To begin, we repeat the following theorem for continuity:

THEOREM 7.4 **Generalized Schur Decomposition.** *Suppose A and E are both real matrices with spectrum $\lambda(A, E)$ and $\lambda(A, E)$ is partitioned into two sets, say $\lambda_u(A, E)$ and $\lambda_s(A, E)$ such that $\lambda_u(A, E) \cap \lambda_s(A, E) = \emptyset$. Then, there exist (non-singular) orthogonal matrices, Q and Z, such that*

$$Q^T EZ = \begin{bmatrix} E_{uu} & E_{us} \\ 0 & E_{ss} \end{bmatrix} \quad \text{and} \quad Q^T AZ = \begin{bmatrix} A_{uu} & A_{us} \\ 0 & A_{ss} \end{bmatrix},$$

where all matrices are real, E_{uu} and E_{ss} are upper triangular, and A_{uu} and A_{ss} are block upper triangular, meaning that their diagonal elements are either 1×1 or 2×2 blocks, depending upon whether the eigenvalues are real or occur in complex conjugate pairs. The row dimensions of E_{uu} and A_{uu} and E_{ss} and A_{ss} are $n_u = \text{card } (\lambda_u(A, E))$ and $n_s = \text{card } (\lambda_s(A, E))$, respectively. □

The generalized Schur decomposition of Theorem 7.4 is carried out efficiently by using the so-called QZ algorithm, which is described in detail in Golub and Van Loan [1996]. In turn, the QZ algorithm is implemented in the LAPACK routine *dgges()* (see Anderson [1999]).

As in the solution procedure for QBD processes, we first define the partitions $\lambda_u(A, E)$ and $\lambda_s(A, E)$ as the unstable and stable sets of generalized eigenvalues of A with respect to E. We then define $y_j = u_j Q^T$. We then substitute $u_j Q^T$ and $u_{j-1} Q^T$ for y_j and y_{j-1} in (7.59) and postmultiply both sides of the result by Z to obtain

$$\begin{bmatrix} u_{j,u} & u_{j,s} \end{bmatrix} \begin{bmatrix} E_{uu} & E_{us} \\ 0 & E_{ss} \end{bmatrix} = \begin{bmatrix} u_{j-1,u} & u_{j-1,s} \end{bmatrix} \begin{bmatrix} A_{uu} & A_{us} \\ 0 & A_{ss} \end{bmatrix}, \tag{7.60}$$

where $u_{j,u}$ and $u_{j,s}$ represent the unstable and stable parts of u_j, respectively, and whose dimensions are n_u and n_s, respectively. We recognize that $u_{j,u}$ must be 0 for all j in order to have a stable solution. Thus, (7.60) implies $u_{j,s} E_{ss} = u_{j-1,s} A_{ss}$. This leads to

$$u_{j,s} = u_{j-1,s} A_{ss} E_{ss}^{-1}. \tag{7.61}$$

Because $u_{j,u} = 0$ for all j, $u_j Q^T = \begin{bmatrix} 0 & u_{j,s} \end{bmatrix} Q^T$. Therefore, it is convenient to partition Q^T so that we may write $y_j = u_{j,s} L_s$, where L_s is the matrix containing the last n_s rows of Q^T. Now substitute $y_{j-1} Q_s$ for $u_{j-1,s}$ into (7.61) and then postmultiply the result by L_s to obtain

$$y_j = y_{j-1} Q_s A_{ss} E_{ss}^{-1} L_s. \tag{7.62}$$

Equivalently,

$$y_j \quad = \quad y_0 \left[Q_s A_{ss} E_{ss}^{-1} L_s \right]^j,$$

$$\begin{aligned} y_j &= y_0 Q_s \left[A_{ss}E_{ss}^{-1}\right]^j L_s, \text{ and} \\ \pi_{C+j} &= g\left[A_{ss}E_{ss}^{-1}\right]^j H \quad \text{for all} \quad j \geq 0, \end{aligned} \tag{7.63}$$

where $g = y_0 Q_s$, H is defined as the first $K+1$ columns of L_s, and the second step of the previous equation results from the fact that $L_s Q_s = I$. Thus, (7.63) specifies all of the level probabilities for levels greater than C in terms of $y_0 Q_s$, while $y_0 = [\, \pi_C \quad \pi_{C+1} \quad \cdots \quad \pi_{C+\nu-1} \,]$. Thus, to complete the solution, it remains only to specify y_0.

From (7.58), we have

$$[\, x_0 \quad y_0 \,] \begin{bmatrix} -N \\ D \end{bmatrix} = 0$$

In addition, because Q is orthogonal, $y_j = u_j Q^T$ implies $u_j = y_j Q$ so that $y_j Q_u = u_{j,u}$, where Q_u is the first n_u columns of Q. Thus, we must have

$$y_j Q_u = 0 \text{ for all } j \geq 0.$$

We thus have from the previous two equations

$$[\, x_0 \quad y_0 \,] \begin{bmatrix} -N & 0 \\ D & Q_u \end{bmatrix} = 0. \tag{7.64}$$

We now define

$$x_0 = k_0 \hat{x}_0 \quad \text{and} \quad y_0 = k_0 \hat{y}_0, \tag{7.65}$$

where $\hat{x}_0 \mathbf{e} = 1$. Equation 7.64 can then be rewritten as

$$[\, \hat{x}_0 \quad \hat{y}_0 \,] \begin{bmatrix} -\hat{N} & \mathbf{e} & 0 \\ \hat{D} & 0 & Q_u \end{bmatrix} = [\, 0 \quad 1 \quad 0 \,], \tag{7.66}$$

where $\hat{N}$ and $\hat{D}$ are the matrices N and D with their last columns deleted. Note that (7.66) yields numerical values for $\hat{x}_0$ and $\hat{y}_0$.

In terms of $\hat{y}_0$, (7.63) becomes

$$\pi_{C+j} = k_0 \hat{y}_0 Q_s \left[A_{ss}E_{ss}^{-1}\right]^j H \quad \text{for all} \quad j \geq 0.$$

Since the individual probabilities must sum to unity, we then have

$$1 = x_0 \mathbf{e} + \sum_{i=0}^{\infty} \pi_{C+j} \mathbf{e},$$

or equivalently,

$$\begin{aligned} 1 &= \hat{k}_0 \left[\hat{x}_0 + \hat{y}_0 Q_s \left\{\sum_{i=0}^{\infty} \left[A_{ss}E_{ss}^{-1}\right]^j\right\} H\right] \mathbf{e} \\ &= \hat{k}_0 \left[\hat{x}_0 + \hat{y}_0 Q_s \left\{I - \left[A_{ss}E_{ss}^{-1}\right]\right\}^{-1} H\right] \mathbf{e}. \end{aligned} \tag{7.67}$$

Thus, we may determine k_0 from (7.67).

After finding k_0, we may substitute its value into (7.65) to determine x_0 and y_0, noting that x_0 contains the vectors $\pi_0, \pi_1, \ldots, \pi_{C-1}$. We may then use y_0 in (7.63) to determine π_{C+i} for $i = 0, 1, \ldots$, which results in a complete solution for the level probabilities.

In summary, the generalized state-space solution to multiple-boundary problems within the M/G/1 paradigm is as follows:

1. From the problem statement, determine $\mathcal{A}(z), \mathcal{B}_j(z)$ for $j = 0, 1, \ldots, C-1$ and express these polynomials in right polynomial fractional form as in (7.52).

2. Compute ν using (7.54).

3. Compute the coefficients of $\mathcal{D}(z)$ and $\mathcal{N}(z)$ using (7.56).

4. Using the results of the previous step, (7.58), and (7.59), determine the matrices D, N, E, and A.

5. Perform a generalized Schur decomposition of A with respect to E according to Theorem 7.4. The LAPACK routine *dgges()* may be used for this purpose. This decomposition yields directly Q, n_u, n_s, E_{ss}, and A_{ss}.

6. Partition Q and Q^T to obtain Q_s, Q_u, and H.

7. Formulate the linear system of equations (7.66) and solve to obtain $\hat{x}_0$ and $\hat{y}_0$.

8. Solve for k_0 using (7.67).

9. Find x_0 and y_0 using (7.65).

10. Partition x_0 to find $\pi_0, \pi_1, \ldots, \pi_{C-1}$.

11. Compute all remaining desired π_j using (7.63).

Once the boundary probabilities are found and the remaining level probabilities are expressed in the form given in (7.63),

$$\pi_{C+j} = g\left[A_{ss}E_{ss}^{-1}\right]^j H,$$

$\mathcal{F}_{\tilde{q}}(z)$ can be written in the following alternative form:

$$\mathcal{F}_{\tilde{q}}(z) = \sum_{i=0}^{\infty} z^i \pi_i = \sum_{i=0}^{C-1} z^i \pi_i + z^C g \left[I - Fz\right]^{-1} H. \tag{7.68}$$

The factorial moments of the queue length distribution can then be determined by applying the standard rules to $\mathcal{F}_{\tilde{q}}(z)\mathbf{e}$; that is, we differentiate $\mathcal{F}_{\tilde{q}}(z)\mathbf{e}$ with respect to z and evaluate the result in the limit as $z \to 1$.

EXERCISE 7.14 Beginning with (7.68), develop an expression for the first and second moments of the queue length distribution.

We turn now to a brief discussion of stability; when does the system $\pi = \pi\mathcal{P}_{MG}$ have an equilibrium distribution? We note that $\mathcal{P}_{MG}$ is the one-step transition probability matrix for a particular type of discrete parameter Markov chain, namely one where a concept of *levels* makes sense. The matrix A_i, $i = 0, 1, \ldots$, represents the probability that the level will increase by i between two consecutive epochs whenever the level at the first epoch is at least C. If $A_0 \neq 0$, then it is clear that the system can drop by up to C levels between two successive epochs. For example, if $C = 1$, then we would find that $\pi_0 = \pi_0 B_0 + \pi_1 A_0$, so that A_0 represents the probability of dropping from level 1 to level 0 between two successive epochs. In a queueing system this probability would be then interpreted as the probability of having zero arrivals during a service interval. Similarly, A_i would represent the probability of having i arrivals during a service interval. Intuitively, we would conclude that if the average number of arrivals during a service interval is less than 1, then the system would be stable.

For general values of C, the interpretation of the A_i would be the same; A_i would represent the probability of having i arrivals during a service interval. Since it is possible for the system to drop by up to C levels during a service interval, we would conclude that if the average number of arrivals during a service interval were less than C, then the system would be stable.

In general, we would expect the matrix

$$A(1) = \sum_{i=0}^{\infty} A_i$$

to be stochastic because some number of arrivals must occur and the matrix $A(1)$ reflects all possible numbers of arrivals. In fact, the matrix $A(1)$ is the one-step probability transition matrix for the phase of the arrival process; that is, its i, j element is the probability that the phase of the arrival process is j at the $(k+1)$-st epoch given that the phase at the k-th epoch was i.

Similarly,

$$\frac{d}{dz}A(z)\bigg|_{z=1} = \sum_{i=0}^{\infty} iA_i$$

would represent, in some sense, the average number of arrivals during a service interval. Again, intuitively, we would expect that the expected number of arrivals that occur during an interval would also depend upon the phase of the process during the interval. In fact, the average number of arrivals that occur during an interval is given by

$$\rho C = \phi \left[\sum_{i=0}^{\infty} i A_i \right] \mathbf{e},$$

where ϕ is the stationary vector of $\mathcal{A}(1)$ and ρ is the traffic intensity.

EXERCISE 7.15 Suppose $C = 2$. Define

$$\hat{A}_i = \begin{bmatrix} A_{2i} & A_{2i+1} \\ 0 & A_{2i} \end{bmatrix}, i = 1, 2, \ldots,$$

$$\hat{\mathcal{A}}(z) = \sum_{i=0}^{\infty} \hat{A}_i z^i.$$

and ϕ_2 to be the stationary vector of $\hat{\mathcal{A}}(1)$. Suppose

$$\rho = \phi_2 \left[\sum_{i=0}^{\infty} i \hat{A}_i \right] \mathbf{e}.$$

Show that

(a) $\phi_2 = [\phi \quad \phi]$, where ϕ is the stationary vector of $\mathcal{A}(1)$.

(b)

$$\phi \left[\sum_{i=0}^{\infty} i A_i \right] \mathbf{e} = 2\rho.$$

EXAMPLE 7.4 Consider the queueing behavior of a frame-oriented cellular transmission system. Suppose that files are to be transferred to mobile units from the network and that the file lengths are drawn from a geometric distribution with a mean of 150 packets.

Assume that the packet arrival process to the cell site is governed by a two-state Markov chain. During the *on* phase, the distribution of the number of packets that arrive is binomial with parameters N_t and p_t, where N_t and p_t are a function of the length of the on period and the prescribed traffic intensity. Specifically, we first choose the desired traffic intensity. Then we choose the average length of the on period in frames. We then choose N_t and p_t such that the average packet arrival rate over all time divided by $N_t p_t$ is equal to 150 packets. We arbitrarily then choose N_t to be the smallest integer such that $p \leq 0.8$.

For the wireless link, suppose the number of packets transmitted during a frame depends upon the state of a Markov chain, $\tilde{\sigma}_k, k = 0, 1, \ldots,$ at the

Table 7.5. Transition probabilities for the system of Example 7.4.

i	c_i	$P_{i,i-1}$	$P_{i,i}$	$P_{i,i+1}$	ϕ_i
0	0	-	0.972689	0.027311	0.037460
1	0	0.028507	0.932727	0.038766	0.035889
2	0	0.076871	0.838562	0.084566	0.018099
3	1	0.030882	0.932360	0.036758	0.049560
4	1	0.070049	0.855469	0.074482	0.026006
5	1	0.027740	0.941352	0.030909	0.069828
6	2	0.011990	0.975068	0.012942	0.180006
7	4	0.012731	0.975884	0.011385	0.182985
8	6	0.014657	0.973850	0.011494	0.142136
9	8	0.007113	0.991620	0.001267	0.229685
10	12	0.011199	0.987589	0.001212	0.025995
11	16	0.013381	0.986619	-	0.002355

beginning of the frame. Denote the state of $\tilde{\sigma}_k, k = 0, 1, \ldots$ at the beginning of an arbitrary frame by i, the transmission capability by c_i , and the transition probability from state i to state j by $P_{i,j}$. Assume further that transitions are possible only to adjacent states and the transmission capabilities and the state transition probabilities are as given in Table 7.5:

We wish to examine the impact of the length of the on period, which is a proxy for the speed of the network, on the queue length at the boundary between the cell site and the wireless link at a fixed overall packet arrival rate. In the process, we wish to observe the degrees of polynomials and the dimensions of various matrices involved in the queue length calculations.

Solution: First, we determine the average service capability by taking the sum of the service rate capacity weighted by the stationary vector of the service process; the resulting service rate is approximately 4.277 packets per frame. We then arbitrarily choose an overall arrival rate of 60% of 4.277.

Table 7.6 shows some of the major characteristics of the solution process as well as the mean queue length obtained from the analysis.

From the table, we see that the number of unstable eigenvalues remains constant as the length of the on time is decreased. On the other hand, the fact that the file is delivered to the cell site from the network over a shorter period of time means that the average number of packet arrivals per frame must be increased. Thus, the potential for larger increases in level must be taken into

Table 7.6. Major characteristics of the solution process for the system of Example 7.4.

on time	N_t	p	$K+1$	ν	n_u	n_s	$E[\tilde{q}]$	CPU time (sec)
0.975	6	0.68376	24	22	384	144	804.58	45
0.650	8	0.76923	24	24	384	192	867.99	66
0.325	16	0.76923	24	32	384	384	961.04	188

account. This potential for larger increases in level is reflected in the increase in N_t from 6 to 8 to 16 as the length of the on period is decreased. In addition, this potential is reflected in the increase in n_s from 144 to 192 to 384 as the length of the on period is decreased.

The above table also lists computation times required for our rough program executing on a 600 MHz Power PC platform. These are not based on any scientific study; they are presented solely for the purpose of giving the reader a feel for the execution times required for a fairly large problem.

The mean queue length, which was computed as

$$E[\tilde{q}] = \frac{d}{dz}\mathcal{F}_{\tilde{q}}(z)\mathbf{e}\Big|_{z=1}$$

does not increase dramatically as the length of the on time is decreased.

Figure 7.4 shows graphs of the survivor function for the three choices of on time. It is seen from the graphs that the queue length distributions do not vary dramatically with on time. On the other hand buffer overflows are substantially greater than would be expected for a system having constant service rate and Poisson arrivals.

7.6 Summary

In this chapter, we have presented a brief description of M/G/1 and G/M/1 paradigms and the solution techniques of Lucantoni,Neuts, and Ramaswami. Again, the interested reader is referred to the two excellent books by Neuts the papers by Ramaswami. For a treatment of the M/G/1 model extended to the M/G/1 with vacations and a broad class of interarrival time distributions, the reader is referred to Lucantoni, Meier-Hellstern, and Neuts [1990]. Latouche and Ramaswami [1999] provide a thorough treatment of the types of systems that can be modeled using the G/M/1 paradigm as well as computational techniques.

Additionally, in this chapter we have presented the use of generalized state space techniques of Akar, Oğuz, and Sohraby [1998] in solving the complex

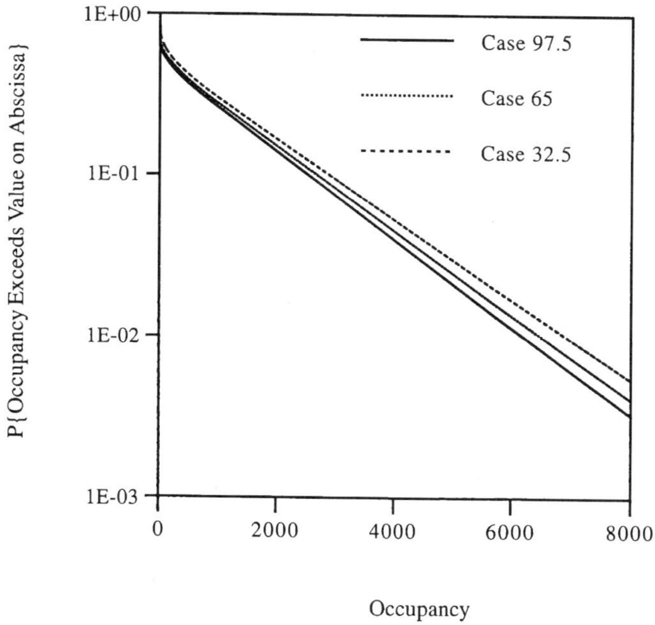

Figure 7.4. Survivor functions for occupancy distributions for wireless communication link with on time as a parameter.

boundary version of queueing problems in the M/G/1 class. We have seen that solving for the stationary probabilities of Markov chains of the M/G/1 type is substantially more complicated than solving QBD models when matrix-analytic techniques are applied. However, when generalized state space techniques are applied, we require only the use of generalized Schur decomposition, and the solution is delivered in matrix-geometric form. We have presented numerical examples to illustrate applications of these techniques. Our own experience is that computational time required to solve complex boundary problems of the M/G/1 type using the generalized state space approach is faster than using matrix-analytic techniques. However, we have not done a thorough investigation into this issue and recommend that readers experiment with a variety of techniques to find the best one for any given situation. The TELPACK queueing analysis package (see Akar, Oğuz, and Sohraby [1998b]), the development of which is an ongoing project, implements numerous alternative computational approaches. This package is client-server based, can be accessed over the internet, and provides a graphical interface.

The phase-dependent arrival and service model of Chapter 3, which is a QBD model, is a special case in which the model is of both the M/G1 and G/M/1 type. As we have seen earlier, when models have a special structure, we can develop special techniques to solve the model efficiently.

It is worth mentioning that sometimes one may obtain a model of the QBD type that approximates the behavior of a system that is naturally modeled by a Markov chain of the M/G/1 type. An example of an analysis based on such an approximation is given in Chipalkatti, Kurose, and Towsley [1989] and the references therein.

We have shown that the methodology described in this chapter is useful in analyzing many types of queueing problems related to computer communications. There are many other applications in addition to the ones described in Sections 7.4 and 7.5. We now discuss some of those briefly.

One rather unusual application that uses M/G/1 matrix-analytic techniques is given in Chandramouli, Neuts, and Ramaswami [1989]. In that paper, the authors analyze a meteor-burst packet-communication system. In meteor-burst communications, successful message transmission is dependent on the level of ionization in the upper atmosphere. The quality of the channel during a given slot is modeled as a discrete-parameter, discrete-valued Markov chain, and the probability of successful message transmission depends on the state of the Markov chain, thus giving rise to a Markov chain of the M/G/1 type.

Li [1990] presents an analysis of a special case in which the arrival process is as described in Daigle, Lee, and Magalhães [1990], but there are multiple servers and there is a one-to-one correspondence between the phase of the arrival process and the number of packets that arrive during a slot; that is, each active source generates one packet during a slot. Li presents mean occupancies for a variety of parameter sets. Although Li's analysis could have been carried out using the methods described here, Li chose to approach the problem by spectral decomposition instead. His methods are quite interesting, and the reader is encouraged to investigate his approach.

Other work of interest along the same lines as presented here are Stavrakakis [1990] and van Arem [1990]. Both consider discrete time queues. Stavrakakis considers a single-server system having an arrival process similar to that of Li's, and van Arem's arrival process has the same form for $\mathcal{F}_{\tilde{a}}(z)$ but an arbitrary form for $\mathcal{P}$. Both present mean values only.

Daigle and Tang [1991] consider the c-server queueing system in which $\mathcal{P}$ has a general form and $\mathcal{F}_{\tilde{a}}$ has the same form as Li [1990]; that is, $\mathcal{F}_{\tilde{a}}(z) = \text{diag}(1, z, z^2, \ldots, z^N)$. Using eigenanalysis along the same lines as presented in Chapter 3, they develop the queue length distribution in terms of partial fraction expansions. Extensions of that work to the case of more general $\mathcal{F}_{\tilde{a}}(z)$ are published in Tang [1995].

Daigle and Moose [1993] and Daigle and Moose [1996] obtain a Markov chain of the M/G/1 type for an ATM multiplexing system where the number of units served during a frame is a random process. The model is similar to that of the example given in Section 7.4, but their solution approach was virtually identical to that of Neuts [1981a].

Since the early 1990s, Li et. al. (see Lau and Li [1993], Sheng and Li [1993], and Li and Hwang [1993]) have developed a spectral analysis approach to understanding queueing systems. The objectives are to develop insights into the behavior of queueing systems by studying their spectral properties, that is the characteristics of the behavior of the power spectral densities. They have developed sets of tools to facilitate queueing analysis in general settings and in more recent years in a network environment (Li and Hwang [1997] and Li, Park, and Arifler [1998]). In Kim and Li [1999], the authors report measurement-based techniques for modeling characteristics of wireless channels as a Markov modulated process, where the service rate of the channel is dependent upon the state of a continuous-time Markov chain. They find that their Markov-chain based modeling techniques yield excellent results when compared to a trace-based analysis. They further show that the dynamics, not just the average statistics, of the propagation environment have a serious effect on the queueing behavior of the system.

More recently, Turin and Zorzi [2002] have analyzed the queue length for a wireless communications system wherein the state of the channel–either good or bad–is modeled as a hidden Markov chain, which turns out to be the special case of either 0 or 1 servers, reminiscent of Towsley's work (Towsley [1980]). Packet arrivals are modeled as on-off processes and queue length distributions are derived. They offer alternative solution techniques for Markov chains of this type.

We point out that some researchers have suggested that the stationary probabilities for the embedded Markov chains could be obtained more efficiently by solving for these probabilities directly; that is, one could simply solve the matrix equations $\pi = \pi\mathcal{P}, \pi\mathbf{e} = 1$ numerically without regard for the structure of $\mathcal{P}$. Related to this topic, a state-of-the-art workshop on the numerical solution of Markov chains was held in January of 1990. The papers presented during the workshop contain many interesting ideas and form the chapters of Stewart [1991]. Additional workshops along the same lines have taken place in 1995 and 1999 (see Stewart [1995] and Stewart [1999]). The papers presented in these publications appear to provide a balanced coverage of the methods available for solving Markov chains of all types.

7.7 Supplementary Problems

7-1 Consider a slotted time, single server queueing system having unit service rate. Suppose the arrival process to the system is a Poisson process modulated by a discrete time Markov chain on $\{0,1\}$ that has the following one-step transition probability matrix:

$$\mathcal{P} = \begin{bmatrix} \beta & 1-\beta \\ 1-\alpha & \alpha \end{bmatrix}.$$

While the arrival process is in phase i, arrivals to the system occur according to a Poisson process with rate λ_i, $i = 0, 1$.

1. Argue that $\mathcal{A}(z) = \mathcal{B}(z)$ for this system.
2. Show that

$$\mathcal{A}(z) = \begin{bmatrix} \beta & 1-\beta \\ 1-\alpha & \alpha \end{bmatrix} \begin{bmatrix} e^{-\lambda_0(1-z)} & 0 \\ 0 & e^{-\lambda_1(1-z)} \end{bmatrix}.$$

3. Determine $\mathcal{A}_i$ for all $i \geq 0$.
4. Suppose that at the end of a time slot, the system occupancy level is zero and the phase of the arrival process is i, $i = 0, 1$. Determine the probability that the system occupancy level at the end of the following time slot will be k, for $k \in \{0, 1, \ldots\}$, and the phase of the arrival process will be j, $j = 0, 1$.
5. Suppose that at the end of a time slot, the system occupancy level is $k > 0$ and the phase of the arrival process is i, $i = 0, 1$. Determine the probability that the system occupancy level at the end of the following time slot will be l, $l = \{k-1, k, k+1\}$ and the phase of the arrival process will be j, $j = 0, 1$.
6. Let $\alpha = 0.5$, $\beta = 0.75$, $\lambda_0 = 1.2$, and $\lambda_1 = 0.3$. Determine the equilibrium probability vector for the phase process and ρ for the system.
7. Write a computer program to determine $\mathcal{K}(1)$ for the parameters given in part (f), and then determine the equilibrium probability vector for the Markov chain for which $\mathcal{K}(1)$ is the one-step transition probability matrix.
8. Compute $E[\tilde{q}]$, the expected number of packets in the system at the end of an arbitrary time slot. Compare the result to the equivalent mean value $E[\tilde{q}]$ for the system in which $\lambda_0 = \lambda_1 = \rho$ as computed in part (f). What can be said about the effect of burstiness on the average system occupancy?

7-2 The objective of this problem is to develop a closed-form expression for the mean queue length, $E[\tilde{q}]$, for the slotted M/D/1 system having phase-dependent arrivals and unit service times. Our point of departure is the expression for the generating function of the occupancy distribution as given in (7.37), which is now repeated for continuity:

$$\mathcal{F}_{\tilde{q}}(z)\left[Iz - \mathcal{P}\mathcal{F}_{\tilde{a}}(z)\right] = \pi_0\left[z-1\right]\mathcal{P}\mathcal{F}_{\tilde{a}}(z). \tag{7.69}$$

1. Differentiate both sides of (7.69) to obtain

$$\begin{aligned}\mathcal{F}_{\tilde{q}}(z)\left[Iz - \mathcal{P}\mathcal{F}_{\tilde{a}}^{(1)}(z)\right] &+ \mathcal{F}_{\tilde{q}}^{(1)}(z)\left[Iz - \mathcal{P}\mathcal{F}_{\tilde{a}}(z)\right] \\ &= \pi_0[z-1]\mathcal{P}\mathcal{F}_{\tilde{a}}^{(1)}(z) + \pi_0\mathcal{P}\mathcal{F}_{\tilde{a}}(z)\end{aligned} \tag{7.70}$$

2. Take limits of both sides of (7.70) as $z \to 1$ to obtain

$$\mathcal{F}_{\tilde{q}}^{(1)}(1)\left[I - \mathcal{P}\right] = \pi_0\mathcal{P} - \mathcal{F}_{\tilde{q}}(1)\left[I - \mathcal{P}\mathcal{F}_{\tilde{a}}^{(1)}(1)\right]. \tag{7.71}$$

3. Define ρ to be the marginal probability that the system is not empty at the end of a slot. Postmultiply both sides of (7.71) by $\mathbf{e}$, and show that $\rho = \mathcal{F}_{\tilde{q}}(1)\mathcal{F}_{\tilde{a}}^{(1)}(1)\mathbf{e}$.

4. Add $\mathcal{F}_{\tilde{q}}^{(1)}(1)\mathbf{e}\mathcal{F}_{\tilde{q}}(1)$ to both sides of (7.71), solve for $\mathcal{F}_{\tilde{q}}^{(1)}(1)$, and then postmultiply by $\mathcal{P}\mathcal{F}_{\tilde{a}}^{(1)}(1)\mathbf{e}$ to obtain

$$\begin{aligned}\mathcal{F}_{\tilde{q}}^{(1)}(1)\mathcal{P}\mathcal{F}_{\tilde{q}}(1) &= \mathcal{F}_{\tilde{q}}^{(1)}(1)\mathbf{e}\rho \\ &\quad+\{\pi_0\mathcal{P} - \mathcal{F}_{\tilde{q}}(1)[I - \mathcal{F}_{\tilde{a}}^{(1)}(1)]\} \\ &\quad[I - \mathcal{P} + \mathbf{e}\mathcal{F}_{\tilde{q}}(1)]^{-1}.\end{aligned} \tag{7.72}$$

Use the fact that $\mathbf{e}\mathcal{F}_{\tilde{q}}(1)\left[I - \mathcal{P} + \mathbf{e}\mathcal{F}_{\tilde{q}}(1)\right] = \mathbf{e}\mathcal{F}_{\tilde{q}}(1)$, as shown in Exercise 7.12 in Section7.3.

5. Differentiate both sides of (7.70) with respect to z, postmultiply both sides by $\mathbf{e}$, take limits on both sides as $z \to 1$, and then rearrange terms to find

$$\begin{aligned}\mathcal{F}_{\tilde{q}}^{(1)}(1)\mathcal{P}\mathcal{F}_{\tilde{a}}^{(1)}(1)\mathbf{e} = \mathcal{F}_{\tilde{q}}^{(1)}(1)\mathbf{e} - \frac{1}{2}\mathcal{F}_{\tilde{q}}^{(1)}(1)\mathcal{F}_{\tilde{a}}^{(1)}(1)\mathbf{e} \\ -\pi_0\mathcal{P}\mathcal{F}_{\tilde{a}}^{(1)}(1)\mathbf{e}.\end{aligned} \tag{7.73}$$

6. Equate right-hand sides of (7.72) and (7.73), and then solve for $\mathcal{F}_{\tilde{q}}^{(1)}\mathbf{e}$ to obtain

$$
\begin{aligned}
E[\tilde{q}] &= \mathcal{F}_{\tilde{q}}(1)\mathbf{e} \\
&= \frac{1}{1-\rho}\Big\{\frac{1}{2}\mathcal{F}_{\tilde{q}}(1)\mathcal{F}_{\tilde{a}}^{(2)}(1)\mathbf{e} + \pi_0\mathcal{F}_{\tilde{a}}^{(1)}(1)\mathbf{e} \\
&+ \left(\pi_0\mathcal{P} - \mathcal{F}_{\tilde{q}}(1)\left[I - \mathcal{F}_{\tilde{a}}^{(1)}\right]\right) \\
&\times \left[I - \mathcal{P} + \mathbf{e}\mathcal{F}_{\tilde{q}}(1)\right]^{-1}\mathcal{P}\mathcal{F}_{\tilde{a}}^{(1)}\mathbf{e}\Big\}.
\end{aligned}
$$

Chapter 8

CLOSING REMARKS

Queueing theory is a vast topic and one can hope to accomplish only a brief introduction in a graduate-level course. As the need arose during the previous chapters, we have suggested reference material to help the reader develop a better understanding of the concepts presented and to obtain a better feel for how the concepts could be used in solving practical problems.

Throughout, our approach has been directed towards the development of an intuitive understanding of how queueing systems work. In most cases, we have carried our discussion far enough so that the reader can obtain numerical results. Our thrust has been to provide the reader with sufficient background to be able to appreciate the major papers currently appearing in the applications literature. In this chapter, we discuss a few additional topics of immediate interest.

An important aspect of defining any queueing problem is the description of its arrival process. A discussion of a very general arrival process, the batch Markovian arrival process (BMAP), is described in detail in the excellent tutorial by Lucantoni (see Lucantoni [1993]). In that tutorial, Lucantoni discusses both the formulation and the solution of the BMAP/G/1 queueing system, addressing lengths of busy periods, virtual waiting times, and both stationary and transient queue length distributions. In Lucantoni [1998] additional transient results concerning the BMAP/G/1 queue are provided. Detailed procedures for obtaining numerical results are not given in those references, but pointers to appropriate references are given.

In recent years, there has been much discussion on whether or not Poisson modeling, or for that matter, any queueing modeling based on Markov chain modeling is useful. Having solved numerous significant problems in the course of my own work, I find the discussion somewhat rhetorical, but useful. Beginning in Chapter 1, we have seen examples of practical systems where very elementary arrival or service processes are simply not sufficient to provide the insights needed for system design. Simply put, use of the wrong models will not yield useful insights into system behavior.

In short, failure to put forth sufficient effort to understand the system under study will almost surely result in modeling efforts that do not yield useful insights. However, when significant thought is put into the understanding of the problem it is sometimes possible to obtain useful results with simple models.

Along these lines, in recent years there have been a number of studies that address formulation of models that are appropriate for analyzing behavior of queueing points within IP-based communication networks. One such study is Cao and Ramanan [2002]. In that reference, the authors examine a queueing system where the input consists of the superposition of a large number, n, of long-range dependent sources and the speed of the server is proportional to n. They demonstrate that $P\{\text{Occupancy} > b\}$ for $b > 0$ is well approximated by a system having Poisson arrivals whenever the level of multiplexing is high. We saw in an example in Chapter 1 that burst length has a significant effect upon queue length distribution. But, in our example, we increased the number of users by spreading their burst out. In other words, our line speed was not increased in proportion to the number of users.

Another example of an analysis of a system having an high level of multiplexing is given in Eun and Shroff [2003]. The authors consider a case where large numbers of flow-controlled sources are multiplexed onto communication lines and outputs from those lines are fed into other communications lines and multiplexed. They show that, as the number of multiplexed sources is increased, the queue length distribution at downstream links behaves more and more like a queueing system in which the upstream queues are transparent. That is, the queueing behavior at the downstream nodes is very much like a queue that is fed directly by the regulated sources, without the interference from the upstream queues. Their analysis is supported by numerous simulations and suggests that it may be possible to gain significant insights from simplified bottleneck analysis in networks.

Significant research has been in progress concerning the behavior of queues whose input is self-similar since the work of Leland, Taqqu, Willinger and Wilson first began to appear in the IEEE INFOCOM conference series and other conferences in the early 1990s (see Leland et. al. [1994]). Much research has been directed towards understanding the queueing behavior of systems that multiplex large numbers of streams where each stream has self-similar traffic. In his book published in 2002, Whitt develops a large number of results that describe the queueing behavior of systems under heavy traffic loads, and consequently, offers insights into the behavior at lower loads. Whitt's book offers materials for beginners as well as experts.

Beginning in the early 1990s, researchers began to use a network calculus (see Cruz [1995] and LeBoudec and Thiran [2001]) to explore the behavior of queueing systems that allocate resources in order to serve regulated flows. The main ideas are that each flow is described by a deterministic arrival curve, which specifies the maximum amount of traffic that may arrive over any period of time. In addition, a service curve that guarantees a minimum amount of service to each flow is specified. Under a broad set of conditions, each flow is guaranteed that its delay will not exceed a given level. A probabilistic version of this is given in Boorstyn et. al. [2000]. The net result of the stochastic version is that the probability that a given flow's delay exceeds a certain level is guaranteed to be below some threshold. These techniques have been used in Duan and Daigle [2004a] and Duan and Daigle [2004b] to address resource allocation in a high speed IP switch having credit-based flow controls.

What should be abundantly clear from the contents of this book is that the tools needed to solve a particular queueing problem will be heavily dependent upon the particular problem under consideration. It is hoped that the limited material provided in this book will provide the reader with sufficient background to be able to solve many interesting problems. In case the coverage is inadequate to solve a particular problem, it is hoped that the reader will be able to select and understand books and articles that will contribute to its solution.

References

Abate, J., and W. Whitt. (1988) Transient Behavior of the M/M/1 Queue via Laplace Transforms. *Adv. Appl. Prob.* **20:1**, 145-178.

Abate, J., and W. Whitt. (1989) Calculating Time-Dependent Performance Measures for the M/M/1 Queue. *IEEE Trans. on Commun.* **37:10**, 1102-1104.

Abboud, N. E., and J. N. Daigle. (1997) A Little's Result Approach to the Service Constrained Spares Provisioning Problem for Repairable Items. *Ops. Res.* **45:4**, 577-583.

Ackroyd, M. H. (1980) Computing the Waiting Time Distribution for the $G/G/1$ Queue by Signal Processing Methods. *IEEE Trans. on Commun.* **28:1**, 52-58.

Akar, N. (2004) Solving the Single Server Queue with State Space Methods and the Matrix Sign Function. Submitted.

Akar, N., and E. Arikan. (1996) A Numerically Efficient Method for the MAP/D/1/K Queue via Rational Approximations. *Queueing Systems: Theory and Applications (QUESTA)* **22**, 97-120.

Akar, N., N. C. Oğuz, and K. Sohraby. (1998a) Matrix-Geometric Solutions of M/G/1-Type Markov Chains: A Unifying Generalized State-Space Approach. *IEEE J. Select. Areas in Commun.* **16:5**, 626-639.

Akar, N., N. C. Oğuz, and K. Sohraby. (1998b) An Overview of TELPACK. *IEEE Communications* **36:8**, 84-87.

Asmussen, S. (2003) *Applied Probability and Queues*. Springer-Verlag, New York.

Anderson, E., Z. Bai, C. Bischof, S. Blackford, J. Demmel, J. Dongarra, J. Du Croz, A. Greenbaum, S. Hammarling, A. McKenney, and D. Sorensen. (1999) *LAPACK Users' Guide,* 3rd. Ed. Society for Industrial and Applied Mathematics (SIAM), Philidelphia.

Bender, P., P. Black, M. Grob, R. Padovani, N. Sindhushyana, and A. Viterbi. (2000) CDMA/HDR: A Bandwidth-Efficient High-Speed Wireless Data Service for Nomadic Users. *IEEE Communications* **38:7**, 70-77.

Bertsekas, D., and R. Gallager. (1987) *Data Networks*. Prentice-Hall, Englewood Cliffs.

Beuerman, S. L., and E. J. Coyle. (1987) The Delay Characteristics of CSMA/CD Networks. *IEEE Trans. on Commun.* **36:5**, 553-563.

Beuerman, S. L., and E. J. Coyle. (1989) State Space Expansions and the Limiting Behavior of Quasi-Birth and Death Processes. *Adv. Appl. Prob.* **21:2**, 284-314.

Bhargava, A., and M. G. Hluchyj. (1990) Frame Losses Due to Buffer Overflows in Fast Packet Networks. *Proc. of IEEE INFOCOM'90*, 132-139, San Francisco.

Blefari-Melazzi, N., J. N. Daigle, and M. Femminella. (2003) Stateless Admission Control for QoS Provisioning to VoIP Traffic in a DiffServ Domain. *Proceedings of International Telecommunications Conference: ITC 18*, Berlin.

Boorstyn, R., A. Burchard, J. Liebeherr, and C. Oottamakorn. (2000) Statistical Service Assurance for Traffic Scheduling Algorithms. *IEEE J. Select. Areas Commun.* **18:12**, 2651-2664.

Bruneel, H. (1988) Queueing Behavior of Statistical Multiplexers with Correlated Inputs. *IEEE Trans. on Commun.* **36:12**, 1339-1341.

Burke, P. J. (1956) The Output of a Queueing System. *Ops. Res.* **4**, 699-704.

Cao, J., and K. Ramanan. (2002) A Poisson Limit for Buffer Overflow Probabilities. *Proceedings of IEEE INFOCOM 2002*, **2**, 994-1003, San Fransico.

Chandramouli, Y., M. F. Neuts, and V. Ramaswami. (1989) A Queueing Model for Meteor-Burst Communication Systems. *IEEE Trans. on Commun.* **37:10**, 1024-1030.

Chandy, K. M., and C. H. Sauer. (1981) *Computer Systems Performance Modelling*. Prentice-Hall, Englewood Cliffs.

Chen, C. -T. (1999) *Linear System Theory and Design,* 3rd. Ed. Oxford University Press, New York.

Chipalkatti, R., J. F. Kurose, and D. Towsley. (1989) Scheduling Policies for Real-Time and Non-Real-Time Traffic in a Statistical Multiplexer. *Proc. IEEE INFOCOM'89*, 774-783, Ottawa.

Churchill, R. V., and J. W. Brown. (1987) *Fourier Series and Boundary Value Problems,* 4th. Ed. McGraw-Hill, New York.

Çinlar, E. (1975) *Introduction to Stochastic Processes*. Prentice-Hall, Englewood Cliffs.

Cohen, J. W. (1969) *The Single Server Queue*. Wiley Interscience, New York.

Cooper, R. B. (1972) *Introduction to Queueing Theory*. Macmillan, New York.

Cooper, R. B. (1981) *Introduction to Queueing Theory,* 2nd. Ed. North-Holland, New York. Republished 1990 by CEEPress, The George Washington University, Washington.

Cooper, R. B. (1990) Queueing Theory. *Stochastic Models: Handbooks of Operations Research and Management Science, Vol. 2.* (D. P. Heyman, and M. J. Sobel, eds.) North-Holland, New York, Chapter 10.

Cooper, R. B., and S. C. Niu. (1986) Beneš's Formula for M/G/1-FIFO Explained by Preemptive-Resume LIFO. *J. Appl. Prob.* **23:2**, 350-54.

Coyle, E., and B. Liu. (1985) A Matrix Representation of CSMA/CD Networks. *IEEE Trans. on Commun.* **33:1**, 53-64.

Cruz, R. L. (1995) Quality of Service Guarantees in Virtual Circuit Switched Networks. *IEEE J. Select. Areas Commun.* **13:6**, 1048-1056.

Daigle, J. N. (1977a) Queueing Analysis of a Packet Switching Node with Markov-Renewal Arrival Process. *Proc. IEEE Int. Commun. Conf.*, **1**, 279-283, Chicago.

Daigle, J. N. (1977b) "Queueing Analysis of a Packet Switching Node in a Data Communication System." Doctoral Dissertation, Columbia University.

Daigle, J. N. (1986) Task Oriented Queueing: A Design Tool for Communications Software Design. *IEEE Trans. on Commun.* **34:12**, 250-256.

Daigle, J. N. (1989) Queue Length Distributions from Probability Generating Functions via Discrete Fourier Transforms. *Ops. Res. Let.* **8**, 229-236.

Daigle, J. N., and J. D. Langford. (1985) Queueing Analysis of a Packet Voice Communication System. *Proc. IEEE INFOCOM'85*, 18-26. Washington.

Daigle, J. N., and J. D. Langford. (1986) Models for Analysis of Packet Voice Communication Systems. *IEEE J. of Select. Areas in Commun.* **4:6**, 847-855.

Daigle, J. N., Y. Lee, and M. N. Magalhães. (1990) Discrete Time Queues with Phase Dependent Arrivals. *Proc. of IEEE INFOCOM'90*, 728-732, San Fransisco.

Daigle, J. N., and D. M. Lucantoni. (1990) Queueing Systems Having Phase-Dependent Arrival and Service Rates. *First International Workshop on the Numerical Solutions of Markov Chains*, 375-395. Raleigh.

Daigle, J. N., and M. N. Magalhães. (1991) Transient Behavior of $M/M^{ij}/1$ Queues. *Queueing Systems and Their Applications (QUESTA)* **8**, 357-378.

Daigle, J. N., and M. N. Magalhães. (2003) Analysis of Packet Networks Having Contention-Based Reservation With Application to GPR. *IEEE/ACM Trans. Networking* **11:4**, 602-615.

Daigle, J. N., and R. L. Moose. (1993) Queue Lengths When Service and Arrival Rates are Markov-Modulated with Application to ATM. MITRE Technical Report 9412.

Daigle, J. N., and R. L. Moose. (1996) Queue Lengths when Service and Arrival Rates are Markov-Modulated with Application to ATM. *IEEE COMSOC Ninth Annual Computer Communications Workshop*, Reston.

Daigle, J. N., and M. Roughan. (1999) Queue-Length Distributions for Multi-Priority Queueing Systems. *Proceedings of IEEE INFOCOM'99*, **1**, 641-648, New York.

Daigle, J. N., and S. Tang. (1991) Numerical Methods for Multiserver Discrete Time Queues with Batch Markovian Arrivals. *Communications in Statistics - Stochastic Models* **8:4**, 665-683.

Daigle, J. N., and S. D. Whitehead. (1985) A Balance Equation Approach to Non-Markovian Queueing Systems. University of Rochester W. E. Simon Graduate School of Business Administration Working Paper Series QM-8629.

Demmel, J. W. (1997) *Applied Numerical Linear Algebra*. Society for Industrial and Applied Mathematics (SIAM), Philadelphia.

Disney, R. L., and P. C. Kiessler. (1987) *Traffic Processes in Queueing Networks: A Markov Renewal Approach*. The Johns Hopkins University Press, Baltimore.

Disney, R. L., D. C. McNickle, and B. Simon. (1980) The M/G/1 Queue with Instantaneous Feedback. *Naval Res. Log. Qrt.* **27**, 635-644.

Doshi, B. T. (1986) Queueing Systems with Vacations - A Survey. *Queueing Systems* **1:1**, 29-66.

Doshi, B. T. (1990) Single Server Queues with Vacations. *Stochastic Analysis of Computer and Communication Systems*. (H. Takagi, ed.) North-Holland, New York, 217-265.

Duan, Q., and J. N. Daigle. (2004) Resource Allocation for Statistical QoS Provision in Buffered Crossbar Switch. Accepted for presentation at IEEE International Conference on Communications (ICC-04), Paris.

Duan, Q., and J. N. Daigle. (2004) Resource Allocation for Quality of Service Provision in Multistage Buffered Crossbar Switches. *Computer Networks Journal* in press.

Evans, R. V. (1967) Geometric Distributions in Some Two Dimensional Queueing Systems. *Ops. Res.* **15:5**, 830-846.

Eun, D. Y., and N. B. Shroff. (2003) Simplification of Network Analysis in Large-Bandwidth Systems. *Proceedings of IEEE INFOCOM 2003*, **1**, 597-607, San Francisco.

Feller, W. (1968) *An Introduction to Probability Theory and Its Applications* **I**, John Wiley, New York.

Feller, W. (1971) *An Introduction to Probability Theory and Its Applications* **II**, John Wiley, New York.

Fuhrmann, S. W., and R. B. Cooper. (1985) Stochastic Decomposition in the M/G/1 Queue with Generalized Vacations. *Ops. Res.* **33:5**, 1091-1099.

Gallager, R. G. (1995) *Discrete Stochastic Processes*. Kluwer Academic Press, Boston.

Gavish, B., and K. Altinkemer. (1990) Backbone Design Tools with Economic Tradeoffs. *ORSA Journal on Computing* **2:3**, 236-251.

Gelenbe, E., and G. Pujolle. (1987) *Introduction to Queueing Networks*. John Wiley, New York.

Giffin, W. C. (1978) *Queueing: Basic Theory and Applications*. Grid, Inc., (Reprinted by Books on Demand, UMI, AnnArbor).

Golub, G. H., and C. H. Van Loan. (1996) *Matrix Computations*, 3rd. Ed. Johns Hopkins, Baltimore.

Gordon, J. J. (1990) The Evaluation of Normalizing Constants in Closed Queueing Networks. *Ops. Res.* **38:5**, 863-869.

Gordon, W. L., and G. F. Newell. (1967) Closed Queuing Systems with Exponential Servers. *Ops. Res.* **15:2**, 254-265.

Grassman, W. F. (1990) Finding Transient Solutions in Markovian Event Systems Through Randomization. *First International Workshop on the Numerical Solutions of Markov Chains*, 179-211, Raleigh.

Green, L., and B. Melamed. (1990) An Anti-PASTA Result for Markovian Systems. *Ops. Res.* **38:1**, 173-175.

Hahne, E. L., A. K. Choudhury, and N. F. Maxemchuk. (1990) Improving the Fairness of Distributed Queue-Dual-Bus Networks. *Proceedings of IEEE INFOCOM'90*, 175-184, San Francisco.

Hammond, J. L., and P. J. P. O'Reilly. (1986) *Performance Analysis of Local Computer Networks*. Addison-Wesley, Reading.

Harrison, M. (1985) On Normalizing Constants in Queueing Networks. *Ops. Res.* **33:2**, 464-468.

Hewitt, E., and K. Stromberg. (1969) *Real and Abstract Analysis*. Springer-Verlag, New York.

Hogg, R. V., and A. T. Craig. (1965) *Introduction to Mathematical Statistics*. Macmillan, New York.

Hunter, J.J. (1983) *Mathematical Techniques of Applied Probability Volume 1, Discrete Time Models: Basic Theory*. Academic Press, New York.

Iliadis, I., and W. E. Denzel. (1990) Performance of Packet Switches with Input and Output Queueing. *Proceedings of IEEE ICC'90*, 747-753, Atlanta.

Jackson, J. R. (1957) Networks of Waiting Lines. *Ops. Res.* **5**, 518-521.

Jackson, J. R. (1963) Jobshop-Like Queueing Systems. *Management Science* **10**, 131-142.

Jensen, A. (1953) Markoff Chains as an Aid in the Study of Markoff Processes. *Skandinavisk Aktuarietidskrift* **36**, 87-91.

Jewell, W. S. (1967) A Simple Proof of: $L = \lambda W$. *Ops. Res.* **15:6**, 1109-1116.

Jiang, I., and J. S. Meditch. (1990) A Queueing Model for ATM-Based Multi-Media Communication Systems. *Proceedings of IEEE ICC'90*, 264-267, Atlanta.

Keilson, J. (1979) *Markov Chain Models - Rarity and Exponentiality*. Springer-Verlag, New York.

Keilson, J., and L. Servi. (1988) A Distributional Form of Little's Law. *Ops. Res. Let.* **7:5**, 223-227.

Keilson, J., and L. Servi. (1989) Blocking Probability for M/G/1 Vacation Systems with Occupancy Level Dependent Schedules. *Ops. Res.* **37:1**, 134-140.

Keilson, J., and D. M. G. Wishart. (1965) A Central Limit Theorem for Process Defined on a Finite Markov Chain. *Proc. Cam. Phil. Soc.* **60**, 547-567.

Kelly, F. P. (1979) *Reversibility and Stochastic Networks*. John Wiley, New York.

Kleinrock, L. (1975) *Queueing Systems: Vol. 1, Theory*. John Wiley, New York.

Kleinrock, L. (1976) *Queueing Systems: Vol. 2, Computer Applications*. John Wiley, New York.

Kim, H. S., and A. Leon-Garcia. (1990) Performance of Self-Routing ATM Switch under Nonuniform Traffic Pattern. *Proceedings of IEEE INFOCOM'90*, 140-145, San Francisco.

Kim, Y. Y., and S.-Q. Li. (1999) Capturing Important Statistics of a Fading/Shadowing Channel for Network Performance Analysis. *IEEE J. Select. Areas Commun.* **17:5**, 888-901.

Kobayashi, H. (1978) *Model and Analysis: An Introduction to System Performance Evaluation Methodology*. Addison-Wesley, Reading.

Lancaster, P. (1966) *Lambda Matrices and Vibrating Systems*. Pergamon Press Ltd., London.

Langford, J. D. (1990) "Queueing Delays in Systems Having Batch Arrivals and Setup Times." Doctoral Dissertation, University of Rochester.

Latouche, G., and V. Ramaswami. (1999) *Introduction to Matrix Analytic Methods in Stochastic Modeling*, Society for Industrial and Applied Mathematics (SIAM), Philidelphia.

Lau, W.-C., and S.-Q. Li. (1993) Traffic Analysis in Large-Scale High-Speed Integrated Networks: Validation of Nodal Decomposition Approach. *Proceedings of IEEE INFOCOM'93*, **3**, 1320-1329, San Francisco.

Lazowska, D. E., J. Zahorjan, G. S. Graham, and K.C. Sevcik. (1984) *Quantitative System Performance: Computer System Analysis Using Queueing Network Models.* Prentice-Hall, Englewood Cliffs.

Lazowska, D. E., J. Zahorjan, and K.C. Sevcik. (1986) Computer System Performance Evaluation Using Queueing Network Models. *Ann. Rev. Comput. Sci.* **1**, 107-137.

LeBoudec, J., and P. Thiran. (2001) *Network Calculus: a Theory of Deterministic Queueing Systems for the Internet*, Springer-Verlag, New York.

Lea, C.-T. (1990) Design and Evaluation of Unbuffered Self-Routing Networks for Wideband Packet Switching. *Proceedings of IEEE INFOCOM'90*, 148-156, San Francisco.

Leland, W. E., M. Taqqu, W. Willinger, and D. V. Wilson. (1994) On the Self-Similar Nature of Ethernet Traffic. *IEEE/ACM Transactions on Networking* **2:1**, 1-15.

Leon-Garcia, A. (1989) *Probability and Random Processes for Electrical Engineering.* Addison-Wesley, Reading.

Levy, H., and L. Kleinrock. (1986) A Queue with Startup and a Queue with Vacations: Delay Analysis by Decomposition. *Ops. Res.* **34:3**, 426-436.

Levy, H., and M. Sidi. (1990) Polling Systems: Applications, Modeling, and Optimization. *IEEE Trans. on Commun.* **38:10**, 1750-1760.

Li, S.-Q. (1990) A General Solution Technique for Discrete Queueing Analysis of Multi-Media Traffic on ATM. *Proceedings of IEEE INFOCOM'90*, 1144-1155, San Francisco.

Li, S.-Q., and C.-L. Hwang. (1997) On the Convergence of Traffic Measurement and Queueing Analysis: a Statistical-Matching and Queueing (SMAQ) Tool. *IEEE/ACM Transactions on Networking* **5:1**, 95-110.

Li, S.-Q., S. Park, and D. Arifler. (1998) SMAQ: A Measurement-Based Tool for Traffic Modeling and Queueing Analysis Part II: Network Applications. *IEEE Communications* **36:8**, 66-77.

Little, J. D. C. (1961) A Proof of the Queuing Formula $L = \lambda W$. *Ops. Res.* **9**, 383-387.

Lucantoni, D. M. (1991) New Results on the Single Server Queue with a Batch Markovian Arrival Process. *Stochastic Models* **7:1**, 1-46.

Lucantoni, D. M. (1993) The BMAP/G/1: Queue: A Tutorial. *Models and Techniques for Performance Evaluation of Computer and Communications Systems.* (L. Donatiello and R. Nelson, eds.) Springer-Verlag, New York, 330-58.

Lucantoni, D. M. (1998) Further Transient Analysis of the BMAP/G/1 Queue. *Stoch. Mod*. **14: 1& 2**, 461-478.

Lucantoni, D. M., K. S. Meier-Hellstern, and M. F. Neuts. (1990) A Single Server Queue with Server Vacations and a Class of Non-Renewal Arrival Processes. *Adv. in Appl. Prob.* **22:3**, 676-705.

Melamed, B., and W. Whitt. (1990) On Arrivals that See Time Averages. *Ops. Res.* **38:1**, 156-172.

Mieni, B. (1997) An Improved FFT-Based Version of Ramaswami's Formula. *Comm. Statist. Stochastic Models* **13**, 223-238.

Moler, C., and C. F. van Loan. (1978) Nineteen Dubious Ways to Compute the Exponential of a Matrix. *SIAM Review* **20:4**, 801-835.

Neuts, M. F. (1981a) *Matrix Geometric Solutions in Stochastic Models.* The Johns Hopkins University Press, Baltimore.

Neuts, M. F. (1981b) The c-Server Queue with Constant Service Times and a Versatile Markovian Arrival Process. *Applied Probability - Computer Science: The Interface* **1**, 31-67.

Neuts, M. F. (1989) *Structured Stochastic Matrices of the M/G/1 Type and Their Applications.* Marcel-Dekker, New York.

Niu, S.-C., and R. B. Cooper. (1993) Transform-Free Results for the M/G/1 Finite-and Related Queues. *Mathematics of Operations Research* **18**, 486-510.

Noble, B., and J. W. Daniel. (1977) *Applied Linear Algebra.* Prentice-Hall, Englewood Cliffs.

Nussbaumer, H. J. (1982) *Fast Fourier Transforms and Convolution Algorithms.* Springer-Verlag, New York.

Platzman, L. K., J. C. Ammons, and J. J. Bartholdi,III. (1988) A Simple and Efficient Algorithm to Compute Tail Probabilities from Transforms. *Ops. Res.* **36:1**, 137-144.

Press, W. H., B. P. Flanner, S. A. Teukolsky, and W. T. Vetterling. (1988) *Numerical Recipes in C - The Art of Scientific Computing.* Cambridge University Press, Cambridge.

Ramaswami, V., and D. M. Lucantoni. (1985) Stationary Waiting Time Distributions in Queues with Phase-Type Service and in Quasi-Birth Death Processes. *Stochastic Models* **1:2**, 125-134.

Ramaswami, V. (1988a) A Stable Recursion for the Steady State Vector in Markov Chains of the M/G/1 Type. *Stochastic Models* **4**, 183-188.

Ramaswami, V. (1988b) Nonlinear Matrix Equations in Applied Probability - Solution Techniques and Open Problems. *SIAM Rev.* **30**, 256-263.

Ross, S. M. (1983) *Stochastic Processes.* John Wiley, New York.

Ross, S. M. (2003) *Introduction to Probability Models,* 8th. Ed. Academic Press, New York.

Saha, A., and M. D. Wagh. (1990) Performance Analysis of Banyan Networks Based on Buffers of Various Sizes. *Proceedings of IEEE INFOCOM'90*, 157-164, San Francisco.

Schwartz, M. (1987) *Telecommunications Networks: Protocols, Modeling and Analysis.* Addison-Wesley, Reading.

Shanthikumar, J. G. (1984) On a Software Reliability Model: A Review. *Microelectronics and Reliability* **23**, 903-93.

Sheng, H.-D., and S.-Q. Li. (1993) Second Order Effect of Binary Sources on Characteristics of Queue and Loss Rate. *Proceedings of IEEE INFOCOM'93*, **1**, 18-27, San Francisco.

Sinha, R. (1990) Baseline Document: T1S1 Technical Sub-Committee on Broadband Aspects of ISDN. Joint Report of AT&T, Bellcore, BellSouth Services, GTE-Telops, and Northern Telecom.

Spragins, J. D. (1991) *Telecommunications: Protocols and Design.* Addison-Wesley, Reading.

Sriram, K., and D. M. Lucantoni. (1989) Traffic Smoothing Effects of Bit Dropping in a Packet Voice Multiplexer. *IEEE Trans. on Commun.* **37:7**, 703-712.

Stallings, W. (1990a) *Handbook of Computer-Communications Standards: The Open Systems Interconnection (OSI) Model and OSI-Related Standards.* Macmillan, New York.

Stallings, W. (1990b) *Handbook of Computer-Communications Standards: Local Network Standards.* Macmillan, New York.

Stallings, W. (1990c) *Handbook of Computer-Communications Standards: Department of Defense (DOD) Protocol Standards.* Macmillan, New York.

Stavrakakis, I. (1990) Analysis of a Statistical Multiplexer under a General Input Traffic Model. *Proceedings of IEEE INFOCOM'90*, 1220-1225, San Francisco.

Stern, T. E. (1979) Approximations of Queue Dynamics and Their Application to Adaptive Routing in Computer Communication Networks. *IEEE Trans. on Commun.* **27:9**, 1331-1335.

Stern, T. E. (1983) A Queueing Analysis of Packet Voice. *Proc. Global Tele. Conf.*, **1**, 71-76. San Diego.

Stewart, W. J. (1990) *Numerical Solution of Markov Chains*. Marcel-Dekker, New York.

Stewart, W. J. (1996) *Computations with Markov Chains*. (W. J. Stewart ed.) Kluwer Academic Publishers, Boston.

Stewart, W. J. (1999) *Numerical Solution of Markov Chains*. (W. J. Stewart ed.) Editado por Prensas Universitarias de Zaragoza.

Stidham, S., Jr. (1974) A Last Word on $L = \lambda W$. *Ops. Res.* **22:2**, 417-421.

Takács, L. (1962) *Introduction to the Theory of Queues*. Oxford University Press, New York.

Takagi, H. (1987a) Queueing Analysis of Vacation Models, Part I: M/G/1 and Part II M/G/1 with Vacations. TRL Report TR87-0032. IBM Tokyo Research Laboratory, Tokyo.

Takagi, H. (1987b) Queueing Analysis of Vacation Models, Part III: M/G/1 with Priorities. TRL Report TR87-0038. IBM Tokyo Research Laboratory, Tokyo.

Takagi, H. (1990) Queueing Analysis of Polling Models, an Update. *Stochastic Analysis of Computer and Communication Systems*. (H. Takagi, ed.) North-Holland, New York, 267-318.

Tang, S. W. (1995) "Discrete-Time Queues with Batch-Markovian Arrival", Ph.D. Dissertation, Virginia Polytechnic Institute and State University.

Tanenbaum, A. S. (1988) *Computer Networks,* 2nd. Ed. Prentice-Hall, Englewood Cliffs.

Tijms, H. C. (1986) *Stochastic Modeling and Analysis: A Computational Approach*. John Wiley, New York.

Towsley, D. (1980) The Analysis of a Statistical Multiplexer with Nonindependent Arrivals and Errors. *IEEE Trans. Commun.*, **28**, 65-72.

Trivedi, K. (1982) *Probability and Statistics with Reliability, Queueing, and Computer Science Applications*. Prentice-Hall, Englewood Cliffs.

Turin, W., and M. Zorzi. (2002) Performance Analysis of Delay-Constrained Communications Over Slow Rayleigh Fading Channels. *IEEE Trans. Wireless Commun.* **1:4**, 801-807.

van Arem, B. (1990) "Queueing Models for Slotted Transmission Systems." Ph. D. Dissertation, Twente University.

Walrand, J. (1988) *An Introduction to Queueing Networks*. Prentice-Hall, Englewood Cliffs.

Weast, R. C., S. M. Selby, and C. D. Hodgman, eds. (1964) *Mathematical Tables from Handbook of Chemistry and Physics,* 12th. Ed. The Chemical Rubber Co., Cleveland.

Welch, P. D. (1964) On a Generalized M/G/1 Queueing Process in which the First Customer Receives Exceptional Service. *Ops. Res.* **12**, 736-762.

Whitt, W. (2002) *Stochastic-Process Limits: An Introduction to Stochastic-Process Limits and Their Applications to Queues*. Springer-Verlag, New York.

Wolff, R. W. (1970) Work Conserving Priorities. *J. App. Prob.* **7:2**, 327-337.

Wolff, R. W. (1982) Poisson Arrivals See Time Averages. *Ops. Res.* **30:2**, 223-231.

Wolff, R. W. (1989) *Stochastic Modeling and the Theory of Queues*. Prentice-Hall, Englewood Cliffs.

Wolff, R. W. (1990) A Note on PASTA and Anti-PASTA for Continuous-Time Markov Chains. *Ops. Res.* **38:1**, 176-177.

Woodside, C. M., and E. D. S. Ho. (1987) Engineering Calculation of the Overflow Probabilities in Buffers with Markov-Interrupted Service. *IEEE Trans. on Commun.* **35:12**, 1272-1277.

Wortman, M. A., and R. L. Disney. (1990) Vacation Queues with Markov Schedules. *Adv. Appl. Prob.* **22**, 730-748.

Zhang, J., and E. J. Coyle. (1989) Transient Analysis of Quasi-Birth-Death Processes. *Commun. Statist., Stochastic Models* **5:3**, 459-496.

Index

About the Author

John N. Daigle is Director of the Center for Wireless Communications and Professor of Electrical Engineering at The University of Mississippi, Oxford. He was formerly a Principal Engineer for the MITRE Corporation in McLean, Virginia, where he was responsible for research direction in the MITRE Washington Networking Technical Center, and an Adjunct Professor of Electrical Engineering at The George Washington University in Washington, D.C. His experience in electrical communications dates back to 1970 and includes a combined eight years at Bell Labs and NCR, twenty years on the faculties of major research universities, and four years in the USAF. He was also a co-op student with NASA, Houston TX, as an undergraduate. More recently, he has been a visiting researcher at the IBM Zurich Research Laboratory and at The University of Perugia.

Prof. Daigle has taught queueing theory and random processes for over a period of about 25 years. He has also taught in a variety of areas including control theory, computer architecture, mobile and wireless communications, analog and digital communications, coding theory, and computer communication systems and protocols. His research results have been published in leading IEEE technical conferences and IEEE and ORSA journals. He is also the author of the text book, *Queueing Theory for Telecommunications*, published by Addison-Wesley in 1992.

Prof. Daigle is a Fellow of the IEEE and is active in the IEEE activities. He serves as Associate Editor-in-Chief of *IEEE Communications Surveys and Tutorials* and as Senior Technical Editor for *IEEE Network*. He has previously served as Editor-in-Chief of *IEEE Communications Surveys and Tutorials* and *IEEE Network* and as an Editor for *IEEE/ACM Transactions on Networking*. He was formerly Director of Education of the IEEE Communications Society and has previously served on that society's Board of Governors. He is a past chairman of the Society's Technical Committee on Computer Communications (1981-83), and he has served on the technical program committees of numerous IEEE conferences and workshops. He also served as an associate editor of the Journal of Operations Research (2000-02). He holds BS and MS degrees in electrical engineering from Louisiana Tech University and Virginia Polytechnic Institute and State University, respectively. His doctorate, from Columbia University, is in operations research.